AF903246

Herbal Insecticides, Repellents and Biomedicines: Effectiveness and Commercialization

Vijay Veer • Reji Gopalakrishnan
Editors

Herbal Insecticides, Repellents and Biomedicines: Effectiveness and Commercialization

Springer

Editors
Vijay Veer
Defence Research Laboratory (Defence Research and Development Organisation)
Tezpur, Assam, India

Reji Gopalakrishnan
Defence Research Laboratory (Defence Research and Development Organisation)
Tezpur, Assam, India

ISBN 978-81-322-2702-1 ISBN 978-81-322-2704-5 (eBook)
DOI 10.1007/978-81-322-2704-5

Library of Congress Control Number: 2015959024

Springer New Delhi Heidelberg New York Dordrecht London
© Springer India 2016
This work is subject to copyright. All rights are reserved by the Publisher, whether the whole or part of the material is concerned, specifically the rights of translation, reprinting, reuse of illustrations, recitation, broadcasting, reproduction on microfilms or in any other physical way, and transmission or information storage and retrieval, electronic adaptation, computer software, or by similar or dissimilar methodology now known or hereafter developed.
The use of general descriptive names, registered names, trademarks, service marks, etc. in this publication does not imply, even in the absence of a specific statement, that such names are exempt from the relevant protective laws and regulations and therefore free for general use.
The publisher, the authors and the editors are safe to assume that the advice and information in this book are believed to be true and accurate at the date of publication. Neither the publisher nor the authors or the editors give a warranty, express or implied, with respect to the material contained herein or for any errors or omissions that may have been made.

Printed on acid-free paper

Springer (India) Pvt. Ltd. is part of Springer Science+Business Media (www.springer.com)

Foreword

Indians have vast inherited knowledge about traditional and herbal medicines as India being one of the megacentres of biodiversity is home to a large number of plant and animal species. The traditional health-care practices in the country are based on the exploration of this rich biodiversity for the benefit of mankind. Most of the traditional practitioners use traditional medicines according to their own way including method of preparations, recipes, etc. But it is true that rationality or claims are perhaps missing when scientifically judged. The traditional medicines are composed of plants, minerals, and organic matter, whereas herbal drugs are prepared from medicinal plants and have phyto-constitution. The advancements in modem medicine have led to a scenario where many of these indigenous traditional knowledge resources were underutilized due to lack of their proper standardization. However, in recent times there is a renewed interest of academic and government institutions and pharmaceutical companies to tap the potential of natural products as sources of biomedicines and biopesticides by utilizing the modern techniques with integrated approach.

Herbal drugs are considered safe by and large as they are not having much side effects and are cost-effective. It is seen during last few decades that the herbal drugs are also taking major role in the Western world particularly in the treatment of metabolic diseases, terminal disease, etc. Similarly, indiscriminate use of chemical insecticides in agriculture has resulted in the development of insecticide resistance and pest resurgence. Thus the focus in crop pest management is now shifting towards plant-derived insecticides. The farmers and the consumers are increasingly becoming aware of the ill effects of chemical insecticides on environment and health. This has created a huge opportunity for exploration and commercialization of safer and effective herbal products as biomedicines, herbal insecticides and herbal repellents.

In this context, it is laudable that Dr. Vijay Veer and Dr. Reji Gopalakrishnan have taken the effort to bring out a book *Herbal Insecticides, Repellents and Biomedicines: Effectiveness and Commercialization*. This book has chapters contributed by distinguished scientists from the country's premiere research laboratories and academic institutions. The topics covered range from plant-derived insecticides to phyto-pharmaceuticals. The latest trends in botanical formulations are discussed along with the emerging field of plant vaccine development. The highlight of this book is the emphasis given to commercialization of herbal products. A whole chapter is devoted to IPR issues on herbal products, which would be tremendously helpful to those involved in herbal product development. The chapter on the guidelines to be followed in the registration and commercialization of herbal products would enlighten the readers with valuable information in bringing their products from the laboratory to the end users. Information contained in this book will be helpful to the R&D/academic institutions in transfer of technology to industry and in commercialization of their herbal products. I believe that this book will be immensely beneficial to researchers, academicians, industrialists, and students alike for exploring, understanding, and utilizing the herbal wealth of our country. I wish the authors and the editors all success in their future endeavours and hope they continue their pursuit of excellence in herbal research and development.

Defence Research & Development Organisation
New Delhi, India

Manas K. Mandal

Preface

The incidence of various diseases is on the rise world over. This may be due to early detection and accessibility to medical care, change in life style and exposure to a large number of environmental carcinogens. However, on the other hand the discovery of new drugs is slow because the new drug development is a complex and lengthy process and quite expensive one. It takes about 10–12 years and costs about 1 billion US dollar for one drug. In this scenario, there is a need to adopt an integrated approach where Ayurvedic wisdom and traditional or tribal knowledge can synergize the drug discovery from plant sources. Thus there is a need of change in paradigm to involve different disciplines like combinatorial chemistry, analytical chemistry, biological assay, microarray analysis and genetic engineering for new chemical entities. New chemical entity or novel bioactive molecule can be isolated from the right kind of plant through sequential or parallel extraction. This will save time and cost and will also enhance the success rate. Phytochemicals including the secondary metabolites are important source for biomedicines for various diseases, due to their safety and low toxicity. Similarly, phytochemicals can be good sources of effective and environmentally safe biopesticides for the control of insect pests, which cause about 40 % crop loss in field and store.

India is blessed with a 5000-year-old Ayurvedic system and one of the 12 mega biodiversity centres having over 45,000 plant species, including about 3000 plant species of medicinal importance and 23,000 fungi, 25,000 algae, 1600 lichens, 1800 bryophytes and 30 million microorganisms. This gives us ample opportunity for drug development and for the discovery of new drugs. Unfortunately, this has not been exploited gainfully and India has lagged behind. Herbal drug industry is one of the fastest growing industries in the world. Many chemotherapeutic drugs for cancer like paclitaxel and vinca alkaloids were developed from plant materials. There are many others like toxoids, flavonoids, saponins, anthracene derivatives, cardiac glycosides quinolone, indole alkaloids, aromatic and laxatives, which are very much in demand in the world market. Herbal medicines worth 3500 US dollar are produced annually, whereas 1200 single ingredient and 10,500 polyherbal formulations are registered in India. Export-Import Bank of India has estimated the trade in medicinal plants at Rs 5500 crore, which is likely to grow to Rs 20,000 crore by the year 2030.

Herbal medicines for various ailments are preferred in India and elsewhere because of their safety and low cost. There is a need to utilize India's herbal heritage in better ways. This requires quality science in herbal medicine based on our traditional knowledge of medicinal plants. For this, we to have to systematically work on the identification of active constituents from the right kind of plant material and ensure correct amount of the active ingredients in the formulation through adopting proper quality control and good manufacturing practices (GMP).

We should comply with the WHO guidelines for quality assessment of plant material, plant preparation and finished products, and documentation on safety assessment through toxicological studies and ensure stability or shelf life of the formulation and efficacy assessment in animal models. We should focus on polyherbal formulations for multi-target therapeutics rather than one disease–one drug approach. Then the selected medicinal plants need to be cultivated in an organized way involving the farmers and using tissue culture technology. This will not only provide substantial business both locally and globally but also make the traditional medicines available to a large number of rural people who have no access to modern medicine. We sincerely hope that this book will be useful for the development and commercialization of high-quality herbal products.

Defence Research Laboratory (Defence Research and Development Organisation)
Tezpur, Assam, India

Vijay Veer
Reji Gopalakrishnan

Acknowledgements

We sincerely thank Dr. Manas K Mandal, Director General, Life Sciences, Defence Research & Development Organisation, and Dr. Lokendra Singh, Director, Defence Research & Development Establishment, Gwalior, for their support, guidance and encouragement during the preparation of this book. The editors would like to thank all the authors who have enriched this book with their valuable contributions. We are thankful to Springer (India) Private Limited and the editorial team for their painstaking efforts in getting this book published in a timely manner.

Dr. Vijay Veer
Dr. Reji Gopalakrishnan
(Editors)

Contents

1 **Advances in Vector Mosquito Control Technologies, with Particular Reference to Herbal Products** 1
B.K. Tyagi

2 **Sustainable and Novel Eco-friendly Approaches Towards Integrated Disease and Vector Management** 11
Jubilee Purkayastha, Rajesh Arora, and Lokendra Singh

3 **Mosquito Larvicidal Potential of Medicinal Plants** 25
M. Govindarajan

4 **Plants of Himalayan Region as Potential Source of Biopesticides for Lepidopteran Insect Pests** 63
S.G. Eswara Reddy, Shudh Kirti Dolma, and Anuja Bhardwaj

5 **Natural Insecticides from Actinomycetes and Other Microbes for Vector Mosquito Control** 85
M. Gabriel Paulraj, P. Saravana Kumar, S. Ignacimuthu, and D. Sukumaran

6 **Extraction, Purification and Characterisation of Insecticidal Compounds from Plants** 101
Shaswat Barua, Reji Gopalakrishnan, and Vijay Veer

7 **Recent Advancements in Bio-botanical Pesticide Formulation Technology Development** 117
Megha Pant, Saurabh Dubey, and P.K. Patanjali

8 **Medicinal Plants and Phytomedicines** 127
Brahma Singh

9 **Prospects for Development of Biomedicines from the Medicinal Plants of Northeastern India** 147
Nongalleima Khumukcham, Dipak Biswas, Nameirakpam Surjit Singh, and Lokesh Deb

10 **Traditional Anti-poison Plants Used by the Monpa Tribe of Arunachal Pradesh** 189
Jambey Tsering, Hui Tag, Baikuntha Jyoti Gogoi, and Vijay Veer

11 Molecular Farming for Production of Biopharmaceuticals and Edible Vaccines in Plants 205
Ajitabh Bora, Hemanta Kumar Gogoi, and Vijay Veer

12 Intellectual Property Rights Issues for Herbal Products 217
Nidhi Sandal and Avinash Kumar

13 Regulatory Approval of Botanical Products Including Herbal Drugs, Food, and Insecticides for Commercialization 243
P. Chattopadhyay and Vijay Veer

Contributors

Rajesh Arora Office of the Director General-Life Sciences, Defence Research and Development Organisation, New Delhi, India

Shaswat Barua Defence Research Laboratory (Defence Research and Development Organisation), Tezpur, Assam, India

Anuja Bhardwaj CSIR-Institute of Himalayan Bioresource Technology (Council of Scientific and Industrial Research), Palampur, Himachal Pradesh, India

Dipak Biswas Pharmacology Laboratory, Institute of Bioresources and Sustainable Development, A National Institute of Department of Biotechnology, Government of India, Takyelpat Institutional Area, Manipur, India

Ajitabh Bora Defence Research Laboratory (Defence Research and Development Organisation), Tezpur, Assam, India

P. Chattopadhyay Defence Research Laboratory (Defence Research and Development Organisation), Tezpur, Assam, India

Lokesh Deb Pharmacology Laboratory, Institute of Bioresources and Sustainable Development, A national Institute of Department of Biotechnology, Government of India, Takyelpat Institutional Area, Manipur, India

Shudh Kirti Dolma CSIR-Institute of Himalayan Bioresource Technology (Council of Scientific and Industrial Research), Palampur, Himachal Pradesh, India

Saurabh Dubey Centre for Rural Development and Technology, Indian Institute of Technology Delhi, New Delhi, India

Baikuntha Jyoti Gogoi Defence Research Laboratory, (Defence Research and Development Organisation), Tezpur, Assam, India

Hemanta Kumar Gogoi Defence Research Laboratory (Defence Research and Development Organisation), Tezpur, Assam, India

Reji Gopalakrishnan Defence Research Laboratory (Defence Research and Development Organisation), Tezpur, Assam, India

M. Govindarajan Unit of Vector Biology/Phytochemistry and Nanotechnology/Department of Zoology, Annamalai University, Annamalainagar, Tamil Nadu, India

S. Ignacimuthu Entomology Research Institute, Loyola College, Chennai, Tamil Nadu, India

P. Saravana Kumar Entomology Research Institute, Loyola College, Chennai, Tamil Nadu, India

Avinash Kumar Defence Research and Development Organisation, New Delhi, India

Nongalleima Khumukcham Pharmacology Laboratory, Institute of Bioresources and Sustainable Development, A national Institute of Department of Biotechnology, Government of India, Takyelpat Institutional Area, Manipur, India

Megha Pant Institute of Pesticide Formulation Technology, Gurgaon, India

P.K. Patanjali Institute of Pesticide Formulation Technology, Gurgaon, India

M. Gabriel Paulraj Entomology Research Institute, Loyola College, Chennai, Tamil Nadu, India

Jubilee Purkayastha Office of the Director General-Life Sciences, Defence Research and Development Organization, New Delhi, India

S.G. Eswara Reddy CSIR-Institute of Himalayan Bioresource Technology (Council of Scientific and Industrial Research), Palampur, Himachal Pradesh, India

Nidhi Sandal Defence Research and Development Organisation, New Delhi, India

Lokendra Singh Defence Research and Development Establishment, Gwalior, India

Nameirakpam Surjit Singh Pharmacology Laboratory, Institute of Bioresources and Sustainable Development, A national Institute of Department of Biotechnology, Government of India, Takyelpat Institutional Area, Manipur, India

Brahma Singh World Noni Research Foundation, Chennai, India

Defence Research and Development Organisation, New Delhi, India

D. Sukumaran Defence Research and Development Establishment, Gwalior, India

Hui Tag Department of Botany, Rajiv Gandhi University, Doimukh, Arunachal Pradesh, India

Jambey Tsering Department of Botany, Rajiv Gandhi University, Doimukh, Arunachal Pradesh, India

B.K. Tyagi Centre for Research in Medical Entomology (ICMR), Madurai, India

Vijay Veer Defence Research Laboratory (Defence Research and Development Organisation), Tezpur, Assam, India

About the Editors

Dr. Vijay Veer is a renowned defence Scientist who is well known for his contributions to insect ecology, taxonomy and medical entomology. He has more than 38 years of research experience and has developed many products and technologies for the benefit of the armed forces. He has developed Defender Net, the first indigenous long-lasting insecticidal net, which offers protection from malaria, Japanese encephalitis, dengue, filariasis and other insect-borne diseases. His studies on pests of defence importance have led to the development of highly effective pest control products, namely, Diethyl Phenyl Acetamide (DEPA – a mosquito repellent), Wool Care, Ratox, Roachtox and ovitraps. He has contributed immensely towards our understanding on forest pests, thrips, dermestid beetles and tabanid flies and has described many new insect genera and species. He has 25 patents, 8 books, several book chapters and over 155 research publications in national and international journals. He was a member of GFAST, a DRDO think-tank, representing Life Sciences. He is presently the Director of Defence Research Laboratory (DRL), Tezpur, India. DRL, under his able leadership, has successfully developed an array of products and technologies including Dentrap, Toxmos, Biotank cum reed bed system, Capsispray (chilli spray for self-defence), Capsigrenade (chilli grenade for mob dispersal) and DrlPure (water purification unit). Currently, he is pursuing the field of semiochemicals and molecular taxonomy.

Dr. Reji Gopalakrishnan is a DRDO Scientist at the Defence Research Laboratory, Tezpur, India. He studied the simulation of insect population dynamics, pest damage on crop plants and pest-weather relationships during his master's and doctoral research at the Indian Agricultural Research Institute, New Delhi. He has more than 12 years of research experience in entomology and has one patent and many research publications in national and international journals. He is presently working on the development of disease vector monitoring tools and insect repellents. He has developed Dentrap, which is a trapping device for dengue, chikungunya, Japanese encephalitis and filariasis vectors and Toxmos, which is an aerosol formulation for application onto fabrics and for personal protection from mosquitoes and other biting insects.

Advances in Vector Mosquito Control Technologies, with Particular Reference to Herbal Products

1

B.K. Tyagi

Abstract

Despite several decades of control effort, vector-borne diseases (VBDs) are still regarded as a major public health problem in the tropical and subtropical regions of the world. They are more common in the developing as well as resource-poor countries and inflict enormous burden in terms of morbidity and mortality. Control of vectors is the primary available intervention for some of the most devastating VBDs, particularly those lacking vaccines such as malaria, dengue, chikungunya, trypanosomiasis, filariasis, leishmaniasis and Chagas disease. Since long decades vector control programme based on chemical agents has been in place but their effectiveness was soon undermined due largely to the prowess of development of resistance amongst most vector species of these diseases. Thus, in turn, there is a growing need for searching of alternative novel interventions and exploration of next-generation vector control strategies, including repellents, attractants, traps, pneumatic/tracheal explosions in larval populations, biocides integrated with nanoparticles, genetically modified vectors (transgenics), paratransgenics, etc. Plant-based repellents have been used for generations in traditional practice as a personal protection measure against host-seeking mosquitoes. Knowledge on traditional repellent plants obtained through ethnobotanical studies is a valuable resource for the development of new natural products. Recently, commercial repellent products containing plant-based ingredients have gained increasing popularity amongst consumers, as these are commonly perceived as 'safe' in comparison to long-established synthetic repellents although this is sometimes a misconception. There is a need for further meticulously evaluating repellent compounds and develop new products that offer high repellency as well as good consumer safety. Such innovative technologies must also

B.K. Tyagi (✉)
Centre for Research in Medical Entomology (ICMR),
4-Sarojini Street, 625002 Chinna Chokkikulam,
Madurai, India
e-mail: abktyagi@gmail.com; abk.tyagi@yahoo.co.in

© Springer India 2016

Vijay Veer, R. Gopalakrishnan (eds.), *Herbal Insecticides, Repellents and Biomedicines: Effectiveness and Commercialization*, DOI 10.1007/978-81-322-2704-5_1

fit into the concept of integrated vector control management (IVM) which derives today a much greater meaning now by bringing within its fold all the available technologies than ever before, for an effective vector control.

1.1 Introduction

Arthropods, or jointed legs, are invertebrate animals belonging to phylum Arthropoda and include insects, spiders, centipedes, shrimp and crayfish which make the group largest amongst animals, accounting for approximately 80 % of extant species (anywhere close to 30 million) estimated to be present in the world. Insects are one of the most abundant groups of arthropods, with phenomenally diverse behaviour and adaptation. Even though an astoundingly large number of insects are beneficial as pollinator or producer of economically important products, some of them are nefariously known as either dreaded pests or transmitters of deadly or debilitating diseases of both public health and veterinary importance. These arthropod vectors (insects, ticks, mites, etc.) transmit protozoa, helminths, bacteria and viruses, spirochaetes, fungi, etc. These vectors and the diseases transmitted by them inflict heavy losses to humans and their livestock, both directly by biting and sucking blood and indirectly by transmitting vector-borne diseases (Table 1.1). In the early days of vector control, inorganic compounds were used, but later they were largely replaced by organic chemical compounds due to their wider and quicker effects, besides large production. Repellents have always been man's fascination in keeping the mosquito at bay and the world has by now seen highly effective repellents such as DEET, DEPA, etc., albeit the list of effective candidate compounds is very long!

Table 1.1 Some important vector groups and the diseases transmitted by them (WHO 2014a)

Vector		Diseases
Mosquitoes	*Aedes aegypti*	Dengue, yellow fever, chikungunya, Zika virus
	Aedes albopictus	Chikungunya, dengue, West Nile virus
	Culex quinquefasciatus	Lymphatic filariasis
	Culex species	Japanese encephalitis
	Anopheles	Malaria, lymphatic filariasis (in Africa)
	Haemagogus	Yellow fever
Flies	Sandflies	Leishmaniasis, sandfly fever (phlebotomus fever)
Bugs	Triatomine bugs	Chagas disease (American trypanosomiasis)
Ticks		Crimean-Congo haemorrhagic fever, tick-borne encephalitis, typhus, Lyme disease, relapsing fever (borreliosis), rickettsial diseases (spotted fever and Q fever),tularaemia
Fleas		Plague, murine typhus, rickettsiosis
Cyclops		Dracunculiasis (guinea worm disease)
Flies	Flies (various species)	Human African trypanosomiasis, onchocerciasis (river blindness)
Aquatic snails		Schistosomiasis (bilharziasis)

1.2 Global Distribution of Vector-Borne Diseases and Need for Indigenously Developed Technologies

The emergence and re-emergence of vector-borne diseases are the most important challenges threatening global public health in the twenty-first century. VBDs account for more than 17 % of the estimated global burden of infectious diseases, causing more than 1 billion cases and over 1 million deaths annually (WHO 2014b). Mosquitoes and ticks account for the majority of transmissions of the most important vector-borne diseases. At present, mosquito-transmitted diseases are present in more than 125 countries worldwide, mainly in the tropical and subtropical regions, posing major risks to half the world's population. Malaria and dengue remain the most prevalent mosquito-borne infections in many parts of the world with over 50 % of world's population – more than 3.3 billion and 2.5 billion people – at risk for malaria and dengue, respectively, with severe impact on economic and social development. Added to it are other vector-borne diseases like West Nile virus in the Americas, chikungunya and Japanese encephalitis in Asia and Oceania and Rift Valley fever in Western and Eastern Africa rapidly emerging (Gould and Solomon 2008).

India, with over 1.21 billion people, is the second most populous country in the world (75 % of South Asia region's people live in India, World Bank 2010) and the tenth largest economy (with a GDP of US$ 1847.9 billion in 2011), but still continues to share as high as 21 % of the world's global burden of diseases in terms of increased morbidity, mortality and disability (WHO 2012; National Commission on Macroeconomics and Health 2005). Overall, out of 4.2 million disability-adjusted life years lost due to vector-borne diseases, malaria alone was responsible for an estimated 1.85 million years lost/annum in India (Table 1.2) (Kumar et al. 2007; Peters et al. 2001). Insecticides formed the main plank of control strategy in the past, and in fact they are invariably deployed even today in the wake of any disease outbreak. Owing to the environmental and animal health concerns, in addition to development of resistance in the vectors of diseases against most of these synthetic pesticides, there has been an urgent need for exploring novel

Table 1.2 Burden of major vector-borne diseases in India (WHO 2012)

Diseases	Number of cases	Percentage of global burden	Reference
Malaria	1.07 million (2012) and 0.88 million (2013)	India contributes about 70 % of malaria in the Southeast Asian Region	Kumar et al. (2007)
Dengue	75,454 (2013)	Asia accounts for 70 %,in which India alone contributed 49 % and 34 % of the Asia and global burden, respectively	Gubler (1998), WHO (2014a)
Visceral leishmaniasis (kala-azar)	13,869 cases and 20 deaths (2013)	67 % of total visceral leishmaniasis (VL) cases reported by Indian subcontinent	WHO (2014a), Lobo et al. (2011)
Lymphatic filariasis	6 million	About 70 % of the infection worldwide contributed by India, Indonesia, Nigeria and Bangladesh	WHO (2014b)
Onchocerciasis	NA	NA	
Schistosomiasis	NA	NA	
JE	1078 JE cases and 199 deaths (2013)	NA	WHO (2014a)
Chikungunya	18,639 cases (2013)	NA	WHO (2014a)

and innovative methodologies that are eco-friendly and less expensive and can be integrated with IVM approach (Patz et al. 2005; Gubler 1998; Monath 1994; Lobo et al. 2011).

1.3 Vector Control Strategies

For many years, much of the medical research community has been focusing on the development of vaccines or drugs against mosquito-borne diseases, and there has been substantial success achieved as well in this direction (e.g. JE, KFD, etc.). Still there are many other serious and deadly infections, such as dengue, chikungunya and malaria, which have so far been eluding the discovery of an effective vaccine. In cases of certain diseases, such as lymphatic filariasis, leishmaniasis and dengue, it has been proven beyond doubt that an effective control of vector could bring about an appreciable depletion in the disease cases as well. The 'vector population suppression' involves various methods such as use of insecticides, pathogens, predators, lure and kill trapping, environment management, etc., while 'vector population replacement' involves vector's genetic manipulation so that vector should either reproduce its nonviable generations or becomes unfit for reproduction or for disease transmission.

1.3.1 Biochemical Strategies

Like in other countries, in India too, insecticides played a key role in effecting vector control under a national disease control programme such as those of National Malaria, Filariasis and Leishmaniasis Eradication/Control Programmes. The Directorate of National Vector Borne Disease Control programme (NVBDCP, earlier known as NMEP) is the nodal department in preventing and control of vector-borne illnesses in India. Since mosquito is the prime arthropod vector in transmitting the vector-borne infection in India, all the vector control strategies are mainly targeted on mosquito control. Currently, India is using insecticides for public health purposes, namely, the organochlorines (DDT), organophosphates (malathion) and certain groups of synthetic pyrethroid (deltamethrin, cyfluthrin, alpha-cypermethrin, lambda-cyhalothrin, etc.) for indoor residual spraying (IRS), fogging and aerial spraying (ultra-low volume spraying). Since organophosphates are highly toxic to mammals and have a short residual life, it is not used for IRS, whereas pyrethroids are widely used for IRS and are the only insecticide currently used to impregnate bed nets.

Larvicidal treatment of water with temephos and Bti is approved by the WHO. The organophosphate temephos (Abate) has been most widely used against the container-breeding malaria vector *Anopheles stephensi* and the dengue/chikungunya vector *Aedes aegypti.* Destroying the larvae, either chemically or biologically, while in the aquatic environment is an effective vector control method. Recently, insect growth regulators (IGR), such as diflubenzuron and methoprene, have been implemented as larvicide which are target specific but may be toxic to immature stages of other aquatic insects. Although insecticides have been effective in bringing disease under control in the initial stage of application, the current disease burden indicates that the strategies deployed are no longer effective due to the development of resistance and that the widespread and long term application is not cost effective and logistically difficult especially in developing and underdeveloped countries. In recent decades, resistance to newer classes of insecticides has also been reported quickly after usage because the formulations used for vector control were originally used for agricultural purposes. Multiple and cross-resistance to insecticides in major mosquito vectors is reported against organochlorines, organophosphates, carbamates and pyrethroid insecticides (Hemingway and Ranson 2000).

The modes of insecticide resistance mechanisms in mosquitoes are through (i) metabolism-based resistance (altered activities of enzyme groups, which inhibit the insecticides from reaching their potential target sites), (ii) behaviour resistance (changing their customary behaviour in a way to exposure to chemical), (iii) target site resistance (changes in the target sites and proteins that insecticides bind to) and (iv) penetration resistance (occurs when the cuticle (outer layer of insects) absorbs the insecticides molecule's much more slowly than the susceptible insects) (Hemingway and Ranson 2000). The development of resistance over various biochemicals makes a shift to insecticide combinations and also searching for new chemical formulations. This has compelled to look for eco-friendly alternative control methods so that the use of insecticides can be minimized. Several environment-friendly methods involving use of insectivorous fishes, biopesticides, pheromones, sterilized males, refractory mosquitoes, endosymbiont, midgut symbionts, etc. are being developed with varying degrees of success. Apart from these, a whole array of effective repellents, sticky traps, lethal traps, etc. is also proposed.

1.3.2 Biological Control Agents

Biocontrol is the use of natural enemies for control of vector population. Biological agents, such as parasites, pathogens and predators, can be used to target various life stages of the mosquito. Bacteria, fungi, viruses, fish, predaceous insects such as dragonflies and notonectids and copepods have been employed to decrease the mosquito larvae populations. These agents are inexpensive to implement, and safe for humans and nontarget organisms. They therefore provide a potentially environmentally friendly option. Fishes feeding on mosquito larvae include the world famous *Gambusia affinis*, some cyprinids (carps and minnows) and killifish which have been used for many decades in malaria endemic regions.

1.3.3 Preventive Strategies

Preventative measures to combat against VBDs include the personal prophylactics like bed nets, repellents and attractants, insecticide-treated nets (ITNs) or long-lasting insecticide nets (LLINs) or insecticide-treated wall linings, environmental management and source reduction measures and the careful design of human settlements (covering the windows and doors of human dwellings with thin muslin).

1.3.3.1 Insecticide-Treated Materials (ITMs), including Long-Lasting Insecticide Nets (LLIN)

It consists of insecticide-treated nets (ITNs) or long-lasting insecticide nets (LLINs), window curtains, sheet cover and wall hangings which have increased in demand in recent decades. The effectiveness of untreated nets could be improved through the use of chemicals, which either killed or repelled insect vectors (Lines and Addington 2001) The ITNs were not very successful because they require reimpregnation every 6 months which was quite laborious and became ineffective after a few washes. This led to the development of LLINs wherein the chemicals were bonded to the fibre of the net (Masum et al. 2010). Insecticide-treated wall linings (ITWLs) or pyrethroid-impregnated indoor linings have been proposed as a safer alternative to indoor residual spraying. Window curtains, screens, doorway or wardrobe curtains, etc. all appear to have promising results in different settings.

1.3.3.2 Repellents

Repellents are made based on chemical products with an offensive smell or taste to mosquitoes. Plant-derived compounds with repellent properties are most likely chemicals that are produced in defence against insects that pose a threat to the plant itself. These chemicals can be grouped into different categories, based on the functional groups present. They include nitrogen-containing compounds, terpenoids, phenolics, proteinase inhibitors and growth regulators. These compounds are generally produced to fight off a broad spectrum of insects including mosquitoes. Plants with better repellent properties fall into distinct families, with the Poaceae family (*citronella* based especially *Cymbopogon* spp.) being the pre-eminent one. Species of Lamiaceae, Fabaceae and Asteraceae also show promising results. Prior to the extensive use of synthetically produced repellents, aromatic/essential oils were commonly used. The military was a significant consumer of these oils, creams containing citronella, camphor and paraffin. Numerous essential oil-producing plants from the Lamiaceae, Poaceae, Rutaceae and Myrtaceae families have very well-known repellent activity. Synthetic products that have been used as repellents include indalone, dimethylphthalate 2-ethyl-1, 3-hexane diol (Rutgers 612) and N,N-diethyl-m-toluamide (DEET). DEET is by far the most effective and widely used repellent. The Indian counterpart of the DEET is N,N-diethyl phenyl acetamide (DEPA) which is now commercially available in different formulations and has been claimed to have the protection rate of about 8 h, against not only the mosquito pests and vectors but also other hematophagous arthropods as well as ticks, mites and even leeches.

1.3.3.3 Insect Traps

The ovitrap or oviposition trap was mostly used and invented for the surveillance of *Aedes* vectors and then modified to render it lethal to adults or larvae of *Ae. aegypti* (Chan 1972; Lok et al. 1977). In principle, ovitraps could kill adult mosquitoes if the ovistrip was treated with insecticide (Zeichner and Perich 1999) or destroy progeny by using fine nylon netting for trapping the larvae (Lok et al. 1977). Lethal ovitraps (which incorporate an insecticide on the oviposition substrate) with deltamethrin-treated ovistrips killed 89 % of *Ae. aegypti* adults and produced more than 99 % larval mortality during 1-month field trials in Brazil (Perich et al. 2003). The autocidal ovitraps (which allow oviposition but prevent adult emergence) and sticky ovitraps (which trap the mosquito when it lands) have been used on a limited basis. The autocidal ovitrap was used in Singapore for the control of Aedes vectors in urban areas with a high density of *Aedes* (Lok et al. 1977). The advantages of lethal ovitraps include their simplicity, their specificity for and effectiveness against container breeders like *Ae. aegypti* and its potential for integration with other chemical or biological control methodologies.

A newer approach to killing mosquitoes in a nontoxic way is to use a device that burns propane, thus generating carbon dioxide, warmth and water vapour which draws the mosquitoes towards the propane flame, where they are then sucked into a net or holder where they are collected. Some newer mosquito traps or known mosquito attractants are also available which are based on the principle of disabling its odour receptor as host-seeking female mosquitoes are guided by attractant odours released by their target. *Anopheles gambiae* is attracted to ammonia, lactic acid and other carboxylic acids naturally present in the body odour and sweat produced by warm-blooded animals. There are traps like black hole mosquito and midge trap and Jakmax insect trap which were developed based on this concept which release heat and CO_2 and captures mosquitoes, sandflies, midges, termites and other flies that are attracted by black light.

1.3.3.4 Strategies Based on Genetic Modification

Genetic modification of vector mosquitoes is one of such technologies, which may be mainly used either to suppress or replace the wild populations of a vector so as to decrease vector populations or reduce the vector's ability to transmit. These applications include the release of reared mosquitoes in the environment to introduce modified genetic traits in wild population. GM encompasses multiple approaches, which are broadly categorized into two types. The first category includes sterilized insect technique (SIT) for population suppression and the second category is the gene drive systems for population replacement or manipulation. The sterilized insect technique (SIT) includes the dominant lethal gene systems (RIDL), *Wolbachia*-mediated cytoplasmic incompatibility (CI) and classical radiation-induced male sterility, while population manipulative technologies include *Medea*-based gene drive, under-dominance gene drive, homing endonuclease, *Wolbachia*-mediated heritable biocontrol or genetically modified midgut bacteria and transposable element like *PiggyBac*. However, most of them are still in laboratory development. There are various methods either based on population suppression or population manipulation; all these techniques are still in the initial stage of development and require different region-based field testing under independent monitoring in order to prove the result to gain the consent of society to implement. These techniques require large volume of mosquitoes to be released into the environment at different period of interval to either suppress or replace mosquito population.

1.4 Plant-Based (Herbal) Mosquito Repellents

Plants offer a great promise for offering molecules that can be exploited to develop effective antimosquito products (Tyagi and Shahi 2001; Tyagi 2002a, b; Tyagi and Shahi 2002; Tyagi 2003a, b; Shahi et al. 2000). Most plants contain compounds, falling under several categories, including repellents, feeding deterrents, toxins and growth regulators, which they use in preventing attack from phytophagous (plant-eating) insects. These compounds may be categorized under (i) nitrogen compounds (primarily alkaloids), (ii) terpenoids, (iii) phenolics, (iv) proteinase inhibitors and (v) growth regulators. Those volatile components released as a consequence of herbivory are now best known for being effective against mosquitoes and other biting insects (Pichersky and Gershenzon 2002). Insects detect odours when volatile odours bind to odorant receptor (OR) proteins on ciliated dendrites of specialized odour receptor neurons (ORNs) often on the antennae and maxillary palps of the insect. Some ORNs, such as OR83b, that is important in olfaction and blocked by the gold standard synthetic repellent DEET (N, N-diethyl-3-methylbenzamide), are highly conserved across insect species. Plants commonly produce volatile 'green leaf volatiles' when leaves are damaged in order to deter herbivores. However, it is most likely that many plant volatiles are deterrent or repellent because they have high vapour toxicity to insects.

In India, a large number of examples are there where the plants, whole or part of it (in dried or fresh and in different forms: leaf, fruit, seed, root, etc.), have been exploited for thousands of years most simply by hanging bruised plants in houses or by burning in fire after preparing a formulation with animal dung (neem leaves are the best example of natural repellent). All over the world, there is now a common knowledge of using these plant-based "natural" smelling repellents because plants are perceived as a safe and trusted means of mosquito bite prevention (Dam et al. 2000a, b).

1.4.1 Some of the Best Known Plants for Mosquito Repellency

1.4.1.1 Eucalyptus

Corymbia citriodora (Myrtaceae), also known as lemon eucalyptus plant, is known for many centuries to have insect repellency properties. Lemon eucalyptus essential oil, comprising 85 % citronellal, is far more effective at repelling mosquitoes for several minutes. However one of its ingredients, para-menthane-3,8-diol, which has a lower vapour pressure than volatile monoterpenes found in most plant oils, provides very high protection from a broad range of insect vectors over several hours. Nanotechnology has of course recently opened new vistas in using effectively the eucalyptus extracts (Sugumar et al. 2014).

1.4.1.2 Citronella

Best known as lemongrass, the essential oils and extracts belonging to plants in the citronella genus (Family:Poaceae) are commonly used as ingredients of plant-based mosquito repellents mainly *Cymbopogon citratus*, *C. nardus*, *C. schoenanthus*, *C. winterianus* and *C. jwarancusa.* Citronella, which mainly contains citronellal, citronellol, geraniol, citral, α pinene and limonene, was used by the Indian Army to repel mosquitoes at the beginning of the twentieth century. Today, citronella is one of the most widely used natural repellents on the market, used at concentrations of 5–10 %. Citronella-based repellents only protect from host-seeking mosquitoes for about 2 h although formulation of the repellent is very important. Recently, the use of nanotechnology, particularly allowing vanillin (5 %) with the essential oil, has allowed higher protection rate with slower release rates of oils. Encapsulated citronella oil nanoemulsion is prepared by high-pressure homogenization of 2.5 % surfactant and 100 % glycerol, to create stable droplets that increase the retention of the oil and slow down release.

1.4.1.3 Neem

Neem is widely appreciated globally for its manifold applications, one of which being the repellency against pests of various kinds including hematophagous insects such as mosquitoes. Several field studies from India have shown very high efficacy of neem-based preparations. It is just unfortunate that it could not be yet exploited fully for its unmatched and wonderful repellency characteristics, albeit considered as a natural alternative to DEET.

1.4.2 Promising Developments in Plant-Based Repellents

Plants-based repellents have a great future, as these are considered much safer and eco-friendly than even the best known synthetic preparations. The field of plant-based repellents is moving forward as consumers demand means of protection from arthropod bites that are safe, pleasant to use and environmentally sustainable. Perhaps the most important consideration is improving the longevity of those repellents that are effective but volatile such as citronella. Several studies looked at improving formulations of plant oils to increase their longevity through development of nano-emulsions, improved formulations and fixatives, while alternate uses such as spatial activity and excito-repellency have also been investigated. *Eucalyptus*, *Citronella* and *Cymbopogon* species have shown huge potential in offering molecules that can be manipulated into highly effective repellents. New developments have also been seen in understanding the function of plant-based repellents in insects.

1.5 Conclusion

Major arthropod vector-borne diseases, particularly malaria, dengue, Japanese encephalitis, African trypanosomiasis, Chagas disease, schistosomiasis, filariasis, etc. are life threatening and infect billions of people throughout the world where the children and the poor are highly susceptible to infection. Even though conventional methodologies and techniques are successful in certain ecosystems and settings, mostly the vectors remain unaffected largely due to their potential to quickly develop resistance against the insecticides in vogue! There is therefore an urgent need to search for alternative products which function through different lethal mechanisms on the vector. Such new systems should innovatively integrate with the IVM since there is no panacea for the whole problem of vector control or the vector-borne disease control.

References

Chan KL (1972) The eradication of Aedes aegypti at the Singapore Paya Lebar international airport. In: Chan Yow-Cheo, Chan Kai-Lok, HoBeng-Chuan (eds) Vector control in Southeast Asia. Proceedings of the first SEAMEO-TROPMED workshop, Singapore, pp 85–88

Dam PK, Ramnath T, Yadav SP et al (2000a) Use of ethnomedicinal practices in the Thar desert, north western India. J Econ Taxon Bot 24:303–318

Dam PK, Yadav SP, Tyagi BK (2000b) Thar Marukschetra mein aushadhiukta vanaspatiyan: ek purvalokana (in Hindi) [The medicinal plants of Thar Desert region: a review]. ICMR Patrika 14(8):61–65

Gould EA, Solomon T (2008) Pathogenic flaviviruses. Lancet 371:500–509

Gubler DJ (1998) Dengue and dengue hemorrhagic fever. Clin Microbiol Rev 11:480–496

Hemingway J, Ranson H (2000) Insecticide resistance in insect vectors of human disease. Annu Rev Entomol 45:371–391

Kumar A, Valecha N, Jain T, Dash AP (2007) Burden of malaria in India: retrospective and prospective view. Am J Trop Med Hyg 77:69–78

Lines J, Addington W (2001) Insecticide-treated nets in Tanzania. Lancet 358:671

Lobo DA, Velayudhan R, Chatterjee P et al (2011) The neglected tropical diseases of India and South Asia: review of heir prevalence, distribution, and control or elimination. PLoS Negl Trop Dis 5:10

Lok CK, Kiat NS, Koh TK (1977) An autocidal ovitrap for the control and possible eradication of Aedes aegypti. Southeast Asian J Trop Med Public Health 8:56–62

Masum H, Shah R, Schroeder K et al (2010) Africa's largest long-lasting insecticide-treated net producer: lessons from A to Z textiles. BMC Int Health Hum Rights 10(Suppl 1):S6

Monath TP (1994) Dengue: the risk to developed and developing countries. Proc Natl Acad Sci U S A 91:2395–2400

National Commission on Macroeconomics and Health (2005) Burden of disease in India. Ministry of Health and Family welfare, Government of India, New Delhi

Patz JA, Campbell-Lendrum D, Holloway T et al (2005) Impact of regional climate change on human health. Nature 438:310–317

Perich MJ, Kardec A, Braga IA et al (2003) Field evaluation of a lethal ovitrap against dengue vectors in Brazil. Med Vet Entomol 17:205–210

Peters D, Yazbeck A, Ramana G et al (2001) Raising the sights: better health systems for India's poor. World Bank, Washington, DC

Pichersky E, Gershenzon J (2002) The formation and function of plant volatiles: perfumes for pollinator attraction and defense. Curr Opin Plant Biol 5:237–243. doi:10.1016/S1369-5266(02)00251-0

Shahi AK, Tyagi BK, Chaudhary RC, Kaul BK (2000) Cymbopogon schoenanthus (L.) Spreng. (Poaceae) – a versatile aromatic and medicinal plant of Indian Thar Desert. Flavours Fragr Assoc India J 1:31–33

Sugumar S, Clarke SK, Nirmala MJ et al (2014) Nanoemulsion of eucalyptus oil and its larvicidal activity against Culex quinquefasciatus. Bull Entomol Res 104(3):393–402. doi:10.1017/S0007485313000710

Tyagi BK (2002a). Future of phytochemicals against vector mosquitoes. Pap. Abstr. VI AZRA Congr. on Adv. in Applied Zoological Researches for Food Production and Environmental Safety, Cuttack, 19–21 Dec 2002

Tyagi BK (2002b) Future of phytochemicals against vector mosquitoes. J Appl Zool Res 14: 101–106

Tyagi BK (2003a) A review of plant-based materials active against vector mosquitoes with particular emphasis on repellent potential of Cymbopogon (Poaceae). J Econ Taxon Bot 28: 489–493

Tyagi BK (2003b) Cymbopogon essential oils: their chemistry and mosquito repellent activities. Pap. Abstr. Natl. Sym. on biochemical science health & environmental aspects, Agra, pp 140–142

Tyagi BK, Shahi AK (2001) Evaluation of repellent potential of two cymbopogon species (Poaceae), endemic to the Thar Desert (India), against major vector mosquitoes. International conference globalization of Indian essential oils, Agra, 21–23 Sep 2001

Tyagi BK, Shahi AK (2002) Genus cymbopogon (Poaceae) – a potent group of mosquito repellent bearing characteristics. Pap. Abstr. XVI Natl. Congr. Parasitology, Bareilly, 31 Oct, 2 Nov 2002, p 91

WHO (2012) Country cooperation strategy at a glance – India. World Health Organisation, Geneva

WHO (2014a) A global brief on vector-borne diseases. World Health Organisation, Geneva

WHO (2014b) Vector borne diseases. World Health Organisation, Geneva Available via DIALOG. http://www.who.int/mediacentre/factsheets/fs387/en/. Accessed 31 Aug 2014

World Bank (2010) World Bank South Asia economic update 2010: moving up, looking east: World Bank. Washington, DC. Available via DIALOG. http://siteresources.worldbank.org/ SOUTHASIAEXT/Resources/223546-1269620455636/6907265-1275784425763/SAREconomicUpdate7June2010.pdf. Accessed 31 Aug 2014

Zeichner BC, Perich MJ (1999) Laboratory testing of a lethal ovitrap for Aedes aegypti. Med Vet Entomol 13:234–238

2 Sustainable and Novel Eco-friendly Approaches Towards Integrated Disease and Vector Management

Jubilee Purkayastha, Rajesh Arora, and Lokendra Singh

Abstract

Herbal medicine has been used since time immemorial for treating a large number of human ailments. Despite the widespread use of western system of medicine, in recent years, most people even in the western world resort to herbal medicines as holistic, inexpensive treatments allude the common man. In addition, western medicine does not offer solutions to chronic incurable diseases and other such problems that confront humanity. Most patients undergoing cancer therapy or treatment for AIDS, rheumatoid arthritis, etc. use herbal medicines in the hope of a cure irrespective of the recommendation by the physician. Many pesticides are known to create significant health risks like birth defects, nervous system breakdown and cancer. The only way out is the use of herbal biomedicines and herbal pesticides.

2.1 Introduction

Presently, there is a paradigm shift from the use of synthetic drugs to the acceptability of herbal biomedicines. This change of trend is due to the realization of harmful effects of synthetic drugs and the searching for alternative ways of cure. The market for herbal biomedicine for human and animal consumption is growing. Similarly, the demand for herbal insecticides is on the rise. Worldwide, herbal biomedicines have an estimated annual growth rate between 5 % and 15 %, and the global total herbal drug market is about US $62 billion. In India, the value of herbal product trade is estimated to be approximately US $10 billion per annum and US $1.1 billion is the annual export value, while for China, it is US $48 billion with an export value of US $3.6 billion. Apart from India, currently, the United States is the largest market for Indian herbal products accounting for about 50 % of the total exports. Traditional Chinese medicine (TCM) and herbal medicines from Japan, Hong Kong, Korea and Singapore find extensive use in the United States (Citarasu 2010).

J. Purkayastha (✉) • R. Arora
Office of the Director General-Life Sciences, Defence Research and Development Organization, Rajaji Marg, New Delhi 110 011, India
e-mail: purkayasthaj@gmail.com

L. Singh
Defence Research and Development Establishment, Gwalior 474 002, India

© Springer India 2016
Vijay Veer, R. Gopalakrishnan (eds.), *Herbal Insecticides, Repellents and Biomedicines: Effectiveness and Commercialization*, DOI 10.1007/978-81-322-2704-5_2

At the global level, there is a shifting from antibiotics over to natural products due to the development of multidrug resistance in pathogens. Herbal medicine is comparatively low in cost, although it is becoming an advanced way of traditional phytotherapy. Plants are the potential store of safer and cheaper chemicals and cure various diseases due to their active principles having antioxidant and anti-microbiological activities. The occurrence of active principles like alkaloids, flavonoids, pigments, phenolics, terpenoids, steroids and essential oils in plant products is reported to promote various activities like antistress, growth promotion, appetite stimulation, tonic and immunostimulation, aphrodisiac and antimicrobial properties (Citarasu 2010).

In addition, serious environmental problems have been encountered as a result of intensive application of pesticides in modern agriculture (Dua et al. 1998; Waliszewski et al. 1999; Singh et al. 2004). Some of the pesticides are highly persistent compounds and pose immense environmental concerns, and some examples are pentachlorophenol (PCPs), polychlorinated biphenyls (PCBs), atrazine (S-triazines), organochlorines (OCs), organophosphates (OPs) and carbamates. The use of these pesticides in agriculture contaminates the adjoining freshwater bodies posing threat to potential freshwater organisms especially important animals, fishes and prawns (Saravanan et al. 2003; Selvarani and Rajamanickam 2003; Park et al. 2004). Recently, the uses of herbal products have got unparalleled momentum throughout the world. Herbal biomedicines are increasingly finding applications for prophylactic and therapeutic purposes, often even for diseases for which treatment modalities are unavailable or ineffective.

On the other hand, the problems associated with the use of synthetic pesticides and insecticides are manifold. Many pesticides are known to create significant, known health risks to people, including birth defects, damage to the nervous system, disruption of hormones and endocrine systems, respiratory disorders, skin and eye irritations and various types of cancer. Also, exposure to persistent organic pollutants through diet has been known to cause breast and other types of cancer, immune system suppression, nervous system disorder, reproductive damage and disruption of hormonal systems (Kristin 2000). An alternate strategy to overcome the problems associated with the use of synthetic insecticides and pesticides is the use of green herbal insecticides for integrated management and vector control.

2.2 Herbal Biomedicines

An overview of herbal biomedicines as antibacterial, antiviral, antifungal and growth-enhancing agents and as immunostimulant, antistress and appetizing agents as per Citarasu (2010) is provided below.

2.2.1 Antibacterial Agents

The herbal compounds responsible for antibacterial activities include mainly phenolics, polysaccharides, proteoglycans and flavonoids present in plants. The antibacterial active principles of the herbal products may act by breaking the cell wall, by blocking DNA and protein synthesis and by inhibiting the enzyme synthesis and then interfering the cellular signalling mechanism. Also, it has been proposed that the means of the antimicrobial effects engage the inhibition of various cellular processes, followed by an increase in plasma membrane permeability and finally ion leakage from the cells. The antagonistic effects of the methanolic extracts of *Coleus aromaticus* of Lamiaceae and *Tabernaemontana divaricata* of Apocynaceae family were found to be most effective against the fish pathogen *Aeromonas hydrophila*. *Ocimum basilicum* extracts, specially ethanol, methanol and hexane extracts, have shown excellent in vitro antimicrobial properties. Another example is of Indian almond, *Terminalia catappa*, where the extract at a concentration of 0.5 mg ml^{-1} has shown the growth inhibition of two strains of *A. hydrophila* (Citarasu 2010).

2.2.2 Antiviral Agents

Many herbs have potent antiviral activities. Oleuropein derived from olive tree leaf (*Olea europaea*) was found to control salmonid rhabdovirus. White spot syndrome virus (WSSV) was effectively suppressed by the methanolic extracts of the herbs *Acalypha indica*, *Cynodon dactylon*, *Picrorhiza kurroa*, *Withania somnifera* and *Zingiber officinalis*. Strong antiviral activity against WSSV was shown by 20 species of Indian traditional medicinal plants such as *Aegle marmelos*, *C. dactylon*, *Lantana camara*, *Momordica charantia and Phyllanthus amarus* in the form of petroleum ether, benzene, diethyl ether, chloroform, ethyl acetate, methanol and ethanol extraction. The mode of action of the herbal active compounds includes inhibition or blocking of the transcription rate of the virus and thereby reducing the replication rate in the host cells (Citarasu 2010).

2.2.3 Antifungal Agents

Fungi are increasingly becoming resistant to conventional drugs, and the limited numbers of antifungals currently used are showing side effects. The fungal infections, whether superficial or systemic, often pose considerable management problems and are still a major concern. Hence, the substitute 'herbal formulations' have become an option of improved interest. Many different plant extracts have been tested for in vitro antifungal activity. For example, the plant *Datura metel* L. is a source of novel antifungal molecule, 2-(3,4-dimethyl-2,5-dihydro-1H-pyrrol-2-yl)-1-methylethyl pentanoate. It is reported to show anti-*Aspergillus* properties and also possess activities against ten clinical isolates of Candida and 19 clinical isolates of *Aspergillus* and also against a few marine fungi. *O. basilicum* extracts were effective in controlling the pathogens *Aspergillus flavus* and *Fusarium oxysporum* (Citarasu 2010). Herbal-based fungicidal products like coconut diethanolamide, *Azadirachta siamensis* and *Melaleuca alternifloria* along with a amalgamation of synthetic fungicide showed prominent fungicidal activities against *Aphanomyces invadans*. The mode of action of the herbal antifungal agent is by altering cell permeability, cell wall lysis, affecting metabolism, and RNA and protein synthesis (Citarasu 2010).

2.2.4 Growth-Enhancing Agents

Many herbal products like those based on *Hygrophila spinosa*, *Withania somnifera*, *Zingiber officinalis*, *Solanum trilobatum*, *Andrographis paniculata*, *Psoralea corylifolia*, *Eclipta erecta*, *Ocimum sanctum*, *Picrorhiza kurroa*, *Phyllanthus niruri*, *Tinospora cordifolia*, etc. have the potential of growth promotion, antistress activities, immunostimulation and antibacterial properties. Livol (IHF-1000), a well-explored commercial herbal growth promoter, drastically improved digestion, and the dietary ginseng herb greatly improved the growth performance, diet consumption efficiency and also haematological indices. Stressol-I and stressol-II, the herbal products enriched with *Artemia nauplii*, increased the growth rate significantly and reduced the osmotic stress in *Penaeus indicus*. Similarly, the herbal product 'Tefroli' improved the growth rate and moulting competence when fed to *Penaeus monodon* postlarvae and 'Trasina', a well-known herbal product, improved the growth and stress tolerance efficiencies radically (Citarasu 2010).

2.2.5 Immunostimulant, Antistress and Appetizing Agents

The methanolic extracts of five herbal medicinal plants, viz., *Cynodon dactylon*, *Aegle marmelos*, *Tinospora cordifolia*, *P. kurroa and Eclipta alba*, showed better performance of haematological, biochemical and immunological parameters when fed to shrimps. Extracts of the herbs *O. sanctum*, *W. somnifera* and *Myristica fragrans* were efficient in improving the immune parameters like bactericidal activity, phagocytic activity and serum albumin–globulin (A/G) ratio against *Vibrio harveyi*. The immunostimulants obtained from herbs like *Emblica officinalis*, *Cynodon dactylon* and *Adhatoda vasica* found to improve the immune system and reduce microbial contamination in the goldfish *Carassius auratus*. Dietary intake of *O. sanctum* also enhanced the immunostimulatory effects, antibody response and disease resistance. Herbs like *Tinospora cordifolia* and *Picrorhiza kurroa* have been used as an antistress agent and are known to have the ability to scavenge free radicals, and the mode of action is similar to superoxide dismutase and metal ion chelators. Herbal drugs also act as appetite stimulators as they improve the performance by increasing gut secretions. Intake of some hot spices increases the salivation and, hence, improves digestibility by increasing enzyme production (Citarasu 2010).

2.3 Herbal Insecticides

Herbal insecticides have an unbeaten track record in insect pest control. However, as of now herbal insecticides are of a small part of the whole insecticide/pesticide industry as large part is covered by synthetic ones. Recently, public concern about the use of synthetic insecticides has grown dramatically resulting in organic agriculture, restricting synthetics and following herbal insect pest control. In many parts of Europe and North America, the use of synthetic pesticides are now excluded in urban areas favouring unconventional control measures such as biopesticides, biocontrol and other methodologies.

Some of the advantages of plant-derived pesticides are that they are selectively toxic, do not bioaccumulate and exhibit relatively short persistence in the environment (Shanker and Solanki 2000) and food chain. There are constant attempts going on for searching new classes of insecticides derived from plants having low toxicity and less persistence in nature (Singh et al. 1996; Kloos and Mc Cullough 1987) like alpha-terthienyl (Nivsarkar et al. 2001). Many plant products like alkaloids, vegetable oils, triterpenoids, rotenone and azadirachtin have been used in the development of alternative pest control agents (Shanker and Solanki 2000). The reasons that have prompted the use of plant products and their encouragement by different countries are because of their properties, like high pesticidal activities, fast availability, easy biodegradability with less danger of environmental contamination, little or no mammalian toxicity, solubility in water, low cost, etc.

The long history of safety of the natural products has provided a lot of confidence for their further exploration. For example, many new botanical pesticides, such as piperamides and alpha-terthienyl, show that they are degraded in the environment in a couple of hours or days. Phytochemicals in botanical extracts are diverse, and the presence of many analogues of one compound increases the efficacy of plant extracts through analogous cooperation and the evolution of pesticide resistance under selection pressure in course of time. Moreover, there is immense scope of deriving many more new novel compounds having insect repellent or killing properties as huge diversity of novel phytocompounds is available. The process of coadaptation is never ending across the plant kingdom for defence against insect pests and diseases, and therefore, plants possibly synthesize a number of such compounds (Arnason et al. 2015).

There are reports of plant-derived compounds as insect behaviour modifying antifeedants, essential oils as repellent as well as insecticides and many more compounds with novel modes of action. Despite these developments, the herbal pesticide market has not as grown as much as for synthetics, even today. The main reasons behind this are the barriers of costs involved in toxicology testing for new herbal products, cost-effective supply of plant product, quality control and lack of stability, competition from other biopesticides and biocontrol agents (Arnason et al. 2015).

Plants, the richest source of renewable bioactive organic compounds, produce around 40,000 compounds, of which around 10,000 are secondary metabolites and responsible for defensive activities (Cooper and Johnson 1984). Many plant-derived defensive chemicals of various categories causing behavioural and physiological changes in pests have been already identified and explored. A few of these important bio-compounds (Singh et al. 2010; Bernhoft 2010) are described below:

(a) *Alkaloids*: Alkaloids are naturally occurring organic bases with at least one nitrogen atom either in the heterocyclic ring or linked to an aliphatic skeleton. These are colourless, crystalline, nonvolatile solids, with little solubility in water but having good solubility in ethanol, ether and chloroform.
(b) *Glycosides*: Various types of secondary metabolites joining to a mono- or oligosac-

charide or to uronic acid make the glycosides, which are composed of two parts called the glycone and aglycone. The major groups of glycosides are cardiac glycosides, cyanogenic glycosides, glucosinolates, saponins and anthraquinone glycosides. Furthermore, flavonoids frequently occur as glycosides.

(c) *Flavonoids*: Flavonoids are C15 compounds (exclusive of O-alkyl groups and secondary substituents), composed of two phenolic nuclei connected by three carbon units.

(d) *Saponins*: Saponins are naturally occurring plant glycosides consisting of a sugar moiety and an aglycone unit. The saponins having toxic activity are the monodesmosidic saponins (sugar moiety only at position C-3); on the other hand, biodesmosidic saponins (sugar moiety both at C-3 and C-28) are inactive.

(e) *Tannins*: Tannins are complex phenolic compounds, which can be divided into two groups, (i) the esters of gallic acid and also glycosides of these esters known as the hydrolysable tannins and (ii) polymers derived from various flavonoids, the condensed tannins.

(f) *Resins*: Resins are complex lipid-soluble mixtures consisting of diterpenoid, triterpenoid, monoterpenoid and sesquiterpenoid. Resins are mostly present in woody plants, but resins to some extent are also present in herbaceous plants. Most resins are having properties like antimicrobial and wound healing, and they are generally safe in handling, but reports of contact allergy are there.

(g) *Lignans*: Lignans are lipophilic, are composed of two phenylpropanoid units forming an 18-carbon skeleton, have various functions related to plant cell membranes and are present in highly concentrated oil seeds.

(h) *Iridoids*: Iridoids are the monocyclic monoterpenoids, which possess a lactone ring as a replacement of the ρ-menthane skeleton.

(i) *Furanocoumarins*: Furanocoumarins are those compounds in which the 1,2-benzopyrene skeleton is fused with a furan ring.

(j) *Diterpenoids*: Diterpenoids have a general molecular formula of C20 H32, are not steam volatile and are usually obtained from plants. A new class of diterpenes which are esters of phorbol possess high toxicity against pests.

(k) *Monoterpenoids*: Monoterpenoids are the major constituents of essential oils and are made up of two isoprene units (C_{10} H_{16}).

(l) *Sesquiterpene lactones*: Sesquiterpenoid lactones possess a sesquiterpene skeleton having an additional lactone ring.

(m) *Proteins and peptides*: There are proteins and peptides with bioactivity which are not hydrolysed in the digestive tract and exert their specific action, for instance, ricin in seeds of *Ricinus communis* (castor bean).

2.4 Conventionally Used Herbal Insecticides

Some of the conventionally used herbal insecticides/pesticides include nicotine, rotenone, ryania, sabadilla and pyrethrum. Nicotine (Fig. 2.1), an alkaloid obtained from the leaves of tobacco plants (*Nicotiana tabacum*), is an effective insecticide/pesticide. The insecticide resulting from extracts of the tropical legumes *Derris* and *Lonchocarpus* is named as rotenone (Isman 2006). The main active principle, the isoflavonoid rotenone (Fig. 2.1), is extremely poisonous to insects and fish, due to its fast uptake and inhibitory activity of respiratory electron transport at site 1 (Arnason et al. 2015).

The extract from the South American shrub *Ryania* sp. is known as ryania and contains the diterpene alkaloid ryanodine (Fig. 2.1). It is a contact and stomach poison against horticultural and ornamental crop pests that acts by blocking Ca^{++} ion channels. The seed extract of the neotropical lily *Schoenocaulon officinale* is known as sabadilla. It contains veratridine alkaloids having a neurotoxic mode of action. The extract is a valuable contact insecticide against a number of agricultural insect pests, viz., lepidoptera, leafhoppers and thrips. However, the market for these products is relatively small (Arnason et al. 2015).

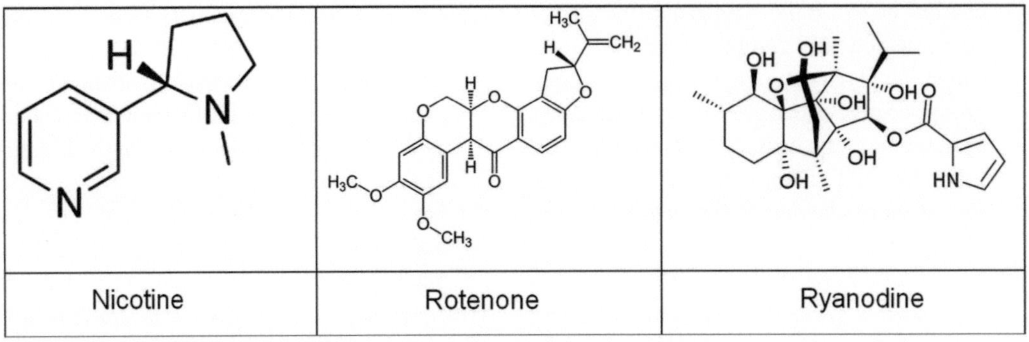

Fig. 2.1 Conventional herbal insecticides/pesticides (Source: Wikipedia 2015)

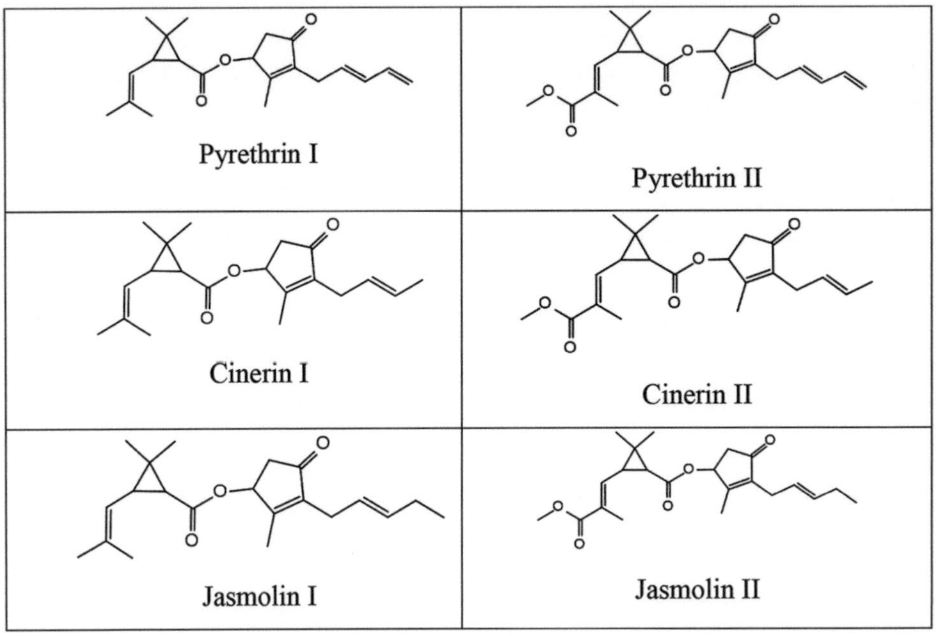

Fig. 2.2 Important conventional herbal insecticide, pyrethrins from *Chrysanthemum cinerariifolium* (Source: Wikipedia 2015)

The most important traditional herbal insecticide is pyrethrum. It is derived from African daisy, *Chrysanthemum cinerariifolium.* The plant produces an insecticidal oleoresin which is extracted in organic phase. The pyrethrum extract contains six major pyrethrin compounds: pyrethrins I and II, jasmolins I and II and cinerins I and II (Fig. 2.2).

Pyrethrin is specially known for its quick knockdown of flying insects as well as a wide variety of home and garden pest insects. It acts via the insect nervous system at the Na^+ channels and has low mammalian toxicity, but does cause significant toxicity to fish and aquatic invertebrates. However, due to the growth of more stable and active synthetic pyrethroids, the market for natural pyrethrum dropped. Now natural pyrethrin is showing more application due to its application in natural pest control methods (Arnason et al. 2015).

Some of the important plants, products and their biological activities are enumerated in Table 2.1.

Table 2.1 Some important plants, active components and their biological activities (Source: Pino et al. 2013 supplemented and modified by the authors)

Plant	Product	Bioactive component(s)	Biological activity	Mode of action	References
Capsicum spp. (*Capsicum frutescens* Mill.)	Capsicum oleoresin	Capsaicin	Repellent, fungicide, nematicide, bactericide	Neurotoxic, repellent	Copping and Duke (2007), Dayan et al. (2009)
Syzygium aromaticum, *Eugenia caryophyllus* Spreng	Clove essential oil	Eugenol (mixture of several predominantly terpenoid compounds)	Insecticide, herbicide	Neurotoxic, interference with the neuromodulator octopamine	Isman (2006), Copping and Duke (2007), Dayan et al. (2009), Fischer et al. (2013), Isman and Machial (2006)
Thymus vulgaris L., *Thymus* spp.	Thyme essential oil	Thymol, carvacrol	Insecticide, fungicide, herbicide	Neurotoxic, interference with GABA-gated chloride channels	Copping and Duke (2007), Dayan et al. (2009), Fischer et al. (2013)
Rosmarinus officinalis	Rosemary essential oil	1,8-cineole (borneol, camphor, monoterpenoids)	Insecticide, acaricide, fungicide	Octopamine antagonist; membrane disruptors, others	Dayan et al. (2009), Fischer et al. (2013), Isman and Machial (2006)
Cinnamomum zeylanicum	Cinnamon essential oil	Cinnamaldehyde	Insecticide, herbicide	Octopamine antagonists; membrane disruptors, others	Dayan et al. (2009), Fischer et al. (2013)
Cymbopogon nardus, *Cymbopogon citratus* Stapf., *Cymbopogon flexuosus* DC	Lemon grass essential oil	Citronellal, citral	Insecticide, herbicide	Octopamine antagonists; membrane disruptors, others	Dayan et al. (2009), Fischer et al. (2013)
Mentha species (mint)	Mint essential oil	Menthol	Insecticide	Octopamine antagonists; membrane disruptors, others	Dayan et al. (2009), Fischer et al. (2013)

(continued)

Table 2.1 (continued)

Plant	Product	Bioactive component(s)	Biological activity	Mode of action	References
Cassia tora L., *Cassia obtusifolia*	Cinnamaldehyde	Cinnamaldehyde	Fungicide, insect attractant	Disruption of the fungal membranes, repellent and attractant	Copping and Duke (2007), Dayan et al. (2009)
Reynoutria sachalinensis (Fr. Schm.) Nakai	Extract of giant knotweed	Physcion, emodin	Fungicide, bactericide	Induction of SAR (phenolic phytoalexins)	Dayan et al. (2009), Regnault-Roger (2012)
Macleaya cordata R. Br.	Pink plume poppy extract	Alkaloids, sanguinarine chloride and chelerythrine chloride	Fungicide	Induction of SAR (phenolic phytoalexins)	Dayan et al. (2009), Regnault-Roger (2012)
Trigonella foenum-graecum L.	Stifenia	Alkaloids	Fungicide	Stimulation of plant defence	Regnault-Roger (2012)
Derris indica (Lam.) Bennet	Karanjin	Karanjin	Insecticide, acaricide	Antifeedant/repellent, insect growth regulator	Copping and Duke (2007)
Mentha piperita L.	Phenethyl propionate	Phenethyl propionate	Insecticide, insect repellent, herbicide	Repellent	Isman (2006), Copping and Duke (2007), Dayan et al. (2009)
Simmondsia californica Nutt., *S. chinensis* Link.	Jojoba essential oil	Straight-chain wax esters	Fungicide, insecticide	B suffocation (eggs and immature life stages), repellent, blocking access to oxygen	Copping and Duke (2007), Dayan et al. (2009)
Tanacetum cinerariifolium (Trevisan) Schultz-Bip	Pyrethrum	Ester of chrysanthemic acid and permethric acid (pyrethrins I and II, cinerins I and II jasmolins I and II)	Insecticide, acaricide	Axonic poisons (sodium channels agonists)	Isman (2006), Copping and Duke (2007), Isman and Paluch (2011), Dayan et al. (2009)

Azadirachta indica A. Juss	Neem (neem oil, medium polarity extracts)	Azadirachtin, dihydroazadirachtin, variety of triterpenoids (nimbin, salannin and others)	Insecticide acaricide, fungicide	Moulting inhibitors (ecdysone antagonists), antifeedant/repellent, physical smothering and desiccation	Isman (2006), Copping and Duke (2007), Isman and Paluch (2011), Dayan et al. (2009)
Derris, *Lonchocarpus* and *Tephrosia* species	Rotenone	Rotenone, deguelin (isoflavonoids)	Insecticide, acaricide	Mitochondrial cytotoxin	Isman (2006), Copping and Duke (2007), Isman and Paluch (2011)
Nicotiana spp.	Nicotine	(S)-isomer, (RS)-isomers and (S)-isomer of nicotine sulphate	Insecticide	Neurotoxin (acetylcholine agonist)	Isman (2006), Copping and Duke (2007), Isman and Paluch (2011), Dayan et al. (2009)
Ryania spp. (*Ryania speciosa Vahl*)	Ryania	Ryanodine, ryania, 9,21-didehydroryanodine (alkaloids)	Insecticide	Neuromuscular poison (calcium channel agonist)	Isman (2006), Copping and Duke (2007), Isman and Paluch (2011), Dayan et al. (2009)
Schoenocaulon spp. (*Schoenocaulon officinale* Grey)	Sabadilla	Mixture of alkaloids (cevadine, veratridine)	Insecticide	Axonic poisons (sodium channels agonists, heart and skeletal muscle cell membranes)	Isman (2006), Copping and Duke (2007), Isman and Paluch (2011), Dayan et al. (2009)
Quassia, Aeschrion, Picrasma	Quassia	Quassin (triterpene lactone)	Insecticide	Unknown	Isman (2006), Isman and Paluch (2011)

2.5 Future Areas of Focus

2.5.1 Photodynamic Killing of Insects

The first scientific documentation showing toxicity of sunlight to biological system was provided by Marcacci (1888) by reporting that the fermentation of plant alkaloids as well as amphibian eggs becomes more important under UV/visible light than in the dark. Many phytochemicals are documented as having photochemically active substances, which can be lethally toxic to insects. The major phytotoxin-producing plant family is known to be the sunflower family, Asteraceae.

Moreover, many photosensitizing agents have been shown to be accumulated in significant amounts by a variety of insects when given in association with suitable attractants. The follow-up exposure of such insects to UV/visible light leads to mortality. Two very important photosynthesizers, known to have very high photoinsecticidal activity, are xanthenes (e.g. phloxin B) and porphyrins (e.g. haematoporphyrin). The very high photobleaching of xanthenes and porphyrins when illuminated by visible light, and no major toxicity of such compounds in the dark, makes these as ideal photoinsecticidal agents due to least environmental impact of such compounds (Amor and Jori 2000).

2.5.2 Genetically Engineered Plants for Pest Management

Genes conferring resistance to insect pests have been inserted into several crops, e.g. *cry* genes from *Bacillus thuringiensis* (Bt) have shown considerable potential for pest management especially in crops like cotton and maize (Sharma et al. 2007). The Bt-transgenic crops have been shown to have better yield as well as insect-resistant trait, but detailed in-depth studies regarding their effect on nontarget species need to be evaluated in detail. Moreover, a number of ecological and economic issues need to be considered while considering the use of transgenic crops for pest management (Sharma and Ortiz 2000). Also, the regulatory issues are of major concern in this regard. Currently, transgenic plants for crop protection have been developed using genes encoding for protease inhibitors, plant lectins, secondary plant metabolites, vegetative insecticidal proteins and RNAi technology.

2.5.3 LLIN with Herbal Insecticides

Long-lasting insecticidal nets (LLINs) are impregnated with synthetic pyrethroids and are wash proof till 20 times. Although pyrethroid use are considered to be very safe for human beings, LLINs with herbal insecticides would be more acceptable, keeping in mind the eco-friendliness and almost negligible toxicity. The LLIN developed by Defence Research and Development Organisation, India, known as ‘Defender Net’, is in very high demand; however, if instead of synthetic pyrethroids, herbal insecticides are used, its user-friendliness would increase manifold. Several essential oils like patchouli, together with citronella, clove and makaen, have been shown to have mosquito repellent activities, which could be of immense use if explored for using in LLINs. Endeavours in this direction need to be focussed.

2.5.4 Herbal Therapeutics for Vector-Borne Diseases

Plant products, especially essential oils, are a potential source of mosquito control due to their insecticidal properties (Benner 1993). Essential oils are naturally occurring, volatile, complex compounds with strong odour and are present in aromatic plants as secondary metabolites (Madhumathy et al. 2007). Amongst higher plants, about 17,500 aromatic plant species and about 3,000 essential oils are known, of which 300 are commercially important for pharmaceuticals, cosmetics and perfume industries and also having insecticide/pesticide potential (Palanisami et al. 2014). Some of the essential oils constitute even better alternatives or complementary to synthetic compounds for vector control without showing the same secondary effects. A better understanding of their mode of action would help in exploring their further potential in controlling vector-borne diseases as well as application in human health, agriculture and the environment (Karmegan et al. 1997).

2.5.5 Biolarvicides

Mosquito larvae can be controlled by bacteria such as *Bacillus thuringiensis* var. *israelensis* (*Bti*) and *Bacillus sphaericus* which act as biolarvicides. *Bacillus thuringiensis* var. *israelensis* (*Bti*) is an aerobic spore-forming entomopathogenic bacterium effective against mosquitoes and black flies. These are highly effective against mosquito larvae at very low doses; safe to other nontarget organisms, environment, man and wild life and also appropriate for community use. These biolarvicides are regarded as the most promising microbial control agent against mosquitoes and black flies for use in integrated vector control programme. Similarly, *B. sphaericus* isolated first time from United State in 1965 are also reported to be highly insecticidal against mosquito larvae (Mittal 2003).

2.5.6 Herbal Repellents

Herbs with repellent properties are known to play an important role by minimizing man-vector contact. Nowadays natural insect repellent products that are preferred as chemical repellents are considered not safe for public use and have unpleasant smell, oily feeling to some users and potential toxicity. Repellents of herbal origin do not show toxicity to human and domestic animals and are easily biodegradable. Also, natural products are safe for human use when compared to synthetic compounds (Das et al. 2003).

Certain plants, viz., *Zanthoxylum armatum* DC syn. *Z. alatum* Roxb. (Rutaceae), *Azadirachta indica* (Meliaceae) and *Curcuma aromatica* (Zingiberaceae), have been investigated for repellent activity against mosquitoes (Das et al. 2000); *Callistemon rigidus* (bottlebrush), *A. indica* (neem) and *Z. armatum* (timur) have been reported to have repellent activity against land leeches also (Nath et al. 2002). Likewise, repellent action of neem oil in the form of mats and neem cream has been evaluated. Essential oil obtained from *Vitex negundo* and *Lantana camara* flowers is active against *Aedes* mosquitoes (Das et al. 2003).

2.5.7 Nanoparticles of Herbs for Vector Control

The activity of herbal medicines is because of the active constituents present in them, which show synergistic action and thus enhance the curative value (Lu et al. 2007; Williamson 2001). However, most of the herbal constituents are poorly water soluble because of their hydrophobic nature, which leads to decreased bioavailability. Nanoparticles are utilized to increase the herbal drug solubility and for helping in localizing the drug specificity in the body. Nanoparticulate formulations such as polymeric nanoparticles, liposomes, proliposomes, solid lipid nanoparticles and microemulsions have the potential to deliver herbal medicines effectively (Thapa et al. 2013). Hence, insecticidal herbal drugs in the form of nanoparticles will be of immense use in controlling disease vector as in the form of nanoparticles; the efficiency of insecticidal drugs will be at its best.

2.6 Conclusion

Our country possesses a rich biodiversity of medicinal plants, which are used for various purposes. The active principles of the herbs have unique bioactivities, e.g. growth-promoting ability, modulating immune system, antimicrobial capability, appetizing and antistress activities. The major active principles of biomedicines include alkaloids, flavonoids, pigments, phenolics, terpenoids, starch, steroids and essential oils. The use of biomedicines will certainly reduce the use of synthetic compounds, reduce the cost and be an eco-friendly approach to solve the menace of pests.

Plants synthesize a number of secondary metabolites due to their co-evolution with insects, and these serve as defence chemicals against pest attack. Researchers have discovered a completely new paradigm for the control of vectors using secondary plant products, which may be toxic to specific vectors while harmless to nontarget organisms. More than 2,000 plant species of different families and genera have been reported to contain toxic principles effective against insects. There are many plant species containing compounds lethal to target as well as nontarget organisms, which are much below those for synthetic pyrethroids. Moreover, such products have the further advantage of biodegradability, as all such compounds are not only confined to the plants in which they are found but also get disseminated in the environment. An important issue about herbal insecticides/pesticides is that there are reports of toxicity. However, very little literature is available on their mode of action and effects on other organisms. But the hope is that they are safe, as these have been used by mankind for pest management without polluting the environment since ages.

It is anticipated that herbaceous products not only can control the plant/animal disease vectors, storage insects, predatory fishes and mosquitoes but are easily available and inexpensive and have easy biodegradability and greater acceptance amongst the users. It is apparent that with the rapid advancements in biotechnology, such valuable plant pesticide products can be sourced from industries using biotechnological methods also, other than only plant source, and thus in a plant and season-independent manner for sustainable vector management.

Acknowledgements The authors would like to thank the Defence Research and Development Organisation (DRDO) for funding and support. Special thanks are due to Dr. Manas K. Mandal, Director General, Life Sciences for his wholehearted support and guidance.

References

Amor TB, Jori G (2000) Sunlight-activated insecticides: historical background and mechanisms of phototoxic activity. Insect Biochem Mol Biol 30:915–925

Arnason JT, Sims SR, Scott IM (2015) Natural products from plants as insecticides. Available via DIALOG. http://www.eolss.net/sample-chapters/c06/e6-151-13.pdf. Accessed 26 Mar 2015

Benner JP (1993) Pesticidal compounds from higher plants. Pestic Sci 39:95–102

Bernhoft A (2010) A brief review on bioactive compounds in plants. In: Bioactive compounds in plants – benefits and risks for man and animals. The Norwegian Academy of Science and Letters, Oslo, pp 11–17

Citarasu T (2010) Herbal biomedicines: a new opportunity for aquaculture industry. Aquacult Int 18:403–414

Cooper MR, Johnson AW (1984) In reference book 161. Ministry of Agriculture, Fisheries and Food, London

Copping LG, Duke SO (2007) Natural products that have been used commercially as crop protection agents. Pest Manag Sci 63:524–554

Das NG, Nath DR, Baruah I, Talukdar PK, Das SC (2000) Field evaluation of herbal mosquito repellents. J Com Dis 31:241–245

Das NG, Baruah I, Talukdar PK, Das SC (2003) Evaluation of botanicals as repellents against mosquitoes. J Vect Borne Dis 40:49–53

Dayan FE, Cantrell CL, Duke SO (2009) Natural products in crop protection. Bioorg Med Chem 17:4022–4034

Dua VK, Pant CS, Sharma VP, Pathak GK (1998) Determination of HCH and DDT in finger-prick whole blood dried on filter papers and its field application for monitoring concentrations in blood. Bull Environ Contam Toxicol 56:50–57

Fischer D, Imholt C, Pelz HJ, Wink M, Prokopc A, Jacoba J (2013) The repelling effect of plant secondary metabolites on water voles, *Arvicola amphibious*. Pest Manag Sci 69:437–443

Isman MB (2006) Botanical insecticides, deterrents, and repellents in modern agriculture and an increasingly regulated world. Annu Rev Entomol 51:45–66

Isman MB, Machial CM (2006) Pesticides based on plant essential oils: from traditional practice to commercialization. In: Rai M, Carpinella MC (eds) Naturally occurring bioactive compounds. Elsevier, Amsterdam, pp 29–44

Isman MB, Paluch G (2011) Needles in the haystack: exploring chemical diversity of botanical insecticides. In: López O, Fernández-Bolaños J (eds) Green trends in insect control. Royal Society of Chemistry, Cambridge, pp 248–265

Karmegan N, Sakthivadivel M, Anuradha V, Daniel T (1997) Indigenous plant extracts as larvicidal agents against *Culex quinquefasciatus* Say. Bioresour Technol 5:137–140

Kloos H, Mc Cullough FS (1987) Plants with recognized molluscicidal activity. In: Mott KE (ed) Plant molluscicides. UNDP/World Bank/WHO, New York, pp 45–108

Kristin S (2000) Nowhere to hide: persistent toxic chemicals in the U.S. Food Supply. Pesticide Action Network North America, San Francisco

Lu C, Lu Y, Chen J, Zhang W, Wu W (2007) Synchronized and sustained release of multiple components in silymarin from erodible glyceryl monostearate matrix system. Eur J Pharm Biopharm 66:210–219

Madhumathy AP, Aivazi AA, Vijayan VA (2007) Larvicidal efficacy of *Capsicum annum* against *Anopheles stephensi* and *Culex quinquefasciatus*. J Vector Borne Dis 44:223–226

Marcacci A (1888) Sur l'action des alcaloides dans le regne vegetale et animal. Arch Ital Biol 9:2–4

Mittal PK (2003) Biolarvicides in vector control: challenges and prospects. J Vect Borne Dis 40:20–32

Nath DR, Das NG, Das SC (2002) Bio-repellents for land leeches. Def Sci J 52:73–76

Nivsarkar M, Cherian B, Padh H (2001) Alpha-terthienyl: a plant-derived new generation insecticide. Current Sci 81:25

Palanisami S, Natarajan E, Rajamma R (2014) Development of eco-friendly herbal mosquito repellent. J Innov Biol 1:132–136

Park D, Minor MD, Propper CR (2004) Toxic response of endosulfan to breeding and non-breeding female mosquito fish. J Environ Biol 25:119–124

Pino O, Sánchez Y, Rojas MM (2013) Plant secondary metabolites as an alternative in pest management. I: Background, research approaches and trends. Rev Protección Veg 28:81–94

Regnault-Roger C (2012) Trends for commercialisation of biocontrol agent (biopesticide) products. In: Mérillon JM, Ramawat KG (eds) Plant defence: biological control, vol 12, Progress in biological control. Springer, Dordrecht, pp 139–160

Saravanan TS, Mohamed MA, Chandrasekar R, Sundramoorthy M (2003) Freshwater fishes as indicators of Kaveri river pollution. J Environ Biol 24:381–389

Selvarani D, Rajamanickam C (2003) Toxicity of PCB 1232 on mitochondria of fish *Arius caelatus* (Valenciennes). Ind J Exp Biol 41:336–340

Shanker C, Solanki K (2000) Botanical insecticides: a historical perspective. Asi Agric Hist 4(3):221–232

Sharma HC, Ortiz R (2000) Transgenics, pest management, and the environment. Curr Sci 79:421–437

Sharma HC, Arora R, Pampapathy G (2007) Influence of transgenic cottons with *Bacillus thuringiensis* cry1Ac gene on the natural enemies of *Helicoverpa armigera*. BioControl 52:469–489

Singh K, Singh A, Singh DK (1996) Molluscicidal activity of neem (*Azadirachta indica*). J Ethanopharmacol 52:35–40

Singh SK, Tripathi PK, Yadav RP, Singh D, Singh A (2004) Toxicity of malathion and carbaryl pesticides: effects on some biochemical profiles of the freshwater fish *Colisa fasciatus*. Bull Environ Contam Toxicol 72:592–599

Singh SK, Yadav RP, Singh A (2010) Molluscicides from some common medicinal plants of eastern Uttar Pradesh, India. J Appl Toxicol 30:1–7

Thapa RK, Khan GM, Parajuli-Baral K, Thapa P (2013) Herbal medicine incorporated nanoparticles: advancements in herbal treatment. Asian J Biomed Pharmaceut Sci 3:7–14

Waliszewski SM, Aguirre AA, Benitez A, Infanzon RM, Infazon R, Rivera J (1999) Organo-pesticides residues in human blood serum of inhabitants of Veracruz, Mexico. Bull Environ Contam Toxicol 62:397–402

Wikipedia (2015) en.wikipedia.org. Accessed 26 Mar 2015

Williamson EM (2001) Synergy and other interactions in phytomedicines. Phytomedicine 8:401–409

Mosquito Larvicidal Potential of Medicinal Plants

3

M. Govindarajan

Abstract

Mosquitoes act as vectors for many life-threatening diseases like malaria, yellow fever, dengue fever, chikungunya, filariasis, encephalitis, West Nile virus infection, etc. Vector control is by far the most successful method for reducing the incidence of mosquito-borne diseases, but the emergence of widespread insecticide resistance and the potential environmental issues associated with some synthetic insecticides has indicated that additional approaches to control the proliferation of mosquito population would be an urgent priority research. Mosquitoes develop genetic resistance to synthetic insecticides and even to biopesticide such as *Bacillus sphaericus*. Also synthetic insecticides adversely affect the environment by contaminating air, water, and soil. There is an urgent need to find alternatives to the synthetic insecticides which are more potent and low cost. Plants are a rich source of alternative agents for control of mosquitoes, because they possess bioactive chemicals, which act against a limited number of species including specific target insects and are eco-friendly. Traditionally, plant-based products have been used in human communities for many centuries for managing insects. Several secondary metabolites present in plants serve as a defense mechanism against insect attacks. These bioactive chemicals may act as insecticides, antifeedants, molting hormones, oviposition deterrents, repellents, juvenile hormone mimics, growth inhibitors, antimolting hormones, as well as attractants. Plant-based pesticides are less toxic, and there is a delay in the development of resistance because of their new structure and easy biodegradability. In present article, the local and traditional uses of plants in mosquito control, current state of knowledge on phytochemical sources, and the mosquitocidal properties of secondary metabolites have been reviewed.

M. Govindarajan (✉)
Unit of Vector Biology/Phytochemistry and Nanotechnology/Department of Zoology, Annamalai University, Annamalainagar 608002, Tamil Nadu, India
e-mail: drgovind1979@gmail.com

© Springer India 2016

Vijay Veer, R. Gopalakrishnan (eds.), *Herbal Insecticides, Repellents and Biomedicines: Effectiveness and Commercialization*, DOI 10.1007/978-81-322-2704-5_3

3.1 Introduction

Mosquitoes are the major vector for the transmission of malaria, dengue, yellow fever, filariasis, and Japanese encephalitis (JE) (Das and Ansari 2003). In India, malaria is one of the most important causes of direct or indirect infant, child, and adult mortality with approximately two to three million new cases arising every year. *Anopheles stephensi* is the primary vector of malaria in India and other West Asian countries, and improved methods of control are urgently needed (Burfield and Reekie 2005). Malaria infects more than 500 million humans each year, killing approximately 1.2–2.7 million/year. About 90 % of all malaria cases occur in Africa, as does approximately 90 % of the world's malaria-related deaths (Breman et al. 2004). Malaria, caused by *Plasmodium falciparum*, is one of the leading causes of human morbidity and mortality from infectious diseases, predominantly in tropical and subtropical countries (Snow et al. 2005).

Dengue, transmitted by *Aedes aegypti*, is one of the most significant viral diseases which afflicts humans worldwide whose symptoms are ranging from mild fever to a severe and potentially life-threatening hemorrhagic disease. *Aedes aegypti* is of supreme concern because of its wide distribution and close association with humans (Ravikumar et al. 2011). *Aedes aegypti* is present in heavy polluted areas like Asia, America, and some Pacific Islands and infested about 2/3 of the world's population (Hahn et al. 2001). *Culex quinquefasciatus* is the principal vector of filariasis, and it is reported to infect more than a hundred million people every year in more than 110 countries in the tropics (WHO 2006). New control methodologies aim at reducing mosquito breeding sites and biting activity by using a combination of chemical-biological methods to reduce the population of mosquito and to reduce the man-vector contact (Service 1983).

3.2 Vector Control

The past century has witnessed intense activities in unraveling the mystery of the transmission of mosquito-borne diseases. Though therapeutic measures have been identified and used for treatment of diseases, sustainable interruption in transmission of these mosquito-borne diseases has been an eluding success. Vector control still remains an integral and significant part in disease control programs. Organized mosquito control is necessary to prevent mosquito-borne diseases, for which new strategies must be added to our armory of control. Vector control may be directed toward preventing the occurrence of disease, suppressing epidemics, or controlling already existing endemic mosquito-borne disease (John Williams 2007).

3.3 Disadvantage of Chemical Insecticides

For several decades, chemical insecticides were preferred for the control of sudden outbreak of the vectors. The repeated use of the chemical insecticides fosters many environmental hazards including development of resistance in the vectors to these chemicals and the disruption of natural biological control systems (Pushpanathan et al. 2008). Moreover, these chemicals persist in the environment for unpredictable period of time. Mosquito control has been becoming increasingly difficult because of the indiscriminate uses of synthetic chemical insecticides which have an adverse impact on the environment and disturb ecological balance. Majority of the chemical pesticides are harmful to man and animals, some of which are not easily degradable and spreading toxic effects. Due to the increased use of these, insecticides may enter into the food chain, and thereby, the internal organs like the liver, kidney, etc., may be irreversibly damaged. They even result in mutation of genes and these changes become prominent only after a few generations (Sivakumar et al. 2011). Chemical insecticides are also very costly. In larval mosquito control, application of insecticides in ponds, wells, and other water bodies may cause health hazards to human and larvivorous fishes.

Nowadays, mosquito coils containing pyrethroids and other synthetic compounds cause so many side effects, such as breathing problem, eye irritation, headache, asthma, itching, and sneezing to the users. With the use of mosquito repellent, people complained of ill health effect and sometimes required medical treatment. In addition, pests were becoming resistant to chemical treatments. Indoor residual spraying of insecticides stains the walls and leaves a long-lasting unpleasant odor. Mosquitoes are still the world's number one vectors of human and animal diseases and are conspicuous nuisance pests as well, even after massive efforts of eradication or control. The extensive use of chemical pesticides or insecticides resulted in inducing resistance by insect pests besides residue contamination of human food, mammalian toxicity, and environmental pollution. These factors have created the need for environmental safe, degradable, and target-specific agents for pest control purposes. Plant extracts have gained importance in insect control, being considered environmentally safe, less hazardous to nontarget biota, and inexpensive, and can be applied effectively by using techniques more suitable for developing countries (Mohamed et al. 2003).

3.4 Advantage of Botanical Insecticides

Phytochemicals are botanicals which are naturally occurring insecticides obtained from floral resources. Applications of phytochemicals in mosquito control were in use since the 1920s (Shahi et al. 2010), but the discovery of synthetic insecticides such as DDT in 1939 sidetracked the application of phytochemicals in mosquito control program. After facing several problems due to injudicious and overapplication of synthetic insecticides in nature, refocus on phytochemicals that are easily biodegradable and have no ill effects on nontarget organisms was appreciated. Since then, the search for new bioactive compounds from the plant kingdom and an effort to determine its structure and commercial production have been initiated. At present phytochemicals make up to 1 % of world's pesticide market (Isman 1997).

Botanicals are basically secondary metabolites that serve as a means of defense mechanism of the plants to withstand the continuous selection pressure from herbivore predators and other environmental factors. Several groups of phytochemicals such as alkaloids, steroids, terpenoids, essential oils, and phenolics from different plants have been reported previously for their insecticidal activities (Shaalan et al. 2005). Insecticidal effects of plant extracts vary not only according to plant species, mosquito species, geographical varieties, and parts used but also due to extraction methodology adopted and the polarity of the solvents used during extraction. A wide selection of plants from herbs, shrubs, and large trees was used for extraction of mosquito toxins. Phytochemicals were extracted either from the whole body of herbs or from various parts like fruits, leaves, stems, barks, roots, etc., of larger plants or trees. In all cases where the most toxic substances were concentrated upon, found and extracted for mosquito control. Plants produce numerous chemicals, many of which have medicinal and pesticidal properties. More than 2000 plant species have been known to produce chemical factors and metabolites of value in pest control programs.

3.5 Medicinal and Insecticidal Properties of Plants

Herbs are staging a comeback and herbal "renaissance" is happening all over the globe. The herbal products today symbolize safety in contrast to the synthetics that are regarded as unsafe to human and environment. Although herbs had been priced for heir medicinal, flavoring, and aromatic qualities for centuries, the synthetic products of the modern age surpassed their importance, for a while. However, the blind dependence on synthetics is over and people are returning to the naturals with hope of safety and security.

Of the 250,000 higher plant species on earth, more than 80,000 are medicinal. India is one of the world's 12 biodiversity centers with the presence of over 45,000 different plant species. India's diversity is unmatched due to the presence of 16 different agroclimatic zones, 10 vegetation zones, 25 biotic provinces, and 426 biomes (habitats of specific species). Of these, about 15,000–20,000 plants have good medicinal value.

However, only 7000–7500 species are used for their medicinal values by traditional communities. In India, drugs of herbal origin have been used in traditional systems of medicines such as *Unani* and *Ayurveda* since ancient times. The *Ayurveda* system of medicine uses about 700 species, *Unani* 700, *Siddha* 600, Amchi 600, and modern medicine around 30 species. The drugs are derived either from the whole plant or from different organs, like leaves, stem, bark, root, flower, seed, etc. Some drugs are prepared from excretory plant product such as gum, resins, and latex. Even the allopathic system of medicine has adopted a number of plant-derived drugs which form an important segment of the modern pharmacopoeia. Some important chemical intermediates needed for manufacturing the modern drugs are also obtained from plants (e.g., diosgenin, solasodine, b-ionone). Not only that plant-derived drug offers a stable market worldwide but also plants continue to be an important source for new drugs.

3.6 Botanical Used for Mosquito Control

A large number of plant extracts have been reported to have mosquitocidal or repellent activity against mosquito vectors (Sukumar et al. 1991) but very few plant products have shown practical utility for mosquito control. Plant products can be obtained either from the solvents. Botanicals are used as larvicides and growth inhibitors; some plants possess repellent properties and they are used as repellents against mosquitoes. Biological control of mosquito was very popular during the early part of the twentieth century, but with the development and availability of chemicals such as the organochlorines and organophosphates, it was replaced by insecticidal control. However, because of problems with insecticide resistance and greater awareness of environmental contamination, there has been renewed interest in biological (biological control) methods. Plants and plant derivatives have tremendous advantages over synthetic insecticides.

3.7 Larvicidal Activity of Plants

Larviciding is a general term for killing immature mosquitoes by applying agents, collectively called larvicides, to control mosquito larvae and/or pupae. Larval source management (LSM) involves both the modification of water habitats, often referred to as source reduction, and the direct application of larvicides to control mosquito production. Most mosquito species spend much of their life cycle in the larval stage when they are highly susceptible to both predation and control efforts. They often are concentrated within defined water boundaries, immobile with little ability to disperse, and accessible. Adult mosquitoes, in contrast, fly in search of mates, blood meals, or water sources for egg laying and are often inaccessible, not concentrated, and widely distributed. Therefore, effective larviciding can reduce the number of adult mosquitoes available to disperse, potentially spread disease, create a nuisance, and lay eggs which leads to more mosquitoes. The effective control of larvae and/or pupae is a basic principle of integrated pest management (IPM). Effective IPM involves understanding the local mosquito ecology and patterns of arbovirus transmission and then selecting the appropriate mosquito control tools. The most common methods of IPM include environmental management or source reduction, larviciding, and adulticiding. Other mosquito control principles include biocontrol, as well as additional methods not discussed here such as herbiciding and hand removal of aquatic plants. These methods may be used to control immature mosquitoes indirectly, usually when there is an obligatory association between the larvae/pupae and specific host plants (Govindarajan et al. 2008).

Several different types of larvicides are available for controlling mosquitoes. Generally, these larvicides are least effective in wastewater systems. The flow-through nature of many wastewater treatment, reuse, and recycling operations rapidly diminishes the effectiveness of many larvicides. Bacteria and other components of wastewater quickly break down or inactivate some larvicides. Increasing the dosage rate and the number of applications or using slow-release formulations may be required to achieve adequate control. At sites where mosquito outbreaks are large and frequent, larvicides may provide only temporary control and may not be cost effective. Larvicide operations must be supported with a quality inspection program. Potential mosquito production sites must be identified and frequently inspected. Larvicide applications should be integrated with other mosquito abatement measures, such as aquatic plant management and water quality improvement. Larvicides should not interfere with the level of mosquito control already provided by natural predators and parasites. Biological control of mosquitoes could be very promising being eco-friendly as well as cost effective. Hence, there is a constant need for developing biologically active plant materials as insecticides, which are expected to reduce the hazards to humans and other organisms by minimizing the accumulation of harmful residues in the environment. Natural products of plant origin are generally preferred because of their less harmful nature to nontarget organisms and their innate biodegradability (Govindarajan 2011) (Table 3.1).

The larvicidal activity of various plant extracts such as *Pedalium murex*, *Cleome icosandra*, and *Dictyota dichotoma* has been found to be promising against *Cx. quinquefasciatus* and *An. stephensi* (Kalyanasundaram and Das 1985). Among many well-known phytochemicals, azadirachtin (a tetranortriterpenoid compound) is one of the most extensively studied biological insecticides extracted from the seed kernel of the neem tree *Azadirachta indica* (Schmutterer 1990). Susceptibility tests were carried out in *Cx. quinquefasciatus* larvae using peel oil extracts of *Citrus aurantium*, *C. sinensis*, and *Citrus limon* (Mwaiko 1992); the extracts of *O. canum* were effective in killing the larvae *Anopheles gambiae* (Lukwa 1994).

The metabolites of *Cucumis sativus* exhibited insecticidal activities to many insect pests, including *Ae. gossypii*, *Tetranychus viennensis*, and larvae of *Pieris rapae* (Zhang et al. 1998). Nirmal Sharma et al. (1998) reported that larvicidal activity of *Gliricidia sepium* crude ethanol extracts of dried leaves, fresh leaves, dried petioles, and stem bark was tested for their activities against third instar larvae of *An. stephensi*, *Ae. aegypti*, and *Cx. quinquefasciatus*. Pushpalatha and Muthukrishnan (1999) evaluated that the methanol seed and leaf extract showed significant larvicidal and growth regulatory activities even at very low concentrations against *An. stephensi*. Murugan and Jeyabalan (1999) studied the effect of some indigenous plants on the larvicide and ovipositional properties on *An. stephensi*.

Rahuman et al. (2000) have reported that a bioassay-guided fractionation of the acetone extract of *Feronia limonia* dried leaves afforded a potent mosquito larvicide, identified as n-hexadecanoic acid, and found to be effective against fourth instar larvae of *An. stephensi*; Vahitha et al. (2002) studied the larvicidal efficacy of *Pavonia zeylanica* and *Acacia ferruginea* against *Cx. quinquefasciatus*. The crude chloroform extract of seeds of *Millettia dura* showed high activity against second instar larvae of *Ae. aegypti* (Yenesew et al. 2003). Essential oils of leaf and bark of *Cryptomeria japonica* demonstrated high larvicidal activity against *Ae. aegypti* larvae (Cheng et al. 2003). The petroleum ether extract showed larvicidal activity against *Ae. aegypti*, *Cx. quinquefasciatus*, *An. dirus*, and *Mansonia uniformis* (Komalamisra et al. 2005). The ethanolic extracts of the orange peel (*C. sinensis*) were tested for the toxicity effect on the larvae of the yellow fever mosquito *Ae. aegypti* (Amusan et al. 2005); the ethanolic and acetone extracts of *Nerium indicum* and *Thuja orientalis* have been studied against third instar larvae of *An. stephensi* and *Cx. quinquefasciatus* (Sharma et al. 2005). The larvicidal activity of methanol extracts of dried root powder of *Chamaecyparis obtusa* was tested against three vector mosquitoes (Jang et al. 2005).

The methanolic extracts of the fresh leaf extract of *Calotropis procera* (Singh et al. 2005), the aqueous extract from the roots of *Hibiscus abelmoschus* (Dua et al. 2006) and the petroleum ether, carbon tetrachloride, and methanol extract of *Artemisia annua*, *Chenopodium album*, and *Sonchus oleraceus* (Sharma et al. 2006) were

Table 3.1 Efficacy of plant extracts against vector mosquitoes

Plants	Family	Plant parts used	Target mosquito species	Lethal concentrations/biological activity	References
Acacia nilotica	Fabaceae	Leaf	*An. stephensi*, *Ae. aegypti*, and *Cx. quinquefasciatus*	LC_{50} value was 55.72 ppm and LC_{90} value was 194.58 ppm	Sakthivadivel and Daniel (2008)
Acalypha alnifolia	Euphorbiaceae	Leaf	*An. stephensi*, *Ae. aegypti*, and *Cx. quinquefasciatus*	LC_{50} values were 125.73, 127.98, and 128.55 ppm against fourth instar larvae of three mosquito species at 24 h	Kovendan et al. (2012)
Acalypha indica	Euphorbiaceae	Leaf	*An. stephensi*	LC $_{50}$ value was 19.25 ppm at 24 h	Govindarajan et al. (2008)
Ageratina adenophora	Asteraceae	Twigs	*Ae. aegypti* and *Cx. quinquefasciatus*	At 24 h, LC_{50} value of the extract was found to be 356.70 ppm for *Ae. aegypti* and 227.20 ppm for *Cx. quinquefasciatus*	Raj Mohan and Ramaswamy (2007)
Ageratina adenophora	Asteraceae	Leaf	*Cx. quinquefasciatus*, *Ae. aegypti*, and *An. stephensi*	The LC_{50} and LC_{90} values of crude methanol extract of leaves of *A. adenophora* on *Cx. quinquefasciatus*, *Ae. aegypti*, and *An. stephensi* larvae in 24 h were 144.86, 132.82, and 113.08 and 250.70, 231.12, and 198.81 mg/l, respectively	Rajeswary and Govindarajan (2013)
Ageratum conyzoides	Asteraceae	Leaf	*Cx. quinquefasciatus*	Potent larvicidal activity was noticed	Saxena et al. (1992)
Aloe barbadensis	Liliaceae	Leaf	*An. stephensi*	LC_{50} values were 29.06 and 22.59 ppm for 24 and 48 h	Maurya et al. (2007)
Annona crassiflora	Annonaceae	Root wood	*Ae. aegypti*	LC_{50} value was 0.71; LC_{90} value was 5.12 μg/ml	Omena et al (2007)
Annona squamosa	Annonaceae	Bark	*Cx. quinquefasciatus* and *An. stephensi*	LC_{50} values of 28.18 and 43.07 ppm against *An. stephensi* and *Cx. quinquefasciatus*, respectively	Kamaraj et al. (2010)
Apium graveolens	Umbelliferae	Seed	*Ae. aegypti*	LD_{50} and LD_{95} values of 81.0 and 176.8 mg/l, respectively, for fourth instar larvae	Choochate et al. (2004)
Argemone mexicana	Papaveraceae	Leaf	*Cx. quinquefasciatus*	Causes 100 % mortality at 250 ppm of each extracts	Karmegan et al. (1997)
Artemisia annua	Asteraceae	Leaf	*Anopheles stephensi*	LC_{50} value was 16.85 ppm after 24 h and 11.45 ppm after 48 h of exposure	Sharma et al. (2006)

Artemisia cina	Compositae	Leaf	*Cx. pipiens*	The EC_{50} for the mosquito at 24 h after treating with extract was 4.0 g/l	Aly and Bardan (1996)
Asparagus racemosus	Asparagaceae	Root	*Culex quinquefasciatus*, *Aedes aegypti*, and *Anopheles stephensi*	The LC_{50} and LC_{90} values were 115.13, 97.71, and 90.97 ppm and 210.96, 179.92, and 168.82 ppm, respectively	Govindarajan and Sivakumar (2013b)
Atlantia monophylla	Rutaceae	Leaf	*An. stephensi*	LC_{50} value of 0.05 mg/l. Insect growth regulating activity with EI_{50} value of 0.065 mg/l	Sivagnaname and Kalyanasundaram (2004)
Atlantia monophylla	Rutaceae	Leaf	*Cx. quinquefasciatus*	Larvae were found susceptible with LC_{50} value of 0.14 mg/l	Sivagnaname and Kalyanasundaram (2004)
Atlantia monophylla	Rutaceae	Leaf	*Ae. aegypti*	Larval growth regulating activity of this extract was found to be pronounced with EI_{50} value of 0.002 mg/l	Sivagnaname and Kalyanasundaram (2004)
Azadirachta indica	Meliaceae	Leaf	*Cx. fatigans*	In comparison with malathion (LC_{50} value was 0.45 ppm), the LC_{50} value of neem fraction (NLX) was found to be higher to the third instar larvae at 390 ppm	Azmi et al. (1998)
Azadirachta indica	Meliaceae	Leaf	*Ae. aegypti*	LC_{50} value is 8.32 mg/ml	Mgbemena (2010)
Caesalpinia pulcherrima	Fabaceae	Leaf	*Cx. tritaeniorhynchus*, *Ae. albopictus*, and *An. subpictus*	The LC_{50} and LC_{90} values were 150.47, 135.24, and 119.27 ppm and 282.57, 261.55, and 243.37 ppm, respectively	Govindarajan et al. (2013a)
Cardiospermum halicacabum	*Sapindaceae*	Leaf	*Culex quinquefasciatus* and *Aedes aegypti*	The LC_{50} values were 174.24, 193.31, 183.36, 150.44, and 154.95 ppm and 182.51, 200.02, 192.31, 156.80, and 164.54 ppm, respectively	Govindarajan (2011)
Carica papaya	Caricaceae	Seed	*Cx. quinquefasciatus*	LC_{50} value of 0.15, 0.11, 0.07, and 0.20 % against first, second, third, and fourth instar larvae	Rawani et al. (2009)
Cassia fistula	Fabaceae	Leaf	*Cx. tritaeniorhynchus* and *Anopheles subpictus*	The LC_{50} and LC_{90} values of *C. fistula* against early third instar of *Cx. tritaeniorhynchus* and *An. subpictus* were 45.57 and 33.76 ppm and 82.05 and 60.63 ppm, respectively	Govindarajan et al. (2011a)

(continued)

Table 3.1 (contniued)

Plants	Family	Plant parts used	Target mosquito species	Lethal concentrations/biological activity	References
Cassia fistula	Fabaceae	Flowers	*Cx. tritaeniorhynchus*, *Ae. albopictus*, and *An. subpictus*	LC_{50} and LC_{90} values of *C. fistula* flower against early third instar of *Cx. tritaeniorhynchus*, *Ae. albopictus*, and *An. subpictus* were 136.59, 118.64, and 96.51 ppm and 243.67, 231.79, and 174.39 ppm, respectively	Govindarajan (2013)
Cassia obtusifolia	Leguminosae	Seed	*Ae. aegypti*, *Ae. togoi*, and *Cx. pipiens* pallens	Showed a strong larvicidal activity of 100 % mortality at 25 mg/l. The biologically active component was emodin. The LC_{50} values of emodin were 1.4, 1.9, and 2.2 ppm, respectively	Yang et al. (2003)
Cassia obtusifolia	Leguminosae	Leaf	*An. stephensi*	LC_{50} and LC_{90} values were 52.2 and 108.7 mg/ l	Rajkumar and Jebanesan (2009)
Cassia tora	Caesalpiniaceae	Seed	*Ae. aegypti* and *Cx. pipiens* pallens	LC_{50} value was 20 mg/l for both the larval species	Jang et al. (2002)
Centella asiatica	Umbelliferae	Leaf	*Cx. quinquefasciatus*	LC_{50} ranged between 6.84 ppm at 19 °C and 1.12 ppm at 31 °C. LC_{90} varied from 9.12 to 3.63 ppm at the two temperatures, respectively	Rajkumar and Jebanesan (2005)
Cestrum diurnum	Solanaceae	Leaf	*An. stephensi*	The LC_{50} value of the active ingredient was determined as 0.70, 0.89, 0.90, and 1.03 mg/100 mL for first, second, third, and fourth instar larva, respectively, in 24 h study period	Ghosh and Chandra (2006)
Chamaecyparis obtusa	Cupressaceae	Leaf	*An. stephensi*	The bioactive component in the leaf extract was characterized as beta-thujaplicin by spectroscopic analyses. The LC_{50} value of beta-thujaplicin was 2.91 ppm	Jang et al. (2005)
Chrysanthemum indicum	Asteraceae	Leaf	*Cx. tritaeniorhynchus*	LC_{50} value was 42.29 mg/ml after 24 h	Kamaraj et al. (2010)

Citrullus vulgaris	Cucurbitaceae	Leaf	*Ae. stephensi*	100 % mortality was exerted at 250 ppm and the corresponding LC_{50} value was 18.56 ppm	Mullai et al. (2008)
Citrus aurantium	Rutaceae	Fruit peel	*Cx. quinquefasciatus*	LC_{90} values were 53.80 and 32.52 ppm after 24 and 48 h of treatment	Kassir et al. (1989)
Citrus reticulata	Rutaceae	Seed	*Cx. quinquefasciatus* and *Ae. aegypti*	LC_{50} value against *Ae. aegypti* and *Cx. quinquefasciatus* larvae was 2,267.71 and 2,639.27 ppm, respectively	Sumroiphon et al. (2006)
Citrus sinensis	Rutaceae	Fruit peel	*An. subpictus*	LC_{50} value was 58.25 and LC_{90} value was 298.31 ppm	Bagavan et al. (2009)
Clausena anisata	Rutaceae	Leaf	*Anopheles subpictus* and *Aedes albopictus*	The LC_{50} values of β-pinene, sabinene, germacrene D, estragole, and linalool appeared to be most effective against *Anopheles subpictus* (LC_{50} – 24.68, 20.93, 17.16, 12.25, 36.26 ppm) followed by *Aedes albopictus* (LC_{50} – 28.10, 22.66, 19.51, 13.01, 39.99 ppm)	Govindarajan and Sivakumar (2013a)
Coccinia indica	Cucurbitaceae	Leaf	*Cx. quinquefasciatus*	The LC_{50} and LC_{90}values of hexane, ethyl acetate, benzene, chloroform, and methanol extracts of *C. indica* against adults of *Cx. quinquefasciatus* were 146.65, 133.68, 122.38, 112.85, and 68.88 ppm and 259.86, 246.34, 225.99, 212.48, and 129.56 ppm, respectively	Sivakumar and Govindarajan (2013a)
Coccinia indica, *Cucumis sativus*, and *Momordica charantia*	Cucurbitaceae	Leaf	*Cx. quinquefasciatus* and *Ae. aegypti*	LC_{50} values of the respective plants were 377.69, 623.80, and 207.61 and 309.46, 492.73, and 199.14 ppm against the two vector species	Rahuman and Venkatesan (2008)
Coleus aromaticus	Lamiaceae	Oil	*Cx. tritaeniorhynchus*, *Ae. albopictus*, and *An. subpictus*	24.83 and 22.06 LC_{50} values of 72.70, 67.98, and 60.31 μg/mL, respectively	Govindarajan et al. (2013b)
Curcuma aromatica	Zingiberaceae	Rhizome	*Ae. aegypti*	LC_{50} value was 36.30 ppm	Choochate et al. (2005)

(continued)

Table 3.1 (contniued)

Plants	Family	Plant parts used	Target mosquito species	Lethal concentrations/biological activity	References
Cybistax antisyphilitica	Bignoniaceae	Stem wood	*Ae. aegypti*	A natural quinone identified as 2-hydroxy-3-(3-methyl-2-butenyl)-1.4-naphthoquinone (lapachol) was quite potent with LC_{50} value of 26.3 μg/ml	Rodrigues et al. (2005)
Aloe turkanensis	Asphodelaceae	Leaf	*An. gambiae*	100 % mortality was achieved at a concentration of 0.2 mg/ml and it had a LC_{50} value of 0.11 mg/ml	Matasyoh et al. (2008)
Delonix elata	Fabaceae	Leaf	*An. stephensi* and *Ae. aegypti*	LC_{50} and LC_{90} values being 93.59 and 111.83 and 163.69 and 202.77 ppm, respectively	Govindarajan et al. (2012a)
Delonix elata	Fabaceae	Leaf, seed	*Cx. quinquefasciatus*	The LC_{50} and LC_{90} values of *D. elata* against early third instar of *Cx. quinquefasciatus* were 124.84 and 147.86 mg/L and 213.88 and 289.43 mg/L, respectively	Govindarajan et al. (2012b)
Dysoxylum malabaricum	Meliaceae	Leaf	*An. stephensi*	4 % concentration of leaf extract killed more than 97 % of first instars, 92 % of fifth instars, 93 % of pupae, and 91 % of adults	Senthil Nathan et al. (2006a)
Eucalyptus globulus	Myrtaceae	Seed, leaf	*Cx. pipiens*	Both the extracts at a dose of 1000 ppm caused 100 % and 80 % mortality to the tested larvae	Sheeren (2006)
Eclipta alba	Asteraceae	Leaf	*Cx .quinquefasciatus*	The methanol extract and the LC_{50} and LC_{90}values were 119.83 and 234.19 ppm, respectively, followed by chloroform, benzene, and ethyl acetate, and hexane extracts with the LC_{50} and LC_{90} values were 134.65, 140.72, 144.11, and 157.97 ppm and 257.29, 260.60, 274.43, and 291.21 ppm, respectively	Sivakumar and Govindarajan (2013b)

Ervatamia coronaria	Apocynaceae	Leaf	*An. stephensi*, *Ae. aegypti*, and *Cx. quinquefasciatus*	The LC_{50} and LC_{90} values were 79.08, 89.59, and 96.15 ppm and 150.47, 166.04, and 174.10 ppm, respectively	Govindarajan et al. (2011b)
Erythrina indica	Fabaceae	Leaf	*An. stephensi*, *Ae. aegypti*, and *Cx. quinquefasciatus*	LC_{50} and LC_{90} values of 69.43, 75.13, and 91.41 ppm and 125.49, 134.31, and 167.14 ppm, respectively	Govindarajan and Sivakumar (2013c)
Eucalyptus citriodora	Myrtaceae	Leaf	*An. stephensi*, *Ae. aegypti*, and *Cx. quinquefasciatus*	The LC_{50} values against fourth instar larvae of three species were 69.86, 81.12, and 91.76 ppm, respectively, after 24 h and 26.7, 29.9, and 38.8 ppm, respectively, after 72 h	Singh et al. (2007)
Eucalyptus globulus	Myrtaceae	leaf, seed	*Culex pipiens*	Both the extracts at a dose of 1000 ppm caused 100 and 80 % mortality to the tested larvae	Sheeren (2006)
Euphorbia hirta	Euphorbiaceae	Stem bark	*Cx. quinquefasciatus*	LC_{50} value was 424.94 and LC_{90} value was 1314.01 ppm	Rahuman et al. (2007)
Euphorbia tirucalli	Euphorbiaceae	Stem bark	*Cx. pipiens* pallens	LC_{50} value was 200.76 and LC_{90} value was 343.515 mg/l	Yadav et al. (2002)
Feronia limonia	Rutaceae	Leaf	*Cx. quinquefasciatus*, *An. stephensi*, and *Ae. aegypti*	LC_{50} values of 129.24, 79.58, and 57.23 ppm for three mosquito species, respectively	Rahuman et al. (2000)
Ficus benghalensis	Moraceae	Leaf	*Culex quinquefasciatus*, *Aedes aegypti*, and *An. stephensi*	LC_{50} values of 41.43, 58.21, and 74.32 ppm; 56.54, 70.29, and 80.85 ppm; and 60.44, 76.41, and 89.55 ppm for three mosquito species, respectively	Govindarajan (2010a)
Ficus benghalensis	Moraceae	Leaf	*Cx. tritaeniorhynchus* and *An. subpictus*	The LC_{50} and LC_{90} values of *F. benghalensis* against early third instar of *Cx. tritaeniorhynchus* and *An. subpictus* were 100.88 and 159.76 ppm and 56.66 and 85.84 ppm, respectively	Govindarajan et al. (2011c)
Hemidesmus indicus	Asclepiadaceae	Root	*Cx. quinquefasciatus*	80 % mortality was observed in 5 % concentration after 1 day of exposure	Khanna and Kannabiran (2007)
Jatropha curcas	Euphorbiaceae	Leaf	*Cx. quinquefasciatus*	LC_{50} value was 11.34 and LC_{90} value was 46.52 ppm	Rahuman et al. (2007)

(continued)

Table 3.1 (contniued)

Plants	Family	Plant parts used	Target mosquito species	Lethal concentrations/biological activity	References
Kaempferia galanga	Zingiberaceae	Rhizome	*Cx. quinquefasciatus*	LC_{50} value was 42.33 ppm	Choochate et al. (1999)
Khaya senegalensis	Meliaceae	Leaf	*Cx. annulirostris*	LC_{50} value was 5.86 mg/l	Shaalan et al. (2005)
Melia azedarach	Meliaceae	Leaf, seeds	*An. stephensi*	The extract showed strong larvicidal activity	Senthil Nathan et al. (2006b)
Mentha spicata	Lamiaceae	Leaf	*Cx. quinquefasciatus*, *Ae. aegypti*, and *An. stephensi*	LC_{50} values of 62.62, 56.08, and 49.71 ppm and LC_{90} values of 118.70, 110.28, and 100.99 ppm, respectively	Govindarajan et al. (2012c)
Millettia dura	Leguminosae	Seed	*Ae. aegypti*	Rotenoids, deguelin, and tephrosin isolated from the seeds of this plant showed potent activities, with LC_{50} values of 1.6 and 1.4 μg /ml at 24 h, respectively	Yenesew et al. (2003)
Millingtonia hortensis	Bignoniaceae	Leaf	*An. stephensi*, *Ae. aegypti*, and *Cx. quinquefasciatus*	LC_{50} values of 104.70, 138, and 83.18 ppm for second instar larvae of three species at 24 h of bioassay	Kaushik and Saini (2008)
Momordica charantia	Cucurbitaceae	Fruit	*An. stephensi*	LC_{50} value was 0.50 and LC90 value was 1.54 % concentration of the extract	Singh et al. (2006)
Momordica charantia	Cucurbitaceae	Leaf	*Cx. quinquefasciatus*	LC_{50} value was 465.85; LC_{90} value was 2421.46 ppm	Prabakar and Jebanesan (2004)
Moringa oleifera	Moringaceae	Bark	*Cx. gelidus*	LC_{50} value was 38.47 μg/ml	Kamaraj and Rahuman (2010)
Myrtus communis	Myrtaceae	Flower and leaf	*Cx. pipiens* molestus	LC_{50} value was 16 mg/l. Thymol, carvacrol, (1R)–(+)- -pinene, and (1S)–(−)- -pinene were the most effective toxic compounds with LC_{50} values of 36–49 mg /l	Traboulsi et al. (2002)

Nyctanthes arbor-tristis	Nyctantheceae	Flower	*Cx. quinquefasciatus*	LC_{50} values were 25.67 and 22.19, 38.60 and 28.95, 53.14 and 42. 14, and 72.60 and 61.82 ppm and for the isolated compound NCS-2 were 73.31 and 65.48, 83.02 and 67.02, 97.26 and 81.84, and 14.68 and 99.02 ppm for first, second, third, and fourth instar larvae, respectively, at 24 and 48 h postexposure	Khatune et al. (2001)
Ocimum sanctum	Lamiaceae	Leaf	*Ae. aegypti* and *Cx. quinquefasciatus*	The LC_{50} value of *O. sanctum* against the larvae of *Ae. aegypti* was 425.94 and against the larvae of *Cx. quinquefasciatus* was 592.60 ppm	Anees (2008)
Ocimum basilicum	Lamiaceae	Leaf	*An. stephensi* and *Cx. quinquefasciatus*	LC_{50} values of 8.29 and 87.68 ppm, respectively	Maurya et al. (2009b)
Ocimum basilicum	Lamiaceae	Oil	*Cx. tritaeniorhynchus*, *Ae. albopictus*, and *An. subpictus*	LC_{50} values of 14.01, 11.97, and 9.75 ppm and LC_{90} values of 23.44, 21.17, and 18.56 ppm, respectively	Govindarajan et al. (2013c)
Ocimum gratissimum	Lamiaceae	Leaf	*Cx. gelidus* and *Cx. quinquefasciatus*	LC_{50} values were 39.31 and 66.28 μg/ml against fourth instar larvae after 24 h	Kamaraj and Rahuman (2010)
Paullinia clavigera	Sapindaceae	Leaf	*An. benarrochi*	LC_{50} (24 h) value was 0.81; LC_{50} (12 h) value was1.19 %	Iannacone and Pérez (2004)
Pavonia zeylanica	Malvaceae	Leaf	*Cx. quinquefasciatus*	After 24 h of treatment, the LC_{50} value was 2214.7 ppm	Vahitha et al. (2002)
Piper longum	Piperaceae	Fruit exocarp	*Ae. aegypti*	LC_{50} value of 2.23 ppm	Chaithong et al. (2006)
Piper nigrum	Piperaceae	Seed	*Cx. pipiens*	LC_{50} value was 2.6 mg/l	Shaalan et al. (2005)
Piper retrofractum	Piperaceae	Unripe and ripe fruit	*Cx. quinquefasciatus* and *Ae. aegypti*	LC_{50} value of 135 against *Cx. quinquefasciatus* and 79 ppm against *Ae. aegypti*	Chansang et al. (2005)
Pithecellobium dulce	Fabaceae	Leaf, seed	*An. stephensi* and *Ae. aegypti*	The LC_{50} and LC_{90} values are145.43 and 155.78 mg/l and 251.23 and 279.73 mg/l, respectively	Govindarajan et al. (2013d)

(continued)

Table 3.1 (contniued)

Plants	Family	Plant parts used	Target mosquito species	Lethal concentrations/biological activity	References
Plumbago zeylanica, P. dawei, and *P. stenophylla*	Plumbaginaceae	Root	*An. gambiae*	LC_{50} values were 4.1, 6.4, and 6.7 mg/ml, respectively. LC_{90} values were 10.6, 26.2, and 15.6 mg/ml, respectively	Maniafu et al. (2009)
Rhizophora mucronata	Rhizophoraceae	Bark, pith, and stem wood	*Ae. aegypti*	LC_{50} values of 157.4, 168.3, and 1003.4 ppm for bark, pith, and stem wood at 48 h, respectively	Kabaru and Gichia (2001)
Solanum xanthocarpum	Solanaceae	Root	*Cx. pipiens* pallens	LC_{50} and LC_{90} values were 64.99 and 252.43 ppm and 59.20 and 186.15 ppm after 24 and 48 h of exposure, respectively	Mohan et al. (2006)
Sida acuta	Malvaceae	Leaf	*Cx. quinquefasciatus, Ae. aegypti*, and *An. stephensi*	In terms of lethal concentrations for 50 % mortality (LC_{50}), crude extract of *S. acuta* appeared to be most effective against *An. stephensi* (LC_{50}=38.64 mg/L) followed by *Ae. aegypti* (LC_{50}=42.08 mg/L) and *Cx. quinquefasciatus* (LC_{50}=47.91 mg/L)	Govindarajan (2010b)
Solanum nigrum	Solanaceae	Dried fruit	*An. Culicifacies. An. stephensi, Cx. quinquefasciatus*, and *Ae. aegypti*	The LC_{50} values against fourth instar larvae of four species were 9.04, 6.25, 12.25, and 17.63 ppm, respectively	Raghavendra et al. (2009)
Solanum nigrum	Solanaceae	Leaf	*Cx. quinquefasciatus*	LC_{50} value was 17.04 ppm against fourth instar larvae after 24 h	Rawani et al. (2010)
Solanum villosum	Solanaceae	Leaf	*An. stephensi, Cx. quinquefasciatus*, and *Ae. aegypti*	The protein compound responsible for mosquitocidal property was isolated with LC_{50} values of 644.75, 645.75, and 747.22 ppm	Chowdhury et al. (2009)

Solenostemma argel	Apocynaceae	Aerial parts	*Cx. pipiens*	LC_{50} values of 0.037, 0.031, 0.009, and 0.007 ppm and the LC_{95} values were found as 0.394, 0.293, 0.065, and 0.030 ppm, after 1, 2, 4, and 7 days against the larva of *Cx. pipiens* under laboratory conditions	Al-Doghairi et al. (2004)
Thymus capitatus	Lamiaceae	Leaf	*Cx. pipiens*	The volatile oil, thymol, and the unsaponifiable portion proved high larvicidal potency (LC_{50} value was 49.0 ppm)	Mansour et al. (2000)
Tridax procumbens	Compositae	Leaf	*An. subpictus*	LC_{50} value of 39.98 mg/l	Kamaraj et al. (2011)
Vitex negundo	Verbenaceae	Leaf	*Cx. quinquefasciatus*	LC_{50} value was 212.57 ppm	Krishnan et al. (2007)
Zingiber officinalis	Zingiberaceae	Rhizome	*Cx. quinquefasciatus*	The LC_{50} value was 50.78 ppm. Skin repellent test at 1.0, 2.0, 3.0, and 4.0 mg/cm2 concentration of *Z. officinalis* gave 100 % protection up to 15, 30, 60, and 120 min	Pushpanathan et al. (2008)

tested against *An. stephensi*. The dried root power methanol extract of *R. nasutus* was tested against the larvae of *Ae. aegypti* and *Cx. quinquefasciatus* (Rongsriyam et al. 2006). Pandey et al. (2007) report extensive investigations of the larvicidal efficacy of the crude extracts of three *Spilanthes* species (*S. acmella*, *S. calva*, and *S. paniculata*) tested against *An. stephensi*, *An. culicifacies* species, and *Cx. quinquefasciatus* Say mosquito larvae and the larvicidal potential of the petroleum ether root extract of *Solanum xanthocarpum* (Mohan and Ramaswamy 2007). Petroleum ether (60–80 °C) extracts of the leaves of *Vitex negundo* were evaluated for larvicidal activity against larval stages of *Cx. tritaeniorhynchus* in the laboratory (Karunamoorthi et al. 2008). The crude acetone, hexane, ethyl acetate, methanol, and petroleum ether extracts of the leaf of *Centella asiatica*, *Datura metel*, *Mukia scabrella*, and *Toddalia asiatica* of whole plant of *Citrullus colocynthis* and *Sphaeranthus indicus* were assayed for their toxicity against the early fourth instar larvae of *Cx. quinquefasciatus* (Rahuman et al. 2008a).

The methanol extract of *Ocimum canum* and the acetone extract of *O. sanctum* were reported to have a toxic effect against the larvae of *Spodoptera litura*, *Ae. aegypti*, and *Cx. quinquefasciatus* (Bagavan et al. 2008). Antifeedant and larvicidal activity of acetone, chloroform, ethyl acetate, hexane, and methanol peel, leaf, and flower extracts of *Citrus sinensis*, *Ocimum canum*, *Ocimum sanctum*, and *Rhinacanthus nasutus* were studied using fourth instar larvae of *Helicoverpa armigera*, *Sylepta derogata*, and *An. stephensi* (Kamaraj et al. 2008). The compound gluanol acetate isolated from the bark acetone extract of *Ficus racemosa* was assayed for their toxicity against the larvae of *An. stephensi* (Rahuman et al. 2008b); the larvicidal potential of the extracts of commonly available medicinal plants *Saraca indica/asoca*, *Nyctanthes arbortristis*, and *Clitoria ternatea* against *An. stephensi*, *Cx. quinquefasciatus*, and *Ae. aegypti* (Mathew et al. 2009); and the larvicidal potential of the various fruit wall extracts of *Momordica charantia* (Cucurbitaceae) against two species of mosquito vectors, *An. stephensi* and *Cx. quinquefasciatus* (Maurya et al. 2009a).

3.8 Plant-derived Insecticides

Plants have been used through human history in cooking, medicines, cosmetics, and insecticides and have a rich source of chemicals and drugs for man. During the twentieth century, a few of these natural compounds like nicotine, rotenone, and pyrethrins have been used commercially as insecticides (Ware 1978). However, plants produce thousands of other compounds that are insecticidal and have diverse modes of action like hormonal, neurological, nutritional, or enzymatic actions (Rosenthal and Janzen 1979). A number of plant families are known to produce alkaloids, phenolics, and oils which have been used for insect control since a long time. They were called as insect killers and were used by Romans and Chinese (Heitz and Downum 1987). Another group of plant products discovered in the last two decades are known as secondary metabolites. These were considered as waste products from plants for a long time, because they were of no nutritional significance to insect. These secondary plant metabolites having insecticidal/pesticidal properties are of several types, viz., repellent, antifeedants, ovicidal, phagostimulants, and toxins. Alkaloids and phenolics are two important groups of secondary plant metabolites that regulate the tolerance level of host-plant acceptance by insects (Table 3.2).

3.9 Conclusion

Whole plants have been demonstrated to be more active than the isolated active compounds as larvicides against mosquitoes. The high bioactivity of plants shows the feasibility of production of a low-cost larvicide without the need to isolate an active compound, which is an expensive process. However, the selection of chemical markers is essential for the quality control of herbal products. Although these natural products have been

Table 3.2 Mosquitocidal properties of active ingredients

Plants	Active ingredient	Mosquito	LC/LD values	Compound structure	References
Abutilon indicum	β-Sitosterol	*Ae. aegypti*, *An. stephensi*, and *Cx. quinquefasciatus*	LC_{50} values of 11.49, 3.58, and 26.67 ppm, respectively	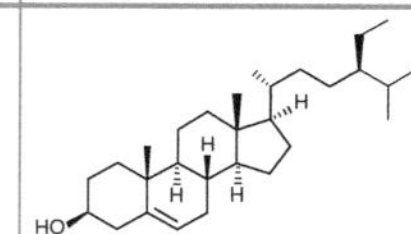 β-sitosterol	Rahuman et al. (2008a)
Aegle marmelos	Marmesin	*An. gambiae*	LC_{50} and LC_{90} values of 0.082 and 0.152 mg/l	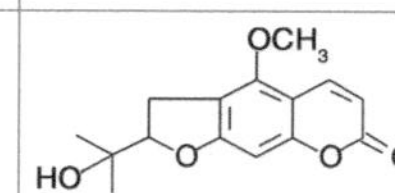 Marmesin	Joseph et al. (2004)
Alpinia purpurata	β-Caryophyllene and β-pinene	*Ae. aegypti*	LC_{50} value of 80.7 mg/L	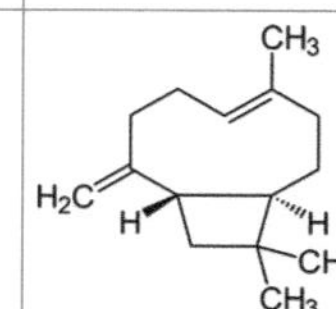 β-Caryophyllene 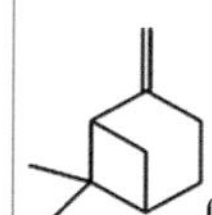 β-pinene	Santos et al. (2012)
Asarum heterotropoides	Methyleugenol and safrole	*Ae. aegypti*	LC_{50} value of 23.82 mg/L	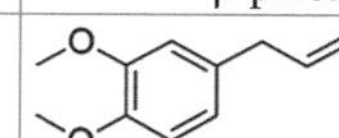 Methyleuge 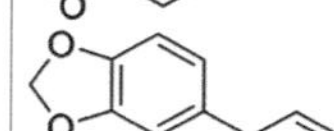 Safrol	Perumalsamy et al. (2010)

(continued)

Table 3.2 (continued)

Plants	Active ingredient	Mosquito	LC/LD values	Compound structure	References
Auxemma glazioviana	α-Bisabolol and α-cadinol	*Ae. aegypti*	LC_{50} value of 2.53 mg/L	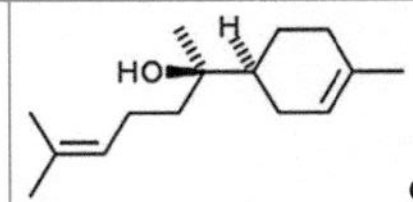 α-Bisabolol 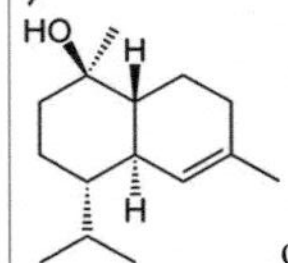 α-cadinol	Costa et al. (2004)
Azadirachta indica	Azadirachtin	*An. stephensi*	EC_{50} values of 0.014, 0.021, 0.028, and 0.034 ppm against first, second, third, and fourth instar larvae, respectively	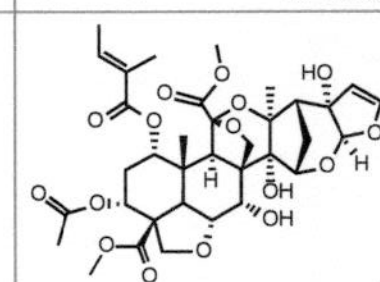 Azadirachtin	Senthil Nathan et al. (2005)
Bauhinia acuruana	epi-α-Cadinol and spathulenol	*Ae. aegypti*	LC_{50} value of 56.22 mg/L	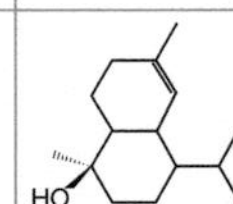 epi-α-Cadinol	Góis (2010)
Capraria biflora	α-Humulene	*Ae. aegypti*	LC_{50} value of 73.39 mg/L	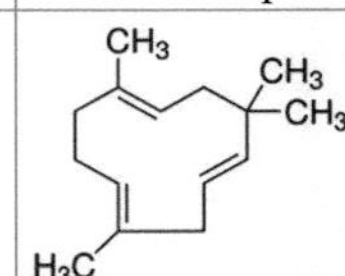 α-Humulene	Souza et al. (2012)
Chenopodium ambrosioides	α-Terpineol	*Ae. aegypti*	LC_{50} value of 35 mg/L	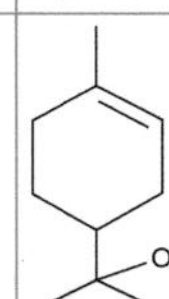 α-Terpineol	Leyva et al. (2009)

Chloroxylon swietenia	Germacrene D	*An. gambiae*, *Cx. quinquefasciatus*, and *Ae. aegypti*	LD_{50} values of 1.8, 2.1, and 2.8×10^{-3}	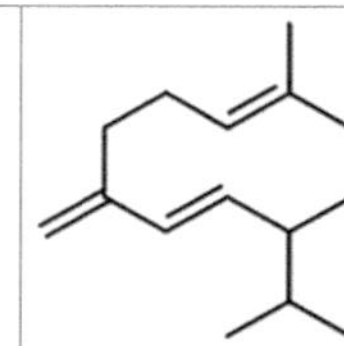 Germacrene D	Kiran and Devi (2007)
Cinnamomum impressicostatum	Benzyl benzoate and α-phellandrene	*Ae. aegypti*	LC_{50} value of 10.7 mg/L	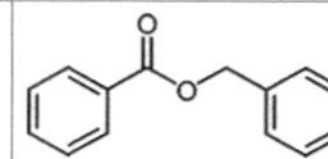 Benzyl benzoate 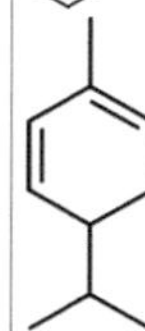 α-phellandrene	Jantan et al. (2005)
Cinnamomum scortechinii	β-Phellandrene and linalool	*Ae. aegypti*	LC_{50} value of 21.5 mg/L	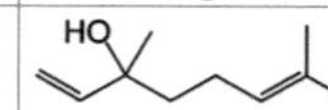 Linalool	Jantan et al. (2005)
Coleus aromaticus	Thymol, terpinen-4-ol, and trans-caryophyllene	*Cx. tritaeniorhynchus*, *Ae. albopictus*, and *An. subpictus*	LC_{50} values of 28.19, 24.83, and 22.06 μg/mL, respectively	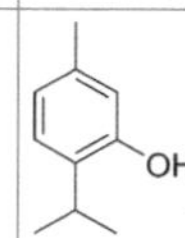 Thymol 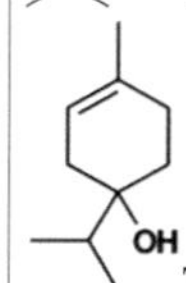 Terpinen-4-ol 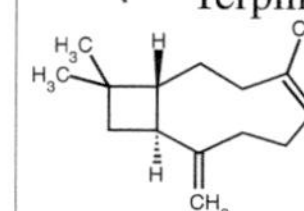 trans-Caryophyllene	Govindarajan et al. (2013b)

(continued)

Table 3.2 (continued)

Plants	Active ingredient	Mosquito	LC/LD values	Compound structure	References
Copaifera multijuga	β-Caryophyllene	*Ae. aegypti*	LC_{50} value of 18 mg/L	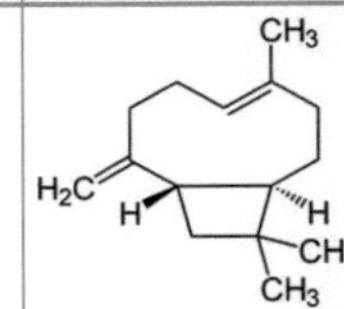 β-Caryophyllene	Trindade et al. (2013)
Cordia leucomalloides	δ-Cadinene and (E)-caryophyllene	*Ae. aegypti*	LC_{50} value of 63.1 mg/L	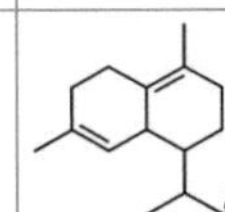 δ-Cadinene 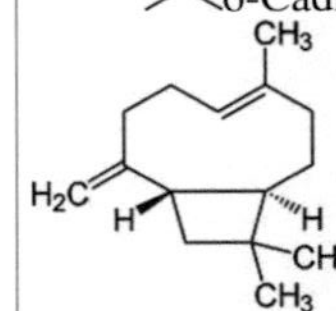 (E)- caryophyllene	Santos et al. (2006)
Croton argyrophylloides	β-trans-Guaiene	*Ae. aegypti*	LC_{50} value of 94.6 mg/L	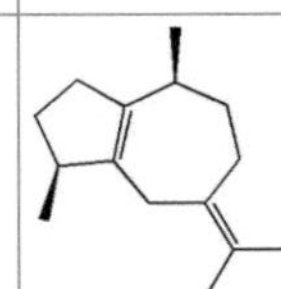 β-trans-Guaiene	Lima et al. (2013)
Croton sonderianus	Spathulenol	*Ae. aegypti*	LC_{50} value of 54.5 mg/L	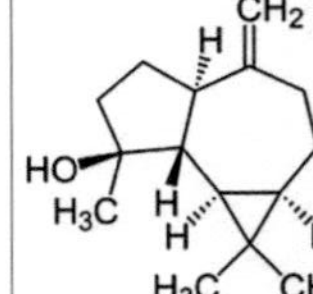 Spathulenol	Lima et al. (2013)

Cryptomeria japonica	16-Kaurene and elemol	*Ae. aegypti*	LC_{50} values of 28.4–56.7 mg/L	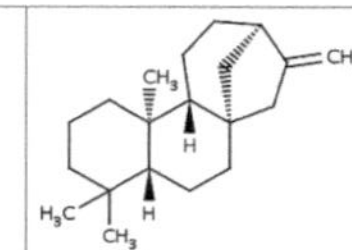 16-Kaurene 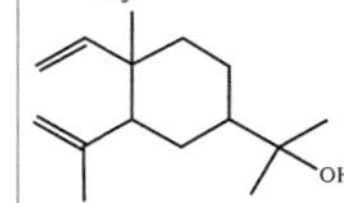 Elemol	Cheng et al. (2009)
Curcuma aromatica	1H-3a,7-Methanoazulene and curcumene	*Ae. aegypti*	LC_{50} value of 36.3 mg/L	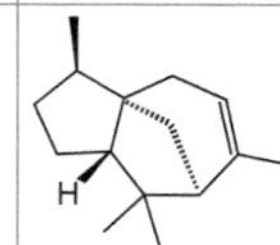 1H-3a,7- Methanoazulene 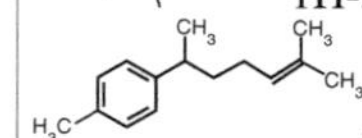 Curcumene	Choochate et al. (2005)
Dendropanax morbifera	γ-Elemene	*Ae. aegypti*	LC_{50} value of 62.32 mg/L	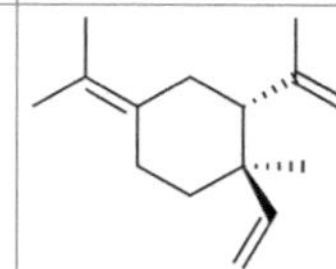 γ-Elemene	Chung et al. (2009)
Eucalyptus camaldulensis	α-Terpinene	*Ae. aegypti*	LC_{50} value of 14.7 μg/mL	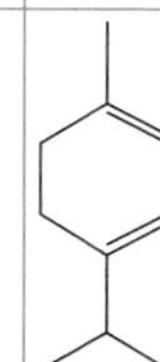 α-terpinene	Lucia et al. (2008)

(continued)

Table 3.2 (continued)

Plants	Active ingredient	Mosquito	LC/LD values	Compound structure	References
Eucalyptus camaldulensis	1,8-Cineol, p-cymene, and β-phellandrene	*Ae. aegypti*	LC_{50} value of 26.75 mg/L	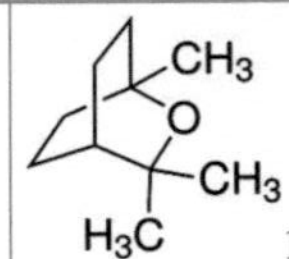 1,8-Cineol 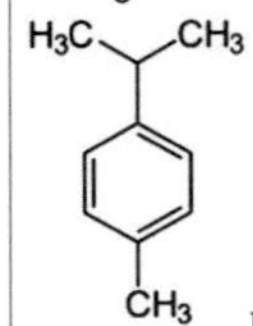 p-cymene 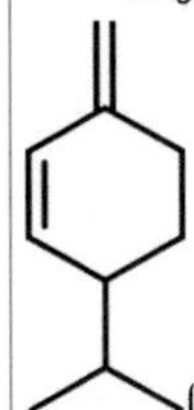 β-phellandrene	Lucia et al. (2008)
Eucalyptus globulus	1,8-Cineol	*Ae. aegypti*	LC_{50} value of 52.9 mg/L	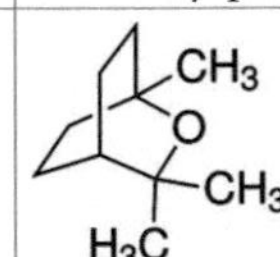 1,8-Cineol	Massebo et al. (2009)
Guarea scabra	cis-Caryophyllene	*Ae. aegypti*	LC_{50} value of 98.6 mg/L	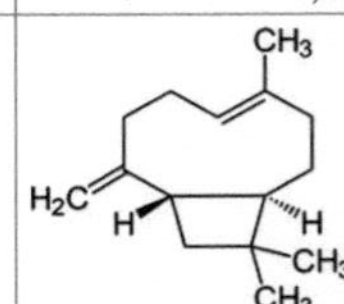 cis-Caryophyllene	Magalhães et al. (2010)

Guatteria friesiana	β-Eudesmol	*Ae. aegypti*	LC_{50} value of 52.6 mg/L	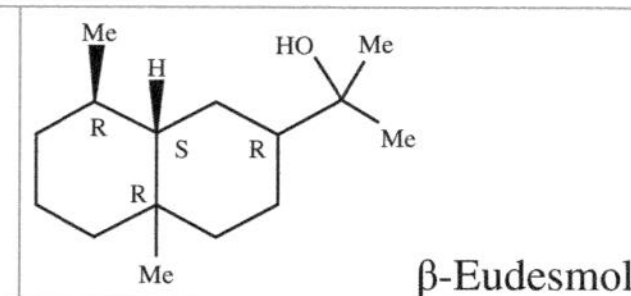	Aciole et al. (2011)
Hugonia castaneifolia	Hugorosenone	*An. gambiae*	LC_{50} values of 0.3028 mg/ml	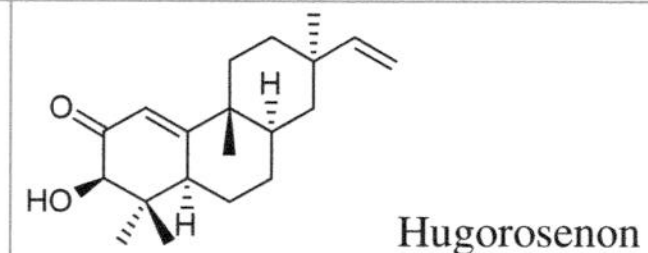	Baraza et al. (2008)
Hymenaea courbaril	α-Copaene and spathulenol	*Ae. aegypti*	LC_{50} value of 14.8 mg/L	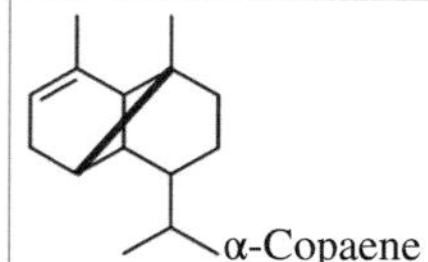	Aguiar et al. (2010)
Hyptis martiusii	δ-3-Carene and 1,8-cineol	*Ae. aegypti*	LC_{50} value of 18.5 mg/L	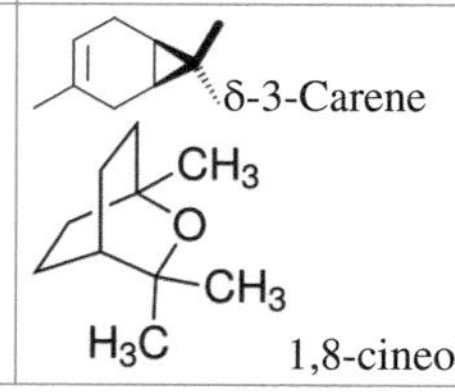	Costa et al. (2005)

(continued)

Table 3.2 (continued)

Plants	Active ingredient	Mosquito	LC/LD values	Compound structure	References
Lavandula gibsoni	α-Terpinolene and thymol	*Ae. aegypti*	LC_{50} value of 48.32 mg/L	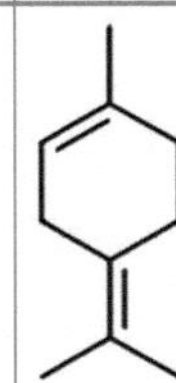 α-Terpinolen 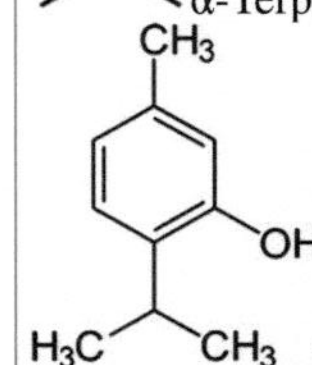 Thymol	Kulkarni et al. (2013)
Lippia sidoides	Thymol	*Ae. aegypti*	LC_{50} value of 19.5 mg/L	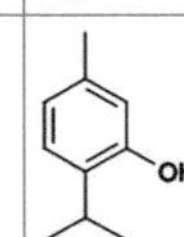 Thymol	Costa et al. (2005)

Melaleuca leucadendron	1,8-Cineol, α-pinene, and α-terpineol	*Ae. aegypti*	LC_{50} value of 41 mg/L	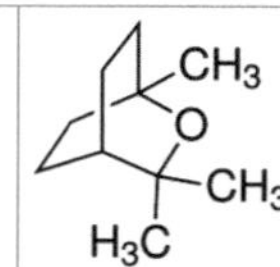 1,8-Cineol 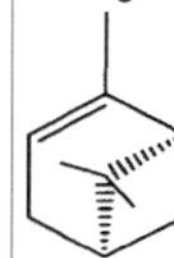 α-pinene 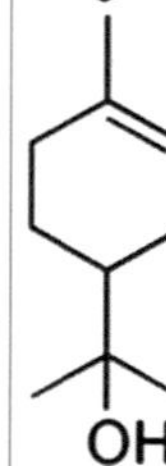 α-terpineol	Leyva et al. (2008)
Mentha spicata	Carvone, *cis*-carveol, and limonene	*Cx. quinquefasciatus*, *Ae. aegypti*, and *An. stephensi*	The LC_{50} values of carvone, cis-carveol, and limonene appeared to be most effective against *An. stephensi* (LC_{50} 19.33, 28.50, and 8.83 ppm) followed by *Ae. aegypti* (LC_{50} 23.69, 32.88, and 12.01 ppm), and *Cx. quinquefasciatus* (LC_{50} 25.47, 35.20, and 14.07 ppm)	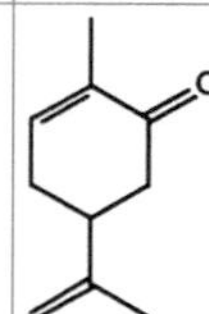 Carvone 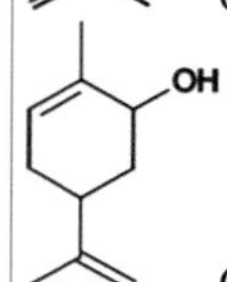 *Cis*-carveol 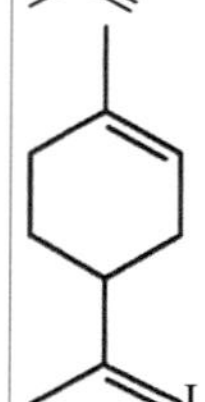 Limonene	Govindarajan et al. (2012c)

(continued)

Table 3.2 (continued)

Plants	Active ingredient	Mosquito	LC/LD values	Compound structure	References
Neorautanenia mitis	Pachyrrhizine	*An. gambiae*	LC_{50} value of 0.007 mg/ml	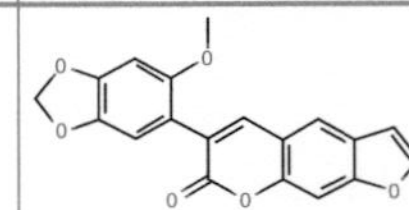 Pachyrrhizine	Joseph et al. (2004)
Nepeta cataria	(E),(Z)-Nepetalactone and (Z),(E)-nepetalactone	*Ae. aegypti*	LC_{50} value of 70 mg/L	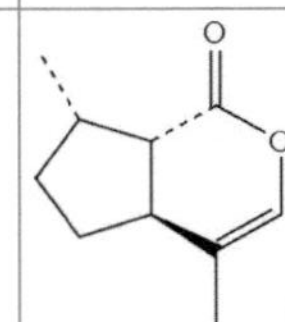 (E),(Z)-Nepeta cataria 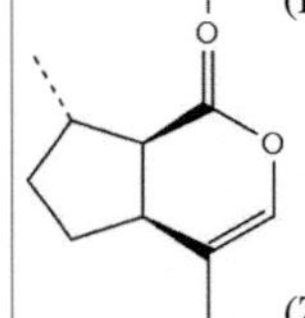 (Z),(E)-nepetalactone	Zhu et al. (2006)
Ocimum americanum	(E)-Methyl-cinnamate	*Ae. aegypti*	LC_{50} value of 67 mg/L	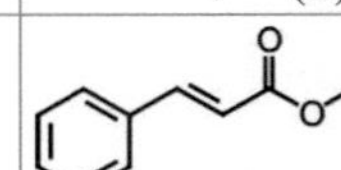 (E)-Methyl-cinnamate	Cavalcanti et al. (2004)
Pinus longifolia	K-terpineol	*Ae. aegypti*	LC_{50} value of 82.1 mg/L	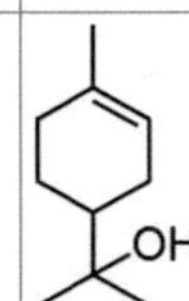 K-terpineol	Ansari et al. (2005)

Piper klotzschianum	1-Butyl-3,4-methylenedioxybenzene, limonene, and α-phellandrene	*Ae. aegypti*	LC_{50} value of 13.27 mg/L	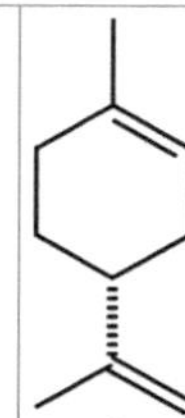 Limonene 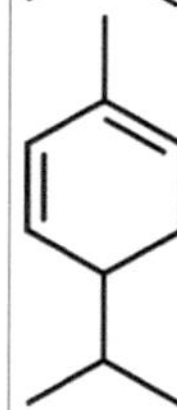 α-phellandrene	Nascimento et al. (2013)
Piper longum	Pipernonaline	*Ae. aegypti* and *Cx. pipiens*	LC_{50} values of 0.25 and 0.21 mg/l, respectively	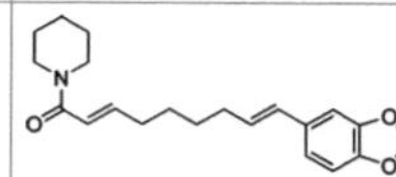 Pipernonalin	Lee (2000)
Piper nigrum	(E)-Caryophyllene and caryophyllene oxide	*Ae. aegypti*	LC_{50} value of 50 mg/L	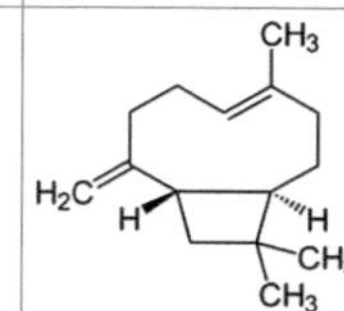 (E)-Caryophyllene 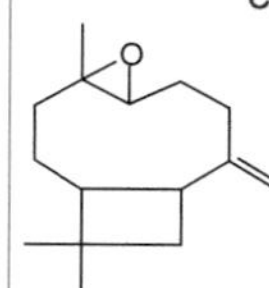 Caryophyllene	Amer and Mehlhorn (2006)
Plumbago zeylanica	Plumbagin	*An. gambiae*	LC_{50} value of 1.9 μg/ml	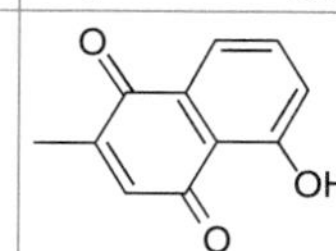 Plumbagin	Maniafu et al. (2009)

(continued)

Table 3.2 (continued)

Plants	Active ingredient	Mosquito	LC/LD values	Compound structure	References
Psidium rotundatum	1,8-Cineol	*Ae. aegypti*	LC_{50} value of 63 mg/L	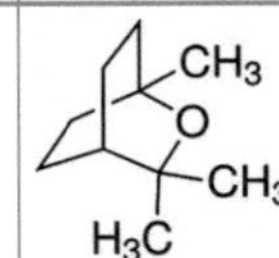 1,8-Cineol	Aguilera et al. (2003)
Spondias purpurea	Caryophyllene oxide and α-cadinol	*Ae. aegypti*	LC_{50} value of 39.7 mg/L	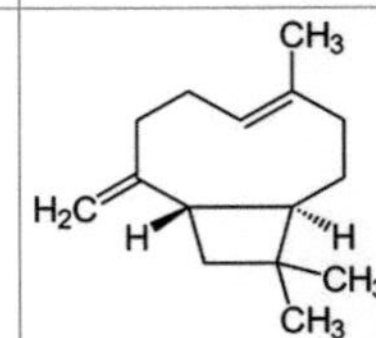 Caryophyllene oxide 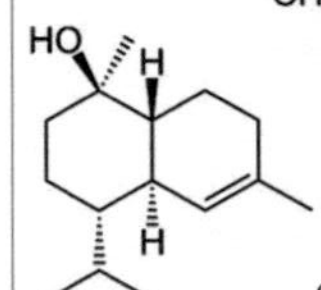 α-cadinol	Lima et al. (2011)
Stemodia maritima	β-Caryophyllene and caryophyllene oxide	*Ae. aegypti*	LC_{50} value of 22.9 mg/L	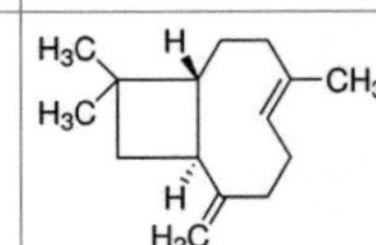 β-Caryophyllene 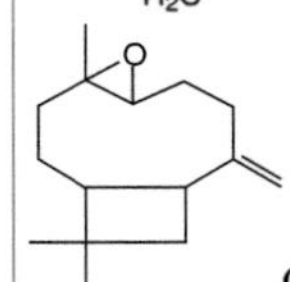 Caryophyllene	Arriaga et al. (2007)

Stemona curtisii	Stemocurtisine, stemocurtisinol, and oxyprotostemonine	*An. minimus*	LC_{50} values of 18, 39, and 4 ppm, respectively	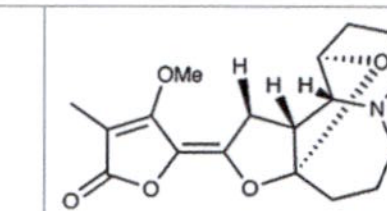 Stemocurtisine 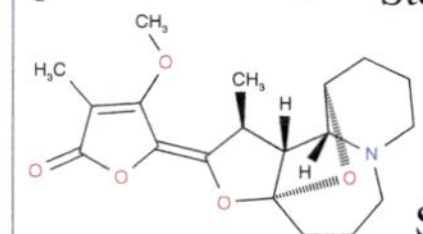 Stemocurtisinol 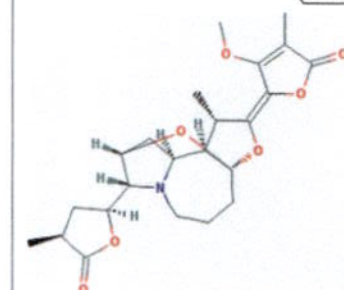 Oxyprotostemonine	Mungkornasawakul et al. (2009)
Tagetes patula	Limonene and terpinolene	*Ae. aegypti*	LC_{50} value of 13.57 mg/L	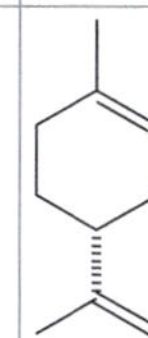 Limonene 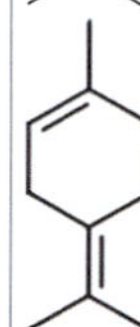 Terpinolene	Dharmagadda et al. (2005)
Triphyophyllum	Dioncophylline-A peltatum	*An. stephensi*	LD_{50} values of 0.5, 1.0, and 2.0 mg/L concentrations at 3.33, 2.66, and 1.92 h	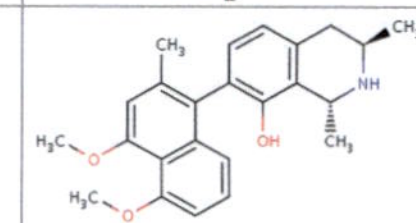 Dioncophylline-A peltatum	Francois et al. (1996)

(continued)

Table 3.2 (continued)

Plants	Active ingredient	Mosquito	LC/LD values	Compound structure	References
Vitex trifolia	Methyl-*p*-hydroxybenzoate	*Cx. quinquefasciatus* and *Ae. aegypti*	LC_{50} values of 5.77 and 4.74 ppm, respectively	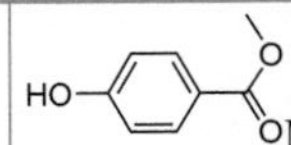 Methyl-*p*- hydroxybenzoate	Kannathasan et al. (2011)
Zanthoxylum limonella	(R)-(+)-Limonene	*Ae. aegypti*	LC_{50} value of 24.6L	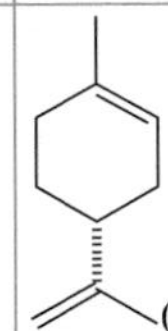 (R)-(+)-Limonene	Pitasawat et al. (2007)
Zanthoxylum oxyphyllum	Methyl heptyl ketone	*Ae. aegypti*	LC_{50} value of 7.52 mg/L	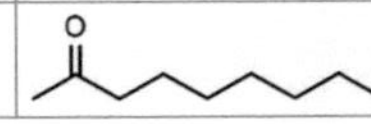 Methyl heptyl	Borah et al. (2012)

widely investigated, less number of patents has been applied regarding the production of formulations for use against mosquito larvae in the field. This review demonstrates the need for standardization of methodologies for the evaluation of larvicides against mosquitoes. The research regarding the search for new larvicides should be performed in a standardized manner. The features from plants (collection, extraction, chemical constitution) and insects (collection, age) and the methodological procedures must be well defined in research. We need to overcome the barriers to commercialization of new botanical insecticides. There has to be a paradigm shift in natural product research, in which studies are conducted with the ultimate goal of producing plant-based larvicides for use in public health programs.

References

Aciole SDG, Piccoli CF, Duquel JE, Costa EV, Navarro-Silva MA, Marques FA, Maia BLNS, Pinheiro MLB, Rebelo MT (2011) Insecticidal activity of three species of *Guatteria* (Annonaceae) against *Aedes aegypti* (Diptera: Culicidae). Rev Colomb Entomol 37:262–268

Aguiar JCD, Santiago GMP, Lavor PL (2010) Chemical constituents and larvicidal activity of *Hymenaea courbaril* fruit peel. Nat Prod Commun 5:1977–1980

Aguilera L, Navarro A, Tacoronte JE, Leyva M, Marquetti MC (2003) Efecto letal de *myrtaceas cubanas* sobre *Aedes aegypti* (Diptera: Culicidae). Rev Cubana Med Trop 55:100–104

Al-Doghairi M, El-Nadi A, El hag E, Al-Ayedh H (2004) Effect of *Solenostemma argel* on oviposition, egg hatchability and viability of *Culex pipiens* L. larvae. Phytother Res 18:335–338

Aly MZY, Badran RAM (1996) Mosquito control with extracts from plants of the Egyptian Eastern Desert. J Herbs Spices Med Plants 3:3540–3580

Amer A, Mehlhorn H (2006) Larvicidal effects of various essential oils against *Aedes, Anopheles*, and *Culex* larvae (Diptera, Culicidae). Parasitol Res 99:466–472

Amusan AA, Idowu AB, Arowolo FS (2005) Comparative toxicity effect of bush tea leaves (*Hyptis suaveolens*) and orange peel (*Citrus sinensis*) oil extract on larvae of the yellow fever mosquito *Aedes aegypti*. Tanzan Health Res Bull 7(3):174–178

Anees AM (2008) Larvicidal activity of *Ocimum sanctum* Linn. (Labiatae) against *Aedes aegypti* (L.) and *Culex quinquefasciatus* (Say). Parasitol Res 101:1451–1453

Ansari MA, Mittal PK, Razdan RK, Sreehari U (2005) Larvicidal and mosquito repellent activities of pine (*Pinus longifolia*, Family: Pinaceae) oil. J Vector Borne Dis 42:95–99

Arriaga AMC, Rodrigues FEA, Lemos TLG, Oliveira MCF, Lima JQ, Santiago GMP, Braz-Filho R, Mafezoli J (2007) Composition and larvicidal activity of essential oil from *Stemodia maritima* L. Nat Prod Commun 2:1237–1239

Azmi MA, Naqvi SNH, Ahmad I, Tabassum R, Anbreen B (1998) Toxicity of neem leaves extract (NLX) compared with malathion (57 E.C.) against late 3rd instar larvae of *Culex fatigans* (Wild Strain) by WHO method. Turk J Zool 22:213–218

Bagavan A, Rahuman AA, Kamaraj C, Geetha K (2008) Larvicidal activity of saponin from *Achyranthes aspera* against *Aedes aegypti* and *Culex quinquefasciatus* (Diptera: Culicidae). Parasitol Res 103:223–229

Bagavan A, Kamaraj C, Rahuman A, Elango G, Zahir AA, Pandiyan G (2009) Evaluation of larvicidal and nymphicidal potential of plant extracts against *Anopheles subpictus* Grassi, *Culex tritaeniorhynchus* Giles and *Aphis gossypii* Glover. Parasitol Res 104:1109–1117

Baraza LD, Joseph CC, Munissi JJE, Nkunya MHH, Arnold N, Porzel A (2008) Antifungal rosane diterpenes and other constituents of *Hugonia castaneifolia*. Phytochemistry 69:200–205

Borah R, Saikia K, Talukdar AK, Kalita MC (2012) Chemical composition and biological activity of the leaf essential oil of *Zanthoxylum oxyphyllum*. Planta Med 78:100

Breman JG, Martin AS, Mills A (2004) Conquering the intolerable burden of malaria: what's new, what's needed: a summary. Am J Trop Med Hyg 71(2):1–15

Burfield T, Reekie SL (2005) Mosquitoes, malaria and essential oils. Int J Aromather 15:30–41

Cavalcanti ESB, Morais SM, Lima MAA, Santana EWP (2004) Larvicidal activity of essential oils from Brazilian plants against *Aedes aegypti* L. Mem Inst Oswaldo Cruz 99:541–544

Chaithong U, Choochote W, Kamsuk K, Jitpakdi A, Tippawangkosol P, Chaiyasit D (2006) Larvicidal effect of pepper plants on *Aedes aegypti* (L.) (Diptera: Culicidae). J Vector Ecol 31:138–144

Chansang U, Zahiri NS, Bansiddhi J, Boonruad T, Thongsrirak P, Mingmuang J (2005) Mosquito larvicidal activity of crude extracts of long pepper (*Piper retrofractum* vahl) Thailand. J Vector Ecol 30:195–200

Cheng SS, Chang HT, Chang ST, Tsai KH, Chen WJ (2003) Bioactivity of selected plant essential oils against the yellow fever mosquito *Aedes aegypti* larvae. Bioresour Technol 89(1):99–102

Cheng SS, Chua MT, Chang EH, Huang CG, Chen WJ, Chang ST (2009) Variations in insecticidal activity and chemical compositions of leaf essential oils from *Cryptomeria japonica* at different ages. Bioresour Technol 100(1):465–470

Choochate W, Kanjanapothi D, Panthong A, Taesotikul T, Jitpakdi A, Chaithong U (1999) Larvicidal, adulticidal and repellent effects of *Kaempferia galangal*.

Southeast Asian J Trop Med Public Health 30:470–476
Choochate W, Tuetun B, Kanjanapothi D, Rattanachanpichai E, Chaithong U, Chaiwong P (2004) Potential of crude seed extract of celery, *Apium graveolens* L., against the mosquito *Aedes aegypti* (L.) (Diptera: Culicidae). J Vector Ecol 29:340–346
Choochate W, Chaiyasit D, Kanjanapothi D, Rattanachanpichai E, Jitpakdi A, Tuetun B, Pitasawat B (2005) Chemical composition and anti-mosquito potential of rhizome extract and volatile oil derived from *Curcuma aromatica* against *Aedes aegypti* (Diptera: Culicidae). J Vector Ecol 30:302–309
Chowdhury N, Chatterjee SK, Laskar S, Chandra G (2009) Larvicidal activity of *Solanum villosum* Mill (Solanaceae: Solanales) leaves to *Anopheles subpictus* Grassi (Diptera: Culicidae) with effect on non-target *Chironomus circumdatus* Kieffer (Diptera: Chironomidae). J Pest Sci 82:13–18
Chung IM, Seo SH, Kang EY, Park SD, Park WH, Moon HI (2009) Chemical composition and larvicidal effects of essential oil of *Dendropanax morbifera* against *Aedes aegypti* L. Biochem Syst 37(4):470–473
Costa JGM, Pessoa ODL, Menezes EA, Santiago GMP, Lemos TLG (2004) Composition and larvicidal activity of essential oils from heartwood of *Auxemma glazioviana* Taub. (Boraginaceae). Flavour Fragr J 19(6):529–531
Costa JGM, Rodrigues FFG, Angélico EC (2005) Chemical-biological study of the essential oils of *Hyptis martiusii, Lippia sidoides* and *Syzigium aromaticum* against larvae of *Aedes aegypti* and *Culex quinquefasciatus*. Braz J Pharmacogn 15:304–309
Das MK, Ansari MA (2003) Evaluation of repellent action of *Cymbopogan martinii martinii* Stapf var Sofia oil against *Anopheles sundaicus* in tribal villages of Car Nicobar Island, Andaman & Nicobar Islands, India. J Vector Borne Dis 40:101–104
de Omena C, Navarro DMAF, de Paula JE, Ferreira de Lima JSMR, Sant'Ana AEG (2007) Larvicidal activities against *Aedes aegypti* of Brazilian medicinal plants. Bioresour Technol 98:2549–2556
Dharmagadda VSS, Naik SN, Mittal PK, Vasudevan P (2005) Larvicidal activity of *Tagetes patula* essential oil against three mosquito species. Bioresour Technol 6(11):1235–1240
Dua VK, Pandey AC, Alam ME, Dash AP (2006) Larvicidal activity of *Hibiscus abelmoschus* Linn. (Malvaceae) against mosquitoes. J Am Mosq Control Assoc 22(1):155–157
Francois G, Looveren MV, Timperman G, Chimanuka B, Assi LA, Holenz J (1996) Larvicidal activity of the naphthylisoquinoline alkaloid dioncophylline-A against the malaria vector *Anopheles stephensi*. J Ethnopharmacol 54:125–130
Ghosh A, Chandra G (2006) Biocontrol efficacy of *Cestrum diurnum* (L.) (Solanales: Solanaceae) against the larval forms of *Anopheles stephensi*. Nat Prod Res 20:371–379
Gois RWS (2010) Estudo fitoquímico e biológico de Bauhinia acuruana Moric. Dissertation, Federal University of Ceara
Govindarajan M (2010a) Larvicidal efficacy of *Ficus benghalensis* L. plant leaf extract against *Culex quinquefasciatus* Say. *Aedes aegypti* L. and *Anopheles stephensi* L. (Diptera: Culicidae). Eur Rev Med Pharmacol Sci 14(2):107–111
Govindarajan M (2010b) Larvicidal and repellent activities of *Sida acuta* Burm. F. (Family: Malvaceae) against three important vector mosquitoes. Asia Pac J Trop Med 3(9):691–695
Govindarajan M (2011) Mosquito larvicidal and ovicidal activity of *Cardiospermum halicacabum* Linn. (Family: Sapindaceae) Leaf extract against *Culex quinquefasciatus* (say.) and *Aedes aegypti* (Linn.) (Diptera: Culicidae). Eur Rev Med Pharmacol Sci 15(7):787–794
Govindarajan M (2013) Larvicidal activity of *Cassia fistula* flower extract against *Culex tritaeniorhynchus* Giles, *Aedes albopictus* Skuse and *Anopheles subpictus* Grassi (Diptera: Culicidae). J Pure Appl Zool 1(2):117–121
Govindarajan M, Sivakumar R (2013a) Larvicidal, ovicidal, and adulticidal efficacy of *Erythrina indica* (Lam.) (Family: Fabaceae) against *Anopheles stephensi, Aedes aegypti*, and *Culex quinquefasciatus* (Diptera: Culicidae). Parasitol Res 113:777–791
Govindarajan M, Sivakumar R (2013b) Chemical composition and larvicidal properties of *Clausena anisata* (Willd.) Hook. F. Ex Benth (Rutaceae) essential oil against *Anopheles subpictus* and *Aedes albopictus* (Diptera: Culicidae). Int J Curr Med Sci 3(1):10–14
Govindarajan M, Sivakumar R (2013c) Ovicidal, larvicidal and adulticidal properties of *Asparagus racemosus* (Willd.) (Family: Asparagaceae) root extracts against filariasis (*Culex quinquefasciatus*), dengue (*Aedes aegypti*) and malaria (*Anopheles stephensi*) vector mosquitoes (Diptera: Culicidae). Parasitol Res 113:1435–1449
Govindarajan M, Jebanesan A, Pushpanathan T, Samidurai K (2008) Studies on effect of *Acalypha indica* L. (Euphorbiaceae) leaf extracts on the malarial vector, *Anopheles stephensi* Liston (Diptera: Culicidae). Parasitol Res 103:691–695
Govindarajan M, Sivakumar R, Amsath A, Niraimathi S (2011a) Mosquito larvicidal properties of *Ficus benghalensis* L. (Family: Moraceae) against *Culex tritaeniorhynchus* Giles and *Anopheles subpictus* Grassi (Diptera: Culicidae). Asia Pac J Trop Med 4(7):505–509
Govindarajan M, Sivakumar R, Rajeswari M (2011b) Larvicidal efficacy of *Cassia fistula* Linn. leaf extract against *Culex tritaeniorhynchus* Giles and *Anopheles subpictus* Grassi (Diptera: Culicidae). Asia Pac J Trop Dis 1(4):295–298
Govindarajan M, Mathivanan T, Elumalai K, Krishnappa K, Anandan A (2011c) Mosquito larvicidal, ovicidal, and repellent properties of botanical extracts against *Anopheles stephensi, Aedes aegypti,* and *Culex quin-*

quefasciatus (Diptera: Culicidae). Parasitol Res 109:353–367

Govindarajan M, Sivakumar R, Rajeswari M, Yogalakshmi K (2012a) Larvicidal and ovicidal properties of leaf and seed extracts of *Delonix elata* (L.) Gamble (Family: Fabaceae) against malaria (*Anopheles stephensi* Liston) and dengue (*Aedes aegypti* Linn.) (Diptera: Culicidae) vector mosquitoes. Parasitol Res 111:65–77

Govindarajan M, Rajeswary M, Sivakumar R (2012b) Mosquito larvicidal and ovicidal activity of *Delonix elata* (L.) Gamble against *Culex quinquefasciatus* Say (Diptera: Culicidae). Asia Pac J Trop Dis 2(2):571–573

Govindarajan M, Sivakumar R, Rajeswari M, Yogalakshmi K (2012c) Chemical composition and larvicidal activity of essential oil from *Mentha spicata* (Linn.) against three mosquito species. Parasitol Res 110:2023–2032

Govindarajan M, Rajeshwary M, Amsath A (2013a) Larvicidal properties of *Caesalpinia pulcherrima* (Family: Fabaceae) against *Culex tritaeniorhynchus*, *Aedes albopictus* and *Anopheles subpictus* (Diptera: Culicidae). Int J Pure Appl Zool 1(1):15–23

Govindarajan M, Sivakumar R, Rajeswary M, Veerakumar K (2013b) Mosquito larvicidal activity of thymol from essential oil of *Coleus aromaticus* Benth. against *Culex tritaeniorhynchus*, *Aedes albopictus*, and *Anopheles subpictus* (Diptera: Culicidae). Parasitol Res 112(11):3713–3721

Govindarajan M, Sivakumar R, Rajeswari M, Yogalakshmi K (2013c) Chemical composition and larvicidal activity of essential oil from *Ocimum basilicum* (L.) against *Culex tritaeniorhynchus*, *Aedes albopictus* and *Anopheles subpictus* (Diptera: Culicidae). Exp Parasitol 134(1):7–11

Govindarajan M, Rajeswary M, Sivakumar R (2013d) Larvicidal and ovicidal efficacy of *Pithecellobium dulce* (Roxb.) Benth. (Fabaceae) against *Anopheles stephensi* Liston and *Aedes aegypti* Linn. (Diptera: Culicidae). Indian J Med Res 138(1):129–134

Hahn CS, French OG, Foley P, Martin EN, Taylor RP (2001) Bispecific Monoclonal Antibodies Mediate Binding of Dengue Virus to Erythrocytes in a Monkey Model of Passive Viremia. J Immunol 66(2):1057–1065

Heitz JR, Downum KR (1987) Light activated pesticides A.C.S. Series. American Chemical Society, Washington, p 339

Iannacone J, Pérez D (2004) Insecticidal effect of *Paullinia clavigera* var. bullata Simpson (Sapindaceae) and *Tradescantia zebrine* Hort ex Bosse (Commelinaceae) in the control of *Anopheles benarrochi* Gabaldon, Cova García & López 1941, main vector of malaria in Ucayali. Peru Ecol Aplicada 3:64–72

Isman MB (1997) Neem and other botanical insecticides: barriers to commercialization. Phytoparasitica 25:339–344

Jang YS, Baek BR, Yang YC, Kim MK, Lee HS (2002) Larvicidal activity of leguminous seeds and grains against *Aedes aegypti* and *Culex pipiens pallens*. J Am Mosq Control Assoc 18:210–213

Jang YS, Jeon JH, Lee HS (2005) Mosquito larvicidal activity of active constituent derived from *Chamaecyparis obtusa* leaves against 3 mosquito species. J Am Mosq Control Assoc 21:400–403

Jantan I, Yalvema MF, Ahmad NW, Jamal JA (2005) Insecticidal activities of the leaf oils of eight *Cinnamomum* species against *Aedes aegypti* and *Aedes albopictus*. Pharm Biol 43(6):526–532

John William S (2007) Mosquito, Man's enemy. In: John William S (ed) Defeating the public enemy, the mosquito: a real challenge. Loyola Publication, Chennai, pp 1–32

Joseph CC, Ndoile MM, Malima RC, Nkunya MHH (2004) Larvicidal and mosquitocidal extracts, a coumarin, isoflavonoids and pterocarpans from *Neorautanenia mitis*. Trans R Soc Trop Med Hyg 98:451–455

Kabaru JM, Gichia L (2001) Insecticidal activity of extracts derived from different parts of the mangrove tree *Rhizophora mucronata* (Rhizophoraceae) Lam. against three arthropods. Afr J Sci Technol 2:44–49

Kalyanasundaram M, Das PK (1985) Larvicidal and synergistic activity of plant extracts for mosquito control. Indian J Med Res 82:19–23

Kamaraj C, Rahuman AA (2010) Larvicidal and adulticidal potential of medicinal plant extracts from south India against vectors. Asia Pac J Trop Med 3:948–953

Kamaraj C, Rahuman AA, Bagavan A (2008) Screening for antifeedant and larvicidal activity of plant extracts against *Helicoverpa armigera* (Hübner), *Sylepta derogata* (F.) and *Anopheles stephensi* (Liston). Parasitol Res 103(6):1361–1368

Kamaraj C, Rahuman AA, Bagavan A, Abduz Zahir A, Elango G, Kandan P (2010) Larvicidal efficacy of medicinal plant extracts against *Anopheles stephensi* and *Culex quinquefasciatus* (Diptera: Culicidae). Trop Biomed 27:211–219

Kamaraj C, Bagavan A, Elango G, Zahir AA, Rajkumar G, Mariamuthu S (2011) Larvicidal activity of medicinal plant extracts against *Anopheles stephensi* and *Culex tritaeniorhynchus*. Indian J Med Res 134:101–106

Kannathasan K, Senthilkumar A, Venkatesalu V (2011) Mosquito larvicidal activity of methyl-*p*-hydroxybenzoate isolated from the leaves of *Vitex trifolia* Linn. Acta Trop 120:115–118

Karmegan N, Sakthivadivel M, Anuradha V, Daniel T (1997) Indigenous plant extracts as larvicidal agents against *Culex quinquefasciatus* Say. Bioresour Technol 5:137–140

Karunamoorthi K, Ramanujam S, Rathinasamy R (2008) Evaluation of leaf extracts of *Vitex negundo* L. (Family: Verbenaceae) against larvae of *Culex tritaeniorhynchus* and repellent activity on adult vector mosquitoes. Parasitol Res 103(3):545–550

Kassir JT, Mohsen ZH, Mehdi NS (1989) Toxic effects of limonene against *Culex quinquefasciatus* Say larvae and its interference with oviposition. J Pest Sci 62:19–21

Kaushik R, Saini P (2008) Larvicidal activity of leaf extract of *Millingtonia hortensis* (Family: Bignoniaceae) against *Anopheles stephensi*, *Culex quinquefasciatus* and *Aedes aegypti*. J Vector Borne Dis 45:66–69

Khanna G, Kannabiran K (2007) Larvicidal effect of *Hemidesmus indicus, Gymnema sylvestre*, and *Eclipta prostrata* against *Culex qinquifaciatus* mosquito larvae. Afr J Biotechnol 6:307–311

Khatune NA, Md Haque E, Mosaddik A (2001) Laboratory evaluation of *Nyctanthes arbor-tristis* Linn. flower extract and its isolated compound against common filarial Vector, *Culex quinquefasciatus* Say (Diptera: culicidae) larvae. Pak J Biol Sci 4:585–587

Kiran SR, Devi PS (2007) Evaluation of mosquitocidal activity of essential oil and sesquiterpenes from leaves of *Chloroxylon swietenia* DC. Parasitol Res 101:413–418

Komalamisra N, Trongtokit Y, Rongsriyam Y, Apiwathnasorn C (2005) Screening for larvicidal activity in some Thai plants against four mosquito vector species. Southeast Asian J Trop Med Public Health 36(6):1412–1422

Kovendan K, Murugan K, Vincent S (2012) Evaluation of larvicidal activity of *Acalypha alnifolia* Klein ex Willd. (Euphorbiaceae) leaf extract against the malarial vector, *Anopheles stephensi*, dengue vector, *Aedes aegypti* and Bancroftian filariasis vector, *Culex quinquefasciatus* (Diptera: Culicidae). Parasitol Res 110:571–581

Krishnan K, Senthilkumar A, Chandrasekaran M, Venkatesalu V (2007) Differential larvicidal efficacy of four species of *Vitex* against *Culex quinquefasciatus* larvae. Parasitol Res 10:1721–1723

Kulkarni RR, Pawar PV, Joseph MP, Akulwad AK, Sen A, Joshi SP (2013) *Lavandula gibsoni* and *Plectranthus mollis* essential oils: chemical analysis and insect control activities against *Aedes aegypti, Anopheles stephensi* and *Culex quinquefasciatus*. J Pest Sci 86:713–718

Lee SE (2000) Mosquito larvicidal activity of pipernonaline, a piperidine alkaloid derived from long pepper, *Piper longum*. J Am Mosq Control Assoc 16:245–247

Leyva M, Tacoronte JE, Marquetti MC, Scull R, Montada D, Rodriguez Y, Bruzón RY (2008) Actividad insecticida de aceites esenciales de plantas en larvas de *Aedes aegypti* (Diptera: Culicidae). Rev Cubana Med Trop 60:78–82

Leyva M, Marquetti MC, Tacoronte JE, Scull R, Tiomno O, Mesa A, Montada D (2009) Actividad larvicida de aceites esenciales de plantas contra *Aedes aegypti* (L.) (Diptera: Culicidae). Rev Biomed 20:5–13

Lima MAA, Oliveira FFM, Gomes GA (2011) Evaluation of larvicidal activity of the essential oils of plants species from Brazil against *Aedes aegypti* (Diptera: Culicidae). Afr J Biotechnol 10:11716–11720

Lima GPG, Souza TM, Freire GP, Farias DF, Cunha AP, Ricardo NMPS, Morais SM, Carvalho AFU (2013) Further insecticidal activities of essential oils from *Lippia sidoides* and Croton species against *Aedes aegypti* L. Parasitol Res 112(5):1953–1958

Lucia A, Licastro S, Zerba E, Masuh H (2008) Yield, chemical composition, and bioactivity of essential oils from 12 species of Eucalyptus on *Aedes aegypti* larvae. Entomol Exp Appl 129(1):107–114

Lukwa N (1994) Do traditional mosquito repellent plants work as mosquito larvicides? Cent Afr J Med 40(11):306–309

Magalhães LAM, Lima MP, Marques MOM, Facanali R, Pinto ACS, Tadei WP (2010) Chemical composition and larvicidal activity against *Aedes aegypti* larvae of essential oils from four Guarea species. Molecules 15(8):5734–5741

Maniafu BM, Wilber L, Ndiege IO, Wanjala CCT, Akenga TA (2009) Larvicidal activity of extracts from three *Plumbago* spp against *Anopheles gambiae*. Mem Inst Oswaldo Cruz 104:813–817

Mansour SA, Messeha SS, EL-Gengaihi SE (2000) Botanical biocides. Mosquitocidal activity of certain Thymus capitatus constituents. J Nat Toxins 9:49–62

Massebo F, Tadesse M, Bekele T, Balkew M, Gebre-Michael T (2009) Evaluation on larvicidal effects of essential oils of some local plants against *Anopheles arabiensis* Patton and *Aedes aegypti* Linnaeus (Diptera, Culicidae) in Ethiopia. Afr J Biotechnol 8:4183–4188

Matasyoh JC, Wathuta EM, Kairuki ST, Chepkorir R, Kavulani J (2008) Aloe plant extracts as alternative larvicides for mosquito control. Afr J Biotechnol 7:912–915

Mathew N, Anitha MG, Bala TSL, Sivakumar SM, Narmadha R, Kalyanasundaram M (2009) Larvicidal activity of *Saraca indica*, *Nyctanthes arbor-tristis*, and *Clitoria ternatea* extracts against three mosquito vector species. Parasitol Res 104:1017–1025

Maurya P, Mohan L, Sharma P, Batabyal L, Srivastava CN (2007) Larvicidal efficacy of *Aloe barbadensis* and *Cannabis sativa* against the malaria vector *Anopheles stephensi* (Diptera: Culicidae). Entomol Res 37:153–156

Maurya P, Preeti S, Lalit M, Lata B, Srivastava CN (2009a) Evaluation of larvicidal nature of fleshy fruit wall of *Momordica charantia* Linn. (Family: cucurbitaceae) in the management of mosquitoes. Parasitol Res 105:1653–1659

Maurya P, Sharma P, Mohan L, Batabyal L, Srivastava CN (2009b) Evaluation of the toxicity of different phytoextracts of *Ocimum basilicum* against *Anopheles stephensi* and *Culex quinquefasciatus*. J Asia-Pac Entomol 12:113–115

Mgbemena IC (2010) Comparative evaluation of larvicidal potentials of three plant extracts on *Aedes aegypti*. J Am Sci 6:435–440

Mohamed MI, El-Mohamady RH, Mohamed HA (2003) Larvicidal activity and biochemical effects of certain plant oil extracts against *Culex pipiens* larvae (Diptera: Culicidae). J Egypt Acad Soc Environ Dev 3(1):75–93

Mohan DR, Ramaswamy M (2007) Evaluation of the larvicidal activity of the leaf extract of a weed plant, *Ageratina adenophora*, against two important species of mosquitoes, *Aedes aegypti* and *Culex quinquefasciatus*. Afr J Biol 6:631–638

Mohan L, Sharma P, Shrivastava CN (2006) Evaluation of *Solanum xanthocarpum* extract as a synergist for cypermethrin against larvae of filarial vector *Culex quinquefasciatus* (Say). Entomol Res 36:220–225

Mullai K, Jebanesan A, Pushpanathan T (2008) Mosquitocidal and repellent activity of the leaf extract of *Citrullus vulgaris* (cucurbitaceae) against the malarial vector, *Anopheles stephensi* liston (Diptera: Culicidae). Euro Rev Med Pharmacol Sci 12:1–7

Mungkornasawakul P, Chaiyong S, Sastraruji T, Jatisatienr A, Jatisatienr C, Pyne SG (2009) Alkaloids from the roots of *Stemona aphylla*. J Nat Prod 72:848–851

Murugan K, Jeyabalan D (1999) Effect of certain plant extracts against the mosquito, *Anopheles stephensi* Liston. Curr Sci 76:631–633

Mwaiko GL (1992) Citrus peel oil extracts as mosquito larvae insecticides. East Afr Med J 69(4):223–226

Nascimento JC, David JM, Barbosa LCA, Paula VF, Demuner AJ, David JP, Conserva LM, Ferreira-Jr JC, Guimarães EF (2013) Larvicidal activities and chemical composition of essential oils from *Piper klotzschianum* (Kunth) CDC. (Piperaceae). Pest Manag Sci 59(2):585–595

Nirmal Sharma, Qadry JS, Subramanium B, Verghese T, Rahman SJ, Sharma SK, Jalees S (1998) Larvicidal activity of *Gliricidia sepium* against mosquito larvae of *Anopheles stephansi*, *Aedes aegypti* and *Culex quinquefasciatus*. Pharm Biol 36(1):3–7

Pandey V, Veena A, Raghavendra K, Dash AP (2007) Strong larvicidal activity of three species of Spilanthes (Akarkara) against malaria (*Anopheles stephensi* Liston, *Anopheles culicifacies*, species C) and filaria vector (*Culex quinquefasciatus* Say). Parasitol Res 102:171–174

Perumalsamy H, Chang KS, Park C, Ahn YJ (2010) Larvicidal activity of *Asarum heterotropoides* root constituents against insecticide-susceptible and -resistant *Culex pipiens* pallens and *Aedes aegypti* and *Ochlerotatus togoi*. J Agric Food Chem 46(6):1420–1423

Pitasawat B, Champakaew D, Choochote W (2007) Aromatic plant derived essential oil: an alternative larvicide for mosquito control. Fitoterapia 78(3):205–210

Prabhakar K, Jebanesa A (2004) Larvicidal efficacy of some cucurbitacious plant leaf extracts against *Culex quinquefasciatus*. Bioresour Technol 95:113–114

Pushpalatha E, Muthukrishnan J (1999) Efficacy of two tropical plant extracts for the control of mosquitoes. J Appl Entomol 123:369–373

Pushpanathan T, Jebanesan A, Govindarajan M (2008) The essential oil of *Zingiber officinalis* Linn (Zingiberaceae) as a mosquito larvicidal and repellent agent against the filarial vector *Culex quinquefasciatus* Say (Diptera: Culicidae). Parasitol Res 102(6):1289–1291

Raghavendra K, Singh SP, Subbarao SK, Dash AP (2009) Laboratory studies on mosquito larvicidal efficacy of aqueous & hexane extracts of dried fruit of *Solanum nigrum* Linn. Indian J Med Res 130:74–77

Rahuman AA, Venkatesan P (2008) Larvicidal efficacy of five cucurbitaceous plant leaf extracts against mosquito species. Parasitol Res 103:133–139

Rahuman AA, Gopalarkrishnan G, Saleem G, Arumugam S, Himalayan B (2000) Effect of *Feronia limonia* on mosquito larvae. Fitoterapia 71:553–555

Rahuman AA, Gopalakrishnan G, Venkatesan P, Geetha K (2007) Larvicidal activity of some Euphorbiaceae plant extracts against *Aedes aegypti* and *Culex quinquefasciatus* (Diptera: Culicidae). Parasitol Res 102:867–873

Rahuman AA, Gopalakrishnan G, Venkatesan P, Geetha K (2008a) Isolation and identification of mosquito larvicidal compound from *Abutilon indicum* (Linn.) Sweet. Parasitol Res 102:981–988

Rahuman AA, Venkatesan P, Geetha K, Gopalakrishnan G, Bagavan A, Kamaraj C (2008b) Mosquito larvicidal activity of gluanol acetate, a tetracyclic triterpenes derived from *Ficus racemosa* Linn. Parasitol Res 103(2):333–339

Rahuman AA, Venkatesan P, Gopalakrishnan G (2008c) Mosquito larvicidal activity of oleic and linoleic acids isolated from *Citrullus colocynthis* (Linn.) Schrad. Parasitol Res 103(6):1383–1390

Raj Mohan D, Ramaswamy M (2007) Evaluation of larvicidal activity of the leaf extract of a weed plant, *Ageratina adenophora*, against two important species of mosquitoes, *Aedes aegypti* and *Culex quinquefasciatus*. Afr J Biotechnol 6:631–638

Rajeswary M, Govindarajan M (2013) Mosquito larvicidal and phytochemical properties of *Ageratina adenophora* (Asteraceae) against three important mosquitoes. J Vector Borne Dis 50:141–143

Rajkumar S, Jebanesan A (2005) Larvicidal and adult emergence inhibition effect of *Centella asiatica* Brahmi (umbelliferae) against mosquito *Culex quinquefasciatus* Say (Diptera: Culicidae). Afr J Biomed Res 8:31–33

Rajkumar S, Jebanesan A (2009) Larvicidal and oviposition activity of *Cassia obtusifolia* Linn (Family: Leguminosae) leaf extract against malarial vector, *Anopheles stephensi* Liston (Diptera: Culicidae). Parasitol Res 104:337–340

Ravikumar S, Ali M, Beula J (2011) Mosquito larvicidal efficacy of seaweed extracts against dengue vector of *Aedes aegypti*. Asian Pac J Trop Biomed 1:143–146

Rawani A, Mallick Haldar K, Ghosh A, Chandra G (2009) Larvicidal activities of three plants against filarial vector *Culex quinquefasciatus* Say (Diptera: Culicidae). Parasitol Res 105:1411–1417

Rawani A, Ghosh A, Chandra G (2010) Mosquito larvicidal activities of *Solanum nigrum* L. leaf extract against *Culex quinquefasciatus* Say. Parasitol Res 107:1235–1240

Rodrigues AMS, Paula JE, Roblot F, Fournet A, Espíndola LS (2005) Larvicidal activity of *Cybistax antisyphilitica* against *Aedes aegypti* larvae. Fitoterapia 76:755–757

Rongsriyam Y, Trongtokit Y, Komalamisra N, Sinchaipanich N, Apiwathnasorn C, Mitrejet A (2006) Formulation of tablets from the crude extract of *Rhinacanthus nasutus* (Thai local plant) against *Aedes aegypti* and *Culex quinquefasciatus* larvae: a preliminary study. Southeast Asian J Trop Med Public Health 37(2):265–271

Rosenthal GA, Janzen DH (1979) Herbivores: their interaction with secondary plant metabolites. Academic, New York, p 718

Sakthivadivel M, Daniel T (2008) Evaluation of certain insecticidal plants for the control of vector mosquitoes viz. *Culex quinquefasciatus*, *Anopheles stephensi* and *Aedes aegypti*. Appl Entomol Zool 43:57–63

Santos RP, Nunes EP, Nascimento RF, Santiago GMP, Menezes GHA, Silveira ER, Pessoa ODL (2006) Chemical composition and larvicidal activity of the essential oils of *Cordia leucomalloides* and *Cordia curassavica* from the northeast of Brazil. J Braz Chem Soc 17:1027–1030

Santos GKN, Dutra KA, Barros RA, Câmara CAG, Lira DD, Gusmão NB, Navarro DMAF (2012) Essential oils from *Alpinia purpurata* (Zingiberaceae): chemical composition, oviposition deterrence, larvicidal and antibacterial activity. Ind Crops Prod 40:254–260

Saxena RC, Dixit OP, Sukumaran P (1992) Laboratory assessment of indigenous plant extracts for anti-juvenile hormone activity in *Culex quinquefasciatus*. Indian J Med Res 95:199–204

Schmutterer H (1990) Properties and potential of natural pesticides from the neem tree, *Azadirachta indica*. Annu Rev Entomol 35:271–297

Senthil Nathan S, Kalaivani K, Murugan K (2005) Effects of neem limonoids on the malaria vector, *Anopheles stephensi* Liston (Diptera: Culicidae). Acta Trop 96:47–55

Senthil Nathan S, Savitha G, George DK, Narmadha A, Suganya L, Chung PG (2006a) Efficacy of *Melia azedarach* L. extract on the malarial vector *Anopheles stephensi* Liston (Diptera: Culicidae). Bioresour Technol 97:1316–1323

Senthil Nathan S, Kalaivani K, Sehoon K (2006b) Effects of *Dysoxylum malabaricum* Bedd. (Meliaceae) extract on the malarial vector *Anopheles stephensi* Liston (Diptera: Culicidae). Bioresour Technol 97:2077–2083

Service MW (1983) Management of vector. In: Youdeowei A, Service N (eds) Pest and vector management in the tropics. Longman Group, London, pp 7–20

Shaalan EAS, Canyonb D, Younesc MWF, Abdel-Wahaba H, Mansoura AH (2005) A review of botanical phytochemicals with mosquitocidal potential. Environ Int 3:1149–1166

Shahi M, Hanafi-Bojd AA, Iranshahi M, Vatandoost H, Hanafi-Bojd MY (2010) Larvicidal efficacy of latex and extract of *Calotropis procera* (Gentianales: Asclepiadaceae) against *Culex quinquefasciatus* and *Anopheles stephensi* (Diptera: Culicidae). J Vector Borne Dis 47:185–188

Sharma P, Mohan L, Srivastava CN (2005) Larvicidal potential of *Nerium indicum* and *Thuja oriertelis* extracts against malaria and Japanese encephalitis vector. J Environ Biol 26(4):657–666

Sharma P, Mohan L, Srivastava CN (2006) Phytoextract-induced developmental deformities in malaria vector. Bioresour Technol 97(14):1599–1604

Sheeren ME (2006) Larvicidal effects of *Eucalyptus* extracts on the larvae of *Culex pipiens* mosquito. Int J Agric Biol 8:896–897

Singh RK, Mittal PK, Dhiman RC (2005) Laboratory study on larvicidal properties of leaf extract of *Calotropis procera* (family-Asclepiadaceae) against mosquito larvae. J Commun Dis 37(2):109–113

Singh RK, Dhiman RC, Mittal PK (2006) Mosquito larvicidal properties of *Momordica charantia* Linn (Family: Cucurbitaceae). J Vector Borne Dis 43:88–91

Singh RK, Dhiman RC, Mittal PK (2007) Studies on mosquito larvicidal properties of *Eucalyptus citriodora* Hook (family-Myrtaceae). J Commun Disord 39:233–236

Sivagnaname S, Kalyanasundaram M (2004) Laboratory evaluation of methanolic extract of *Atlanta monophylla* (Family: Rutaceae) against immature stage of mosquitoes and nontarget organisms. Mem Inst Oswaldo Cruz 99:115–118

Sivakumar R, Govindarajan M (2013a) Mosquito adulticidal activity of *Coccinia indica* Wight and Arn. (Family: Cucurbitaceae) against *Culex quinquefasciatus* Say (Diptera: Culicidae). Int J Curr Sci Technol 2(1):4–8

Sivakumar R, Govindarajan M (2013b) Mosquito larvicidal efficacy of *Eclipta alba* (L.) Hassk. against filariasis vector, *Culex quinquefasciatus* Say (Diptera: Culicidae). Int J Curr Biochem Biotechnol 2(1):1–5

Sivakumar R, Jebanesan A, Govindarajan M, Rajasekar P (2011) Larvicidal and repellent activity of tetradecanoic acid against *Aedes aegypti* (Linn.) and *Culex quinquefasciatus* (Say.) (Diptera:Culicidae). Asian Pac J Trop Med 4:706–710

Snow RW, Guerra CA, Noor AM, Myint HY, Hay SI (2005) The global distribution of clinical episodes of *Plasmodium falciparum* malaria. Nature 434(7030):214–217

Souza LGS, Almeida MCS, Monte FJQ, Santiago GMP, Braz-Filho R, Lemos TLG, Gomes CL, Nascimento RF (2012) Chemical constituents of *Capraria biflora* (Scrophulariaceae) and larvicidal activity of essential oil. Quim Nova 35:2258–2262

Sukumar K, Perich MJ, Boobar LR (1991) Botanical derivatives in mosquito control: a review. J Am Mosq Control Assoc 7(2):210–237

Sumroiphon S, Yuwaree C, Arunlertaree C, Komalamisra N, Rongsriyam Y (2006) Bioactivity of citrus seed for mosquito-borne diseases larval control. Southeast Asian J Trop Med Public Health 37(3):123–127

Traboulsi AF, Taoubi K, El-Haj S, Bessiere JM, Salma R (2002) Insecticidal properties of essential plant oils against the mosquito *Culex pipiens molestus* (Diptera: Culicidae). Pest Manag Sci 58:491–495

Trindade FTT, Stabeli RG, Pereira AA, Facundo VA, Silva AA (2013) *Copaifera multijuga* ethanolic extracts, oil-resin, and its derivatives display larvicidal activity against *Anopheles darling* and *Aedes aegypti* (Diptera: Culicidae). Braz J Pharmacogn 23(3):464–470

Vahitha R, Venkatachalam MR, Murugan K, Jebanesan A (2002) Larvicidal efficacy of *Pavonia zeylanica* L. and *Acacia ferruginea* D.C. against *Culex quinquefasciatus* Say. Bioresour Technol 82:203–204

Ware GW (1978) In pesticides. Theory and application. Freeman, New York, p 718

World Health Organization (2006) Global program to eliminate lymphatic filariasis. Wkly Epidemiol Rec 81:221–232

Yadav R, Srivastava VK, Chandra R, Singh A (2002) Larvicidal activity of Latex and stem bark of *Euphorbia tirucalli* plant on the mosquito *Culex quinquefasciatus*. J Commun Dis 34:264–269

Yang YC, Lim MY, Lee HS (2003) Emodin isolated from *Cassia obtusifolia* (Leguminosae) seed shows larvicidal activity against three mosquito specie. J Agri Food Chem 51:7629–7631

Yenesew A, Derese S, Midiwo JO, Heydenreich M, Peter MG (2003) Effect of rotenoids from the seeds of *Millettia dura* on larvae of Aedes aegypti. Pest Manag Sci 59(10):1159–1161

Zhang Q, Cai Q, Luo K (1998) Study on the physiological activities of the metabolites from *Paecilomyces arovirens*. Wei Sheng Wu Xue Bao 38(1):74–77

Zhu J, Zeng X, O'neal M, Schultz G, Tucker B, Coats J, Bartholomay L, Xue RD (2006) Mosquito larvicidal activity of botanical-based mosquito repellents. J Am Mosq Control Assoc 24(1):161–168

4 Plants of Himalayan Region as Potential Source of Biopesticides for Lepidopteran Insect Pests

S.G. Eswara Reddy, Shudh Kirti Dolma, and Anuja Bhardwaj

Abstract

The plant-based insecticides have received much attention in insect pest management due to their broad spectrum of activities and low toxicity against nontarget organisms. The plant products contain complex mixtures of monoterpenes, sesquiterpenes, and phenylpropanoid compounds that show multiple anti-insect properties such as toxicity, larvicidal, antifeedant, repellent, insect growth regulator, ovicidal, and fumigant activities. Due to these properties, botanical/plant products are gaining more attention as an alternative source for the management of insect pests. Besides, plant products are biodegradable into nontoxic products and potentially suitable for integrated pest management (IPM) programs. In this chapter, insecticidal activities of crude extracts/fractions/essential oils/pure compounds of major Himalayan plants against Lepidopteran insect pests are reviewed. The extracts/fractions/essential oils of some Himalayan plants screened in our laboratory for insecticidal activities against diamondback moth, *Plutella xylostella* (Lepidoptera: Yponomeutidae), are presented. Similarly, the chemical constituents (composition) of few selected plants with high potential for practical use in pest control are also discussed.

4.1 Introduction

The negative impacts of pesticides and increasing pesticide resistance have increased the interest in alternative control methods, with emphasis being placed on biological control, plant resistance, cultural control, botanicals, and other nonpolluting methods (Cheng 1988; Lim et al. 1996). Due to harmful effects of synthetic pesticides to health, environment, and resistance development in pests, there is a need for the development of safer and effective alternate strategies to contain the pests. The natural plant products can be an excellent alternative source of novel insecticides. Plants are a virtually inexhaustible source of structurally diverse and biologically active substances; approximately 1800 plants have been

S.G.E. Reddy (✉) • S.K. Dolma • A. Bhardwaj
CSIR-Institute of Himalayan Bioresource Technology (Council of Scientific and Industrial Research), Palampur, Himachal Pradesh 176 061, India
e-mail: ereddy2001@yahoo.com; ereddy@ihbt.res.in

© Springer India 2016

Vijay Veer, R. Gopalakrishnan (eds.), *Herbal Insecticides, Repellents and Biomedicines: Effectiveness and Commercialization*, DOI 10.1007/978-81-322-2704-5_4

reported to possess insecticidal properties (Jacobson 1982; Grainge et al. 1984). Plants are a good reservoir of eco-friendly allelochemicals.

The crude extracts/fractions/bioactive compounds/essential oils from plant source gain much interest due to their broad spectrum of activities and low toxicity to nontarget organisms, and these exert multiple anti-insect properties such as larvicidal, antifeedant, repellent, ovicidal, insect growth regulator, and fumigant activities. Due to these properties, plant products are gaining more attention as alternative sources for the control of insect pests and vectors (Prajapati et al. 2005; Kim Soon et al. 2003). Besides, these are biodegradable into nontoxic products and potentially suitable for integrated pest management programs. In addition to possessing insecticidal properties, they may also act as antifungal (Wang et al. 2007; Reichling et al. 2009), antitumor (Dung et al. 2009), and antimicrobial agents (Sacchetti et al. 2005; Reichling et al. 2009).

Insecticidal activity of plant products/fractions/essential oils has been reported against a number of insect pests, viz., cabbage looper, *Trichoplusia ni* (Jiang et al. 2012), and armyworm, *Pseudaletia unipuncta* (Akhtar et al. 2003). Extensive work has been done on bioactivity evaluation of extract/fractions/bioactive compounds/essential oil from various plants against important agricultural insect pests worldwide. Within the rich biodiversity of the Himalayas, there are abundant plant species, many of which are valued for their unique natural products and their biological and insecticidal properties (Tewary et al. 2005). In this chapter insecticidal properties of Western Himalayan plants (restricted to Himachal Pradesh) to Lepidopteran insect pests were reviewed. Similarly, some extracts/fractions/essential oils of Himalayan plants were screened for insecticidal activities, viz., larvicidal (toxicity to larvae), antifeedant, repellent, growth inhibition, and reduction in larval weight against diamondback moth, *Plutella xylostella* (Lepidoptera: Yponomeutidae) in the Entomology Laboratory, CSIR-Institute of Himalayan Bioresource Technology, Palampur, Himachal Pradesh.

In this chapter, preliminary description of plants with common name and place of occurrence was given in Table 4.1. Similarly, common name of insect pest, scientific name, plants damaged, and areas affected are also given in Table 4.2 for the benefit of general readers.

4.2 Mode of Action of Plant Products against Lepidopteran Insect Pests

The biological activities, viz., larvicidal (toxicity to larvae), antifeedant, repellent, and growth inhibition activities of different extracts/fractions/essential oils of some Himalayan plants and its concentration/dose, mortality, lethal concentration, and lethal dose to kill 50 % test population against Lepidopteran insect pests are presented in Tables 4.3 and 4.4.

4.2.1 Physiological Effects

4.2.1.1 Toxicity to Insect Larvae (Larvicidal Activity)

Toxicity of any plant product depends upon the stage and type of insect. Early-instar insects (first and second instars) are more susceptible than late instars (third to fifth instars). The insect larvae will be killed due to the toxicant that enters through cuticle or by ingestion. As per the available scientific reports, methanol extract of Kasmal, *Berberis lycium* roots (Tewary et al. 2005), and whole plant extract of Himalayan mayapple, *Podophyllum hexandrum* (Thakur et al. 2007), showed larvicidal activity to gram caterpillar, *Helicoverpa armigera* (44 % mortality), and large white butterfly, *Pieris brassicae* (LC_{50} = 2.42 %), respectively. Aqueous extract of Indian horse chestnut, *Aesculus indica* (Anwar et al. 1992), and pigweed, *Amaranthus viridis* (Rachokarn et al. 2008), showed toxicity to yellow stem borer, *Scirpophaga incertulas*, and beet armyworm, *Spodoptera exigua*, respectively. *Aspidium*, *Dryopteris filix-mas*, was effective

Table 4.1 Preliminary description of plants used for insecticidal activities

Plant name	Common name	Place of occurrence	Common uses	References
Achyranthes aspera	Prickly chaff flower	Grasslands, waste places, shaded or unshaded areas of Himachal up to 1500 m elevation	Antiperiodic, diuretic, purgative, laxative, antiasthmatic, hepatoprotective, and antiallergic	Chauhan (1999)
Acorus calamus	Sweet flag	Kinnaur, Kullu, Mandi, Shimla, Kangra, Sirmour and Chamba riverbeds, water courses, and lakes up to 2600 m	Sedative, laxative, diuretic, carminative, antioxidant, antimicrobial, and insecticidal activities	Chauhan (1999)
Aesculus indica	Indian horse chestnut	Kullu, Shimla, and Chamba up to elevation between 2000 and 3000 m	Skin diseases, rheumatism, as an astringent, acrid, narcotic, and relief of headache	Sagar and Kaur (2010)
Ageratum conyzoides	Billy goat weed	Kangra, Dharamsala, Chamba, Nahan, and Mandi	Insecticide and nematicide	http://hpforest.nic.in/pages/display/ZjY1c2RhhXU0c2RmYQ==-invasive-alien-species
Amaranthus viridis	Pigweed	Kangra	Inflammations, boils, abscesses, eczema, psoriasis, rashes, snakebites, and scorpion stings	Chowdhery and Wadhwa (1984)
Artemisia annua	Sweet wormwood	Lahaul-Spiti and Solan	Antimalarial drugs, treat fever, cytotoxicity, antibacterial, and antifungal activity	http://www.drugs.com/npp/sweet-wormwood.html
Azadirachta indica	Neem	Nalagarh and Nahan forest region	Anthelmintic, antifungal, antidiabetic, antibacterial, antiviral, contraceptive, sedative, antifeedant, repellent, egg-laying deterrent, and protecting the crop from damage	Chauhan (1999)

(continued)

Table 4.1 (continued)

Plant name	Common name	Place of occurrence	Common uses	References
Berberis lycium	Kasmal	Kinnaur, Kullu, Chamba, Shimla, Kangra, and Sirmour valley areas extending up to 2600 m	Swollen and sore eyes, broken bones, wounds, gonorrhea, curative piles, unhealthy ulcers, acute conjunctivitis, chronic ophthalmia used as bitter tonic astringent, diaphoretic, and febrifuge	Chauhan (1999)
Blumea mollis	Soft blumea	Kangra, Chamba, Kinnaur, and Sirmour	Antibacterial activity, hepatoprotective activity, and anti-inflammatory	Chowdhery and Wadhwa (1984)
Bupleurum falcatum	Chinese thoroughwax	Subalpine region of Himachal between 2801 and 3800 m	Anti-inflammatory, adaptogenic, hepatoprotective, antitussive, a mild sedative, liver stasis, hepatomegaly, splenomegaly, and indigestion	Samant et al. (2007)
Capparis spinosa	Common caper bush	Lahaul-Spiti and Kinnaur	Rheumatism and anticarcinogenic effects	Chauhan (1999)
Cedrus deodara	Deodar	Western Himalayas at altitudes of 1200–3000 m	Antifungal and insect repellent	Chaudhary et al. (2011)
Dryopteris filix-mas	Aspidium	Forests, talus, and rocky slopes	Anthelmintics against tapeworm, good ointment, anodyne, antibacterial, antiviral, astringent, febrifuge, and antirheumatic	http://www.pfaf.org/user/Plant.aspx?LatinName=Dryopteris+filix-mas
Murraya koenigii	Curry tree	Kangra and Hamirpur districts, areas lying between 800 and 1450 m	Antidiabetic, cooking, antiperiodic, leukoderma, and blood disorders	https://www.hort.purdue.edu/newcrop/parmar/12.html
Phytolacca acinosa	Red-ink plant	Chamba, Kullu, Mandi, Sirmour, Lahaul-Spiti, and Shimla	Antiasthmatic, antibacterial, antidote, antifungal, antitussive, diuretic, expectorant, laxative, and vermifuge	Chowdhery and Wadhwa (1984)
Podophyllum hexandrum	Himalayan mayapple	Chamba, Kangra, Kinnaur, Kullu, Lahaul-Spiti, and Shimla	Liver problems, cancer, leukoplakia, rheumatoid arthritis, and uterine cancer	Chowdhery and Wadhwa (1984)

Solanum pseudocapsicum	Capsicum weed	Chamba, Kinnaur, Kullu, Kangra, and Shimla	Antimicrobial, antiviral, antispasmodic, antihypertensive, antioxidant, and hepatoprotective	Chowdhery and Wadhwa (1984)
Swertia chirata	Chirayita	Shimla	Fever, malaria, constipation, worm infestations, upset stomach, loss of appetite, skin diseases, and cancer	Chowdhery and Wadhwa (1984)
Zanthoxylum armatum	Prickly ash	Chamba, Kinnaur, Kullu, Kangra, Lahaul-Spiti, Mandi, and Shimla	Antilarvicidal, antifungal, hepatoprotective, and allopathic	Chowdhery and Wadhwa (1984)
Aegle marmelos	Bael fruit	Una, Kangra, Nalagarh, Bilaspur, Mandi, Kunihar, Nahan, and Paonta Sahib forest ranges	Salads and juice, dyspepsia, sinusitis, constipation, diarrhea, skin carcinogen, antibacterial and natural food preservative	Chauhan (1999)
Cinnamomum camphora	Camphor tree	Kangra	Analgesic, antidepressant, anti-inflammatory, antiseptic, cardiac, carminative, diuretic, febrifuge, hypertensive, insecticide, laxative, rubefacient, stimulant, sudorific, vermifuge, and vulnerary	Chowdhery and Wadhwa (1984)
Curcuma aromatica	Wild turmeric	Kangra	Cosmetic herbal as a culinary ingredient, antibiotic properties remove excessive lipids from the blood and reduce aggregation of platelets	Chowdhery and Wadhwa (1984)
Cymbopogon flexuosus	Lemongrass	–	Antiseptics, antidepressant, astringent, preservative, cough, and nasal congestion keep insects away	http://www.pureorigines.com/#!product/zoom56o/331422161/lemongrass-(cymbopogon-flexuosus)

(continued)

Table 4.1 (continued)

Plant name	Common name	Place of occurrence	Common uses	References
Eupatorium adenophorum	Crofton weed	Roadsides, wastelands, barren lands, culturable wastes and fallow; Chamba, Kullu, and Nahan area	Antifungal, antibacterial, and anti-inflammatory	http://hpforest.nic.in/pages/display/ZjY1c2RhhXU0c2RmYQ==–invasive-alien-species
Hedychium spicatum	Spiked ginger lily	Shimla	Nausea, bronchial asthma, halitosis, vomiting, diminished appetite, hiccups, and local inflammation	Chowdhery and Wadhwa (1984)
Mentha spicata	Mint	Kangra, Chamba, Kullu, Lahaul-Spiti, and Shimla	Strong flavoring agent, fevers, bronchitis, chills, cramps, chronic gastritis, common cold, headaches, indigestion, morning sickness, motion sickness, nasal congestion, nausea, halitosis, and painful menstruation	Chowdhery and Wadhwa (1984)
Tagetes minuta	Marigold	Kinnaur	Herbal tea, colds, respiratory inflammations, stomach problems, and organic dye	Chowdhery and Wadhwa (1984)

Table 4.2 Infestation of insect pests causing damage to different parts of the plants

Name of insect pest	Common name	Plants damaged	Areas affected
Helicoverpa armigera	Gram caterpillar/cotton bollworm/pod borer	Polyphagous (cotton, red gram, bengal gram, cotton, tomato, tobacco, beans, peas, etc.)	Pod, leaves, bolls, fruits, etc.
Scirpophaga incertulas	Yellow stem borer	Monophagous	Paddy
Spodoptera litura	Tobacco caterpillar	Polyphagous (tobacco, cotton, castor, beans, mustard, chrysanthemum, etc.)	Pod, leaves, bolls, fruits, etc.
Plutella xylostella	Diamondback moth	Polyphagous (cabbage, broccoli, radish, turnip, cauliflower, mustard, etc.)	Leaves, head, flower
Glyphodes pyloalis	Lesser mulberry Pyralid	Monophagous (mulberry)	Leaves
Diaphania hyalinata	Melonworm	Polyphagous (cucumber, melon, and squash)	Fruit, leaves, flowers
Corcyra cephalonica	Rice moth	Stored grains (rice)	Grains
Spodoptera littoralis	African cotton leafworm	Polyphagous (tobacco, cotton, castor, beans, mustard, chrysanthemum, etc.)	Pod, leaves, bolls, fruits, etc.
Spodoptera exigua	Beet armyworm	Polyphagous (tobacco, cotton, castor, beans, mustard, chrysanthemum, etc.)	Pod, leaves, bolls, fruits, etc.
Pieris brassicae	Large white butterfly	Polyphagous (cabbage, broccoli, radish, turnip, cauliflower, mustard, etc.)	Leaves, head, flower

against rice moth, *Corcyra cephalonica* (Shukla and Tiwari 2011).

Hexane extract of chirata, *Swertia chirata*, to *H. armigera* and tobacco caterpillar, *Spodoptera litura* (Balaraju et al. 2011); ethyl acetate extract of capsicum weed, *Solanum pseudocapsicum*, against *S. litura* and *H. armigera* (Jeyasankar et al. 2012); hexane and chloroform extract of rhizomes against *Plutella xylostella* (Singh et al. 2012); hexane and methanol extract of common caper bush, *Capparis spinosa*, to *S. littoralis* (Ladhari et al. 2013); and hexane extract of soft blumea, *Blumea mollis*, against *H. armigera* (Baskar et al. 2014) showed good larvicidal activities to insects. However, the pentane fraction and crude oils of deodar, *Cedrus deodara*, showed larvicidal activity to second-instar larvae of *P. xylostella* (Chaudhary et al. 2011), whereas the compound coumarin extracted from hexane extract of billy goat weed, *Ageratum conyzoides*, showed larvicidal activity to melonworm, *Diaphania hyalinata* (Moreira et al. 2007).

Gahloth et al. (2011) reported curry tree, *Murraya koenigii* miraculin-like protein (MKMLP), inhibits the trypsin-like activity and total protease activity of *H. armigera* gut proteases (HGPs) by 78.5 and 40 %, respectively, whereas *S. litura* gut proteases (SGPs) by 81 and 48 %, respectively. The inhibitor was stable and actively inhibited the proteolysis of both HGP and SGP enzymes for up to 72 h. Incorporation of MKMLP into artificial diet adversely affected the growth and development of pests in a dose-dependent manner. After 10 days of feeding on diets containing 200 mM MKMLP, larval weight was reduced to 69 and 44.8 % and larval mortality was increased to 40 and 43.3 % for *H. armigera* and *S. litura*, respectively. The LC_{50} of MKMLP was 0.34 and 0.22 % of the diet for *H. armigera* and *S. litura*, respectively. These results demonstrate the efficacy of MKMLP as a potential plant defense agent against *H. armigera* and *S. litura* (Table 4.3).

Table 4.3 Insecticidal activities of Himalayan plants against Lepidopteran insect pests

Name of the plant and family	Plant part used	Insecticidal properties	Fractions/extracts/oils and its concn.	LC_{50}/LD_{50}/mortality	Target pest	References
Prickly chaff flower, *Achyranthes aspera*, (Amaranthaceae)	Leaves	Larvicidal	AE 1 %	17 % (24 HAT)	*Helicoverpa armigera* (Noctuidae)	Ramya et al. (2008)
Indian horse chestnut, *Aesculus indica* (Sapindaceae)	Leaf and flower	Larvicidal	AE	60–70 % mortality	*Scirpophaga incertulas* (Crambidae)	Anwar et al. (1992)
Chirayita, *Swertia chirata* (Gentianaceae)	Whole plant	Antifeedant	EAE 5 %	44 %	*H. armigera*	Balaraju et al. (2011)
			ME 5 %	69 %	*H. armigera*	Balaraju et al. (2011)
			EAE 5 %	37 %	*S. litura*	Balaraju et al. (2011)
			ME 5 %	57 %	*H. armigera*	Balaraju et al. (2011)
		Larvicidal	ME 5 %	78 %	*H. armigera*	Balaraju et al. (2011)
			ME 5 %	46 %	*S. litura*	Balaraju et al. (2011)
Soft *Blumea*, *Blumea mollis* (Asteraceae)	Plant material	Antifeedant	HE	LC_{50}=4.56 %	*H. armigera*	Baskar et al. (2014)
			CE	LC_{50}=8.35 %	*H. armigera*	Baskar et al. (2014)
		Larvicidal	HE	LC_{50}=5.34 %	*H. armigera*	Baskar et al. (2014)
			CE	LC_{50}=7.87 %	*H. armigera*	Baskar et al. (2014)
		Pupicidal	HE 5 %	54 %	*H. armigera*	Baskar et al. (2014)
			CE 5 %	27 %	*H. armigera*	Baskar et al. (2014)

Deodar, *Cedrus deodara* (Pinaceae)	Woodchips	Larvicidal	PF (1250–10,000 μg/ml)	LC_{50}=287 μg/ml (48 HAT)	*Plutella xylostella*	Chaudhary et al. (2011)
			HF	LC_{50}=361 μg/ml (48 HAT)	*P. xylostella*	Chaudhary et al. (2011)
			AEF	LC_{50}=365 μg/ml (48 HAT)	*P. xylostella*	Chaudhary et al. (2011)
		Larvicidal	Crude oil	LC_{50}=424 μg/ml (48 HAT)	*P. xylostella*	Chaudhary et al. (2011)
Curry tree, *Murraya koenigii* (Rutaceae)	Seeds	Trypsin-like activity	GE (1 mg in 50 mM Tris–HCl)	78 % (72 HAT)	*H. armigera*	Gahloth et al. (2011)
		Trypsin-like activity	GE (1 mg in 50 mM Tris–HCl)	81 % (72 HAT)	*S. litura*	Gahloth et al. (2011)
		Protease activity	GE (1 mg in 50 mM Tris–HCl)	40 % (72 HAT)	*H. armigera*	Gahloth et al. (2011)
		Protease activity	GE (1 mg in 50 mM Tris–HCl)	48 % (72 HAT)	*S. litura*	Gahloth et al. (2011)
		Larvicidal	GE 200 μM	LC_{50}=0.34 %; Mortality=40 % (10 DAT)	*H. armigera*	Gahloth et al. (2011)
			GE 200 μM	LC_{50}=0.22 % Mortality=43 % (10 DAT)	*S. litura*	Gahloth et al. (2011)

(continued)

Table 4.3 (continued)

Name of the plant and family	Plant part used	Insecticidal properties	Fractions/extracts/oils and its concn	LC_{50}/LD_{50}/mortality	Target pest	References
Capsicum weed, *Solanum pseudocapsicum* (Solanaceae)	Seed	Antifeedant	HE (5 mg/L)	41 %	*S. litura*	Jeyasankar et al. (2012)
			DEE (5 mg/L)	57 %	*S. litura*	Jeyasankar et al. (2012)
			DME (5 mg/L)	66 %	*S. litura*	Jeyasankar et al. (2012)
			EAE (5 mg/L)	80 %	*S. litura*	Jeyasankar et al. (2012)
			DEE (5 mg/L)	47 %	*H. armigera*	Jeyasankar et al. (2012)
			DME (5 mg/L)	51 %	*H. armigera*	Jeyasankar et al. (2012)
			EAE (5 mg/L)	82 %	*H. armigera*	Jeyasankar et al. (2012)
	Seed	Larvicidal	DME (5 mg/L)	46 %	*S. litura*	Jeyasankar et al. (2012)
			EAE (5 mg/L)	75 %	*S. litura*	Jeyasankar et al. (2012)
			DME (5 mg/L)	42 %	*H. armigera*	Jeyasankar et al. (2012)
			EAE (5 mg/L)	66 %	*H. armigera*	Jeyasankar et al. (2012)
		IGR	EAE (5 mg/L)	21 & 16 %	*H. armigera* and *S. litura*	Jeyasankar et al. (2012)
Sweet wormwood, *Artemisia annua* (Asteraceae)	Leaves	Acute toxicity	Extract (0.21–0.38 g/ml)	LC_{50}=0.33 % (24 HAT)	*Glyphodes pyloalis* (Pyralidae)	Khosravi et al. (2011)
Common caper bush, *Capparis spinosa* (Capparidaceae)	Leaves	Larvicidal	HE and ME 1 %	100 % (7 DAT)	*S. littoralis*	Ladhari et al. (2013)
		Antifeedant	HE and ME 1 %	71 & 91 % (24 HAT)	*S. littoralis*	Ladhari et al. (2013)

Billy goat weed, *Ageratum conyzoides* (Compositae)	Leaves	Larvicidal	Coumarin (10 mg/g)	LD_{50} = 2.21 mg/g (24 HAT)	*Diaphania hyalinata* (Crambidae)	Moreira et al. (2007)
Neem, *Azadirachta indica* (Meliaceae)	Seed	Larvicidal	Acetone extract (0.01–0.16 %)	LC_{50} = 0.076 μg/larva	*Corcyra cephalonica* (Pyralidae)	Pathak and Tiwari (2012)
Chinese thoroughwax, *Bupleurum falcatum* (Apiaceae)	Stem	Larvicidal	AE (15 mg/g)	43 % (5 DAT)	*S. littoralis* (Noctuidae)	Pavela (2011)
		LGI	AE	25–50 %	*S. littoralis*	Pavela (2011)
Pigweed, *Amaranthus viridis* (Amaranthaceae)	Leaves	Larvicidal	AE	LC_{50} = 50.57 mg/ml (24 HAT)	*S. exigua* (Noctuidae)	Rachokarn et al. (2008)
Aspidium, *Dryopteris filix-mas* (Polypodiaceae)	Root and rhizome	Larvicidal	AE 0.14 %	88 % (48 HAT)	*C. cephalonica* (Pyralidae)	Shukla and Tiwari (2011)
		Pupation	AE 0.14 %	12 %	*C. cephalonica*	Shukla and Tiwari (2011)
		Pupal death	AE 0.14 %	100 %	*C. cephalonica*	Shukla and Tiwari (2011)
		Adult emergence	AE 0.14 %	0 %	*C. cephalonica*	Shukla and Tiwari (2011)
Sweet flag, *Acorus calamus* (Acoraceae)	Rhizome	Larvicidal	HE	88 %	*P. xylostella*	Singh et al. (2012)
			CE	82 %	*P. xylostella*	Singh et al. (2012)
Kasmal, *Berberis lycium* (Berberidaceae)	Roots	Larvicidal	AME 1 %	44 %	*H. armigera* (Noctuidae)	Tewary et al. (2005)
Sweet flag, *Acorus calamus* (Acoraceae)	Rhizome	Larvicidal	EO 1 %	88 %	*H. armigera*	Tewary et al. (2005)
			EO 1 %	30 %	*P. xylostella*	Tewary et al. (2005)
			EO 1 %	65 %	*S. litura*	Tewary et al. (2005)

(continued)

Table 4.3 (continued)

Name of the plant and family	Plant part used	Insecticidal properties	Fractions/extracts/oils and its concn.	LC_{50}/LD_{50}/mortality	Target pest	References
Prickly ash, *Zanthoxylum armatum*	Leaves	Larvicidal	EO 1 %	46 %	*H. armigera*	Tewary et al. (2005)
			EO 1 %	42 %	*P. xylostella*	Tewary et al. (2005)
Himalayan mayapple, *Podophyllum hexandrum* (Berberidaceae)	Rhizomes	Larvicidal	ME 0.5–10 %	LC_{50}=2.34 % (48 HAT)	*Pieris brassicae*	Thakur et al. (2007)
	Roots	Larvicidal	ME 0.5–10 %	LC_{50}=4.36 % (48 HAT)	*P. brassicae*	Thakur et al. (2007)
	Whole plant	Larvicidal	ME 0.5–10 %	LC_{50}=2.42 % (48 HAT)	*P. brassicae*	Thakur et al. (2007)
Red-ink plant, *Phytolacca acinosa* (Phytolaccaceae)	–	Antifeedant	Extract (0.05 g/ml)	82 % (48 HAT)	*P. brassicae*	Wang et al. (2010)

HAT hours after treatment, *DAT* days after treatment, *AE* aqueous extract, *HE* hexane extract, *ME* methanol extract, *EAE* ethyl acetate extract, *DME* dichloromethane extract, *CE* chloroform extract, *IGR* insect growth regulator, *DME* dichloromethane extract, *DEE* diethyl ether extract, *GE* gut extract, *AME* aqueous methanol extract, *AEF* atlantone enriched fraction, *PF* pentane fraction, *HF* himachalene fraction, *EO* essential oil, LC_{50} lethal concentration to kill 50 % test population, LD_{50} lethal dose to kill 50 % test population

Table 4.4 Insecticidal activities of plant extracts/fractions/essential oils of Himalayan plants against larvae of diamondback moth, *Plutella xylostella*

Name of the plant	Family	Plant part used	Insecticidal properties	Fractions//extracts/ oils and its concn.	Mortality (%)
Prickly ash, *Zanthoxylum armatum*	Rutaceae	Leaves	Larvicidal	HE 1 %	86 % (48 HAT)
			FDI	EO 1 %	23 % (48 HAT)
			LGI	EO 1 %	93 % (48 HAT)
			Larvicidal	EO 1 %	60 % (48 HAT)
Curry tree, *Murraya koenigii*	Rutaceae	Leaves	Larvicidal	EO 1 %	90 % (48 HAT)
			PRW	EO 1 %	63 % (48 HAT)
			FDI	EO 1 %	31 % (48 HAT)
Bael fruit, *Aegle marmelos*	Rutaceae	Leaves	Larvicidal	EO 1 %	60 % (48 HAT)
			FDI	EO 1 %	25 % (48 HAT)
Wormwood, *Artemisia maritima*	Asteraceae	Leaves	Larvicidal	AE 1 %	93 % (96 HAT)
			Repellent	EO 1 %	80 %
			FDI	EO 1 %	33 % (48 HAT)
Flossflower, *Ageratum houstonianum*	Asteraceae	Leaves	Larvicidal	EO 1 %	100 % (24 HAT)
			Larvicidal	HE 1 %	100 % (48 HAT)
Crofton weed, *Eupatorium adenophorum*	Asteraceae	Leaves	Larvicidal	EO 0.5 %	100 % (48 HAT)
			Larvicidal	ME 1 %	73 % (48 HAT)
Marigold, *Tagetes minuta*	Asteraceae	Flowers	Larvicidal	EO 1 %	50 % (48 HAT)
			FDI	EO 1 %	51 % (48 HAT)
			LGI	EO 1 %	34 % (48 HAT)
Mint, *Mentha spicata*	Lamiaceae	Leaves	Larvicidal	EO 1 %	100 % (24 HAT)
			Repellent	EO 1 %	79 %
			LGI	EO 1 %	57 % (48 HAT)
			PRW	EO 1 %	104 % (48 HAT)
			FDI	EO 1 %	52 % (48 HAT)

(continued)

Table 4.4 (continued)

Name of the plant	Family	Plant part used	Insecticidal properties	Fractions//extracts/ oils and its concn.	Mortality (%)
Lemongrass, *Cymbopogon flexuosus*	Poaceae	Leaves	Larvicidal	EO 1 %	90 % (24 HAT)
			PRW	EO 1 %	100 % (48 HAT)
			FDI	EO 1 %	26 % (48 HAT)
Camphor tree, *Cinnamomum camphora*	Lauraceae	Leaves	Larvicidal	EO 1 %	33 % (24 HAT)
			PRW	EO 1 %	112 % (48 HAT)
			FDI	EO 1 %	31 % (48 HAT)
Wild turmeric, *Curcuma aromatica*	Zingiberaceae	Rhizome	Larvicidal	EO 1 %	100 % (24 HAT)
			PRW	EO 1 %	117 % (48 HAT)
Spiked ginger lily, *Hedychium spicatum*	Zingiberaceae	Rhizome	FDI	EO 1 %	48 % (48 HAT)
			Larvicidal	EO 1 %	100 % (24 HAT)
			PRW	EO 1 %	140 % (48 HAT)
Sweet flag, *Acorus calamus*	Araceae	Rhizome	Larvicidal	EO 1 %	77 % (48 HAT)
			PRW	EO 1 %	101 % (48 HAT)
Deodar, *Cedrus deodara*	Pinaceae	Woodchips	Larvicidal	EO 1 %	100 % (48 HAT)
			PRW	EO 1 %	122 % (48 HAT)

FDI feeding deterrent index, *EO* essential oil, *LGI* larval growth inhibition, *PRW* percent reduction in weight, *HAT* hours after treatment, *HE* hexane extract, *ME* methanol extract

Our studies report the essential oils and crude extracts isolated from different plants for the insecticidal activities, viz., toxicity, larvicidal, repellent, and antifeedant activities against *P. xylostella* under laboratory conditions. Results indicate that essential oils of flossflower (*Ageratum houstonianum*), mint (*Mentha spicata*), wild turmeric (*Curcuma aromatica*), and spiked ginger lily (*Hedychium spicatum*) showed excellent larvicidal activity (100 % mortality) after 24 h of treatment followed by crofton weed (*Eupatorium adenophorum*), lemongrass (*Cymbopogon flexuosus*), and deodar (*Cedrus deodara*) which recorded 100 % mortality after 48 h of treatment against second-instar *P. xylostella*. Mint (*M. spicata*) and wormwood (*Artemisia maritima*) showed 79–80 % repellence, whereas marigold (*Tagetes minuta*) and *M. spicata* showed 50–51 % feeding deterrent activity to third-instar *P. xylostella* (Table 4.4). Similarly, in the present study, hexane extract (1 %) of *A. houstonianum* showed 100 % mortality to second-instar larvae of *P. xylostella* after 48 h followed by prickly ash (*Zanthoxylum armatum*) which caused 86 % larval mortality. The aqueous extract of *A. maritima* and methanol extract of *E. adenophorum* showed 93 and 73 %

mortality after 96 and 72 h, respectively (Table 4.4).

4.2.1.2 Insect Growth Regulation

Insect growth regulators are defined as those substances which interfere with normal growth and development and molting of insects. Interference can result from substances that serve as agonists or antagonists of insect juvenile hormones or molting hormones or that prevent the synthesis of chitin, a key component of insect exoskeleton. A few examples from plants as natural products have been reported that affect insects through these types of mechanism, the best examples of which are azadirachtin from the seeds of Indian neem tree, *Azadirachta indica*, and precocenes from *Ageratum houstonianum* (Bowers 1982).

4.2.1.3 Larval Growth Inhibition

Majority of plant substances that affect insects in laboratory bioassay by inhibiting the growth of immature insects, which results from reduced consumption, which in turn lead to feeding deterrence or reduction in larval weight gain. Such growth inhibitory substance or their extracts could be efficacious as crop protectant (Wheeler and Isman 2001; Leatemia and Isman 2004). Under field conditions, any product/formulation which slowdown the growth rate of pests would broaden the window of opportunity for natural enemies or a biotic factor to act as mortality factors. Any plant-based product does not kill insects directly within a particular time period which would not be popular with farmers/growers who are adapted to conventional pesticides which kill the pest populations within hours. The results from our studies shows that essential oils extracted from *Z. armatum*, *T. minuta*, and *M. spicata* inhibit the larval growth of third-instar larvae of *P. xylostella* (Table 4.2).

4.2.2 Insect Behavior Modification

The plant products act on sensory system and cause behavioral effects on insects. The use of insect pheromones for mass trapping, mating disruption, and bait and kill applications demonstrates the utility of insect behavior modifying substances in pest management. Based on the experimental observation, different plant compounds act on insects as antifeedant, repellent, ovipositional deterrent, and mating disruption.

4.2.2.1 Antifeedant/Feeding Deterrent

It is defined as inhibition of feeding on the treated surface without killing or repelling the insects. The input causing from taste receptors, after biting the food, decides attraction of repellency. Azadirachtin and salannin suppress the action of taste receptor and hence there is no input, i.e., due to inhibition of gustatory receptors. However, the degree of antifeedant activity depends on the insect species.

Plant compounds/fractions/oils deter insect feeding than affect insects in any other behavioral or physiological manner. There is more scientific literature related to insect feeding deterrence from plant extracts/ isolated compounds/oils from different plants. In spite of sufficient literature, there is not a single commercial antifeedant used in crop protection program (Isman 2002). However in India, neem-based formulations containing azadirachtin is mainly used for the control of sucking/leaf feeders either singly or as one of the components in integrated pest management programs.

There are two reasons why antifeedants have not proven successful as on date. Firstly, insects quickly habituate to substances that deter feeding; when the same insects are subsequently exposed to the same antifeedant, the deterrent activity decreases. Under field conditions, an antifeedant would appear to lose efficacy rapidly within days, if not hours of application. This phenomenon of habituation in caterpillars has been shown to extend to a wide range of chemicals as well as to complex mixture (i.e., plant extract) (Akhtar et al. 2003). Secondly, insect species, even closely related ones, vary widely in their response to a specific antifeedant; a suite of pests on a crop cannot be expected to respond to an antifeedant in the same manner. In contrast, most insecticides have a broad spectrum of activity against pests, enhancing their utility and value as crop protectants.

Among different plant products screened, methanol extract of *S. chirata* against *H. armigera* and *S. litura* (Balaraju et al. 2011); ethyl acetate extract of *S. pseudocapsicum* (Jeyasankar et al. 2012); hexane extract of *B. mollis* against *H. armigera* (Baskar et al. 2014); extract of Chinese thoroughwax, *Bupleurum falcatum* (Pavela 2011), against *S. littoralis*; hexane and methanol extract of common caper bush, *Capparis spinosa*, to *S. littoralis* (Ladhari et al. 2013); and chloroform and ethyl acetate extract of *B. mollis* against *H. armigera* (Baskar et al. 2014) showed promising antifeedant activity (Table 4.1). The results from our studies show that the essential oils of *Z. armatum*, *M. koenigii*, *A. marmelos*, *A. annua*, *T. minuta*, *M. spicata*, *C. flexuosus*, *C. aromatica*, *H. spicatum*, and *A. calamus* reported promising antifeedant activity against third-instar larvae of *P. xylostella* (Table 4.2).

4.2.2.2 Repellents

The repellent is a substance or stimulus that causes an organism to make an oriented movement away from the source due to olfactory sensilla. The term "insect repellent" has become synonymous with personal protectants against biting flies, mosquitoes, and other blood-feeding insects and arthropods, even though it remains unclear as to whether most of these products actually "repel" pests or simply mask the attractiveness of host, resulting in nonacceptance by the pest.

At present, DEET (N,N-diethyl-m-toluamide) is the most widely used insect repellent, but biologist and chemists are engaged in finding safe and effective alternatives or replacements. Many plant oils or their constituents have been commercialized as insect "repellents" in the past decade. These include oils obtained from soybean, lemongrass, cinnamon, and the compounds from lemon eucalyptus; citronellal (from lemongrass) is effective against mosquitoes based on short-term tests with humans (Fradin and Day 2002).

d-Limonene is widely used as an alternative to conventional pesticides for structural pest control (i.e., termite management) in California, and other plant oils (clove, peppermint, etc.) are used in the USA by professional pest control operators as "flushing agents" for cockroach control and for "perimeter treatments" of homes against ants and termites, suggesting that repellence makes a strong contribution to the efficacy of these products. Repellent activity may also motivate the application of these oils in long-term protection of foods and products through their incorporation into packaging materials (Wong et al. 2005).

Based on our results, the essential oils of *A. annua*, *T. minuta*, *M. spicata*, *C. flexuosus*, *C. aromatica*, *H. spicatum*, and *A. calamus* reported promising repellent activity to third-instar larvae of *P. xylostella* under laboratory conditions (Table 4.2).

4.3 Development and Commercialization

For any pesticide, the route from the laboratory to commercial application can be a long one, with many hurdles to clear along the route. Even a natural product with outstanding bioactivity against insect pests in the laboratory may not reach the market unless a long list of practical criteria is met. For the most part, natural plant products are less effective as compared to synthetic products. Therefore, proper formulation, providing good coverage, and delivering the active ingredients to the target pest can make the difference between the product success and failure. In this section plants are selected with high potential for practical use in pest control based on their bio-efficacy/mortality against target pests. The chemical constituents (composition) of few selected plants and their compounds/fractions/oils against some insect pests are also discussed and presented in this section.

4.3.1 Wormwood, *Artemisia annua*

About 60 compounds were identified from *Artemisia annua* extract (Haghighian et al. 2008) from which the major compounds were erythritol (50.33 %) followed by camphor (7.25 %), cineol (1.84 %), coumarin (1.70 %), pinocarveol

(4.13 %), *p*-cymene (0.89 %), furfuryl alcohol (0.93 %), and artemisinin (0.89 %) (Table 4.5).

4.3.1.1 Effects of Leaf Extracts of *A. annua*

Several studies have evaluated the effects of mainly methanolic *A. annua* leaf extracts. Insect development was prolonged in fourth-instar larvae of the lesser mulberry pyralid, *Glyphodes pyloalis*, by an *A. annua* leaf extract applied topically at a concentration equivalent to 0.33 g leaf/L methanol, suggesting hormone-like activity of *A. annua* extract. This concentration killed half of the population (LC_{50} concentration). In the same treatment, fecundity and egg hatchability also were decreased, and significant effects were measured on extracted α-amylase, protease, lipase, esterase, and glutathione S-transferase activities of the insect (Khosravi et al. 2011). If the mortality observed in this study were due solely to the activity of artemisinin, the LC_{50} would correspond to 0.033–1.32 mg artemisinin/L, with the average artemisinin content of 0.01–0.4 %.

Table 4.5 Chemical constituents of *Artemisia annua* extract (Haghighian et al. 2008)

Compound	Percent (%)
Benzophenone	0.63
Artemisinin ($C_{15}H_{18}O_4$)	0.89
Azulene	0.40
Coumarin	1.70
Bicyclo[3.1.1] heptan-3- one,6,6-dimethyl-2- methylene	0.61
p-Cymene	0.89
2-Pinen-4-one	0.71
Valeraldehyde	1.09
Pinocarveol	4.13
Camphor	7.25
2-Phenylethanol	1.82
Furfuryl alcohol	0.83
Erythritol	50.30
α-Butylene glycol	1.52
Benzenemethanol	1.59
Cineol (Eucalyptole)	1.84
1,2,Dimethoxyethane	21.18

4.3.1.2 Effects of Essential Oil from *A. annua*

Topical application of *A. annua* essential oil to the red cotton bug (*Dysdercus koenigii*) resulted in a LC_{50} of 0.48 μl oil/nymph after 48 h of exposure (Rao et al. 1999). Application of 0.125 μl oil/nymph delayed the development of the nymphs by 2 days, hemolymph protein concentration decreased, and the electrophoretic protein pattern was altered. The adults emerging from treated nymphs had poor ovary development at an application of 0.25 μl oil/nymph. An essential oil steam distilled from dried *A. annua* leaf material was tested on two economically important stored product beetles *Callosobruchus maculatus* L. and *Tribolium castaneum* (Herbst) (Tripathi et al. 2000).

4.3.1.3 Effects of Pure Artemisinin

A few studies have focused on testing the insecticidal effects of pure artemisinin. Durden et al. (2011) found either 10 g/L artemisinin or 100 g/L 1,8-cineole was enough to give an effect at the same level as the leaf extract on fruit infestation by codling moth, *Cydia pomonella*. So, either 1,8-cineole or artemisinin alone could explain the repellency effect observed. In contrast Maggi et al. (2005) found that the observed deterrent activity on *Epilachna paenulata* and *Spodoptera eridania* could be explained by artemisinin alone.

4.3.2 Sweet Flag, *Acorus calamus*

Beta-asarone (33.36 %), cis-beta-terpineol (23.44 %), limonene (13.08 %), carvone (5.64 %), and amyl isovalerate (4.92 %) were identified as major chemical compounds (Senthilkumar and Venkatesalu 2012) in the essential oil of *A. calamus* (Table 4.6). Asarones, isolated from the essential oil of *A. calamus* rhizomes, showed potent growth inhibitors and antifeedants to the variegated cutworm, *Peridroma saucia*. The larvicidal activity of *A. calamus* rhizome essential oil may be due to the presence of the major chemical compounds beta-asarone and limonene (Koul et al. 1990). In a similar study, the essential oil of *A. calamus* had promising larvicidal effect

Table 4.6 Chemical constituents of *Acorus calamus* essential oil (Senthilkumar and Venkatesalu 2012)

Compounds	Percent (%)
Alpha-thujene	0.77
Alpha-pinene	0.20
2-Methylbutyl acetate	1.50
Limonene	13.08
Amyl isovalerate	4.92
Perillen	0.89
Alpha-cubebene	1.30
Octyl acetate	2.30
Linalool	2.48
Beta-copaene	2.53
Beta-asarone	33.36
Cis-beta-terpineol	23.44
Alpha-humulene	0.49
Gamma-muurolene	0.92
Carvone	5.64
Cis-calamenene	0.83
Globulol	3.28
Gamma-eudesmol	0.68
Tegretol	0.72
Trans-alpha-bergamotene	0.66

against the early fourth-instar larvae of filarial vector mosquito, *Culex quinquefasciatus*, with LC_{50}=63.43 ml/L. The activity is due to the presence of the compounds beta-asarone (33.36 %) and cis-beta-terpineol (23.44 %) reported by Senthilkumar and Venkatesalu (2012). *A. calamus* oil (70 μL) also showed 44.70 % morality to grubs of *Trogoderma granarium* after 7 days of exposure (Hasan et al. 2006).

4.3.3 Marigold, *Tagetes minuta*

Among the identified compounds, dihydrotagetone (30 %), (Z)-p-ocimene (23.60 %), and (Z)-tagetone (14.10 %) were major constituents (Chisowa et al. 1998). Other compounds were also identified in the oil in lesser amounts (Table 4.7).

Table 4.7 Chemical constituents of Tagetes minuta oil (Chisowa et al. 1998)

Compound	Percent (%)
Limonene	8.30
(Z)-β-ocimene	23.60
(E)-β-ocimene	0.40
Dihydrotagetone	30.00
(E)-tagetone	6.30
(Z)-tagetone	14.10
Linalool	0.50
Isocaryophyllene	1.20
β-Caryophyllene	2.20
Germacrene B	0.50
Cis-carvyl acetate	0.80

Floral, foliar, and root extracts of *T. minuta* were evaluated against Mexican bean weevil, *Zabrotes subfasciatus*, to determine speed of action and toxicities. The 24-h LC_{50} values ranged from 138 μg/cm^2 for males exposed to the root extract (most susceptible) to 803 μg/cm^2 for females exposed to the foliar extract (least susceptible). Increasing the duration of exposure to 48 h decreased all LC_{50} values 20–30 μg/cm^2. The males were more susceptible than the females. The time to incapacitation for 50 % of the test insects (IT_{50}) for floral and foliar extracts indicated fast-acting, volatile components, whereas the root extract data indicated slower-acting components, likely a result of the interaction of photophase with time-dependent efficacy. Floral and foliar extracts of *T. minuta* may be useful as insecticides for controlling stored product pests (Weaver et al. 1994). Larvicidal activity of *Tagetes patula* essential oil was tested against the fourth-instar larvae of *Aedes aegypti*, *Anopheles stephensi*, and *Culex quinquefasciatus*. Five different concentrations of essential oil were studied and the results were compared with that of synthetic insecticide, malathion. *A. aegypti* (LC_{50}=13.57) was the most susceptible followed by *An. stephensi* (LC_{50}=12.08) and *C. quinquefasciatus* (LC_{50}=22.33).

4.3.4 Billy Goat Weed, *Ageratum conyzoides*

The essential oil of *A. conyzoides* was characterized by the presence of high percentages of chromene derivatives (59.20 %), viz., ageratochromene (precocene II, 42.50 %), β-caryophyllene (20.70 %), and demethoxy ager-

Table 4.8 Chemical constituents of Ageratum conyzoides essential oil (Padalia et al. 2010)

Compounds	Percent (%)
Camphene	0.10
Sabinene	0.20
p-Cymene	3.30
(*E*)-β-ocimene	0.10
Linalool	0.20
Borneol	0.10
β-Caryophyllene	20.70
α-Humulene	6.60
Demethoxy ageratochromene (precocene I)	16.70
Germacrene D	1.00
Germacrene D-4-ol	0.20
Ageratochromene (precocene II)	42.50
α-Cadinol	0.10

atochromene (precocene I, 16.70 %) along with α-humulene (6.60 %) and p-cymene (3.30 %) as major constituents (Padalia et al. 2010) (Table 4.8).

The mode of action of these plant products varies with insect type; however, the extracts/fractions/essential oils/pure compound isolated from the plant material show contact toxicity (larvicidal), antifeedant, repellent, ovicidal, and growth inhibition activities against different insects.

4.4 Availability of Potential Plant Resources

Production of botanical insecticides requires continuous availability of plant materials on commercial scale. The few selected plants discussed in this chapter are abundantly available in different parts of Himachal Pradesh and their distribution/place of occurrence is given in Table 4.1. Some of the plants (e.g., marigold, *Tagetes* spp.) can be cultivated in agricultural land for commercial use in pest control, whereas *Ageratum conyzoides* is growing as weed in different ecosystems, road sides, forest, and other wastelands in different parts of Himachal Pradesh (20,343 ha). The raw material of some medicinal plants (root, rhizome, tuber, leaf, whole plant, fruit, flower, seed) will be available in different markets (Delhi, Saharanpur, Dehradun, Amritsar, and Solan) of the country (Samant et al. 2007).

4.5 Conclusion

Due to the harmful effects of synthetic pesticides to health, environment, and development of resistance in pests, there is a need for the development of safer and effective alternate strategies to contain the pests. The natural plant products can be an excellent alternative source of novel insecticides. With some exceptions, botanical insecticides are considered to be less toxic to nontarget species and more environment friendly because of their biodegradable nature.

References

Akhtar Y, Rankin CH, Isman MB (2003) Decreased response to feeding deterrents following prolonged exposure in the larvae of a generalist herbivore, *Trichoplusia ni* (Lepidoptera: Noctuidae). J Insect Behav 16:811–831

Anwar T, Jabbar A, Khalique F, Tahir S, Shakeel MA (1992) Plants with insecticidal activities against four major insect pests in Pakistan. Trop Pest Manage 38:431–437

Balaraju K, Vendan SE, Ignacimuthu S, Park K (2011) Antifeedant and larvicidal activities of *Swertia chirata* Buch-Ham.ex wall against *Helicoverpa armigera* Hubner and *Spodoptera litura* Fab. Elix Soc Sci 31:1902–1905

Baskar K, Kumar AS, Raj GA, Packiam SM, Ignacimuthu S (2014) Bioefficacy of *Blumea mollis* (D. Don) Merr and *Hygrophila schulli* (Buch.-Ham.) (Syn. *H. auriculata*) against *Helicoverpa armigera* (Hübner). Arch Phytopathol Pfl 47:1–12

Bowers WS (1982) Toxicology of precocenes. In: Coast JR (ed) Insecticide mode of action. Academic, New York, pp 403–427

Chaudhary A, Sharma P, Nadda G, Tewary DK, Singh B (2011) Chemical composition and larvicidal activities of the Himalayan cedar, *Cedrus deodara* essential oil and its fractions against the diamondback moth, *Plutella xylostella*. J Insect Sci 11:1–10

Chauhan NS (1999) Medicinal and aromatic plants of Himachal Pradesh. Indus Publishing Company, New Delhi

Cheng EY (1988) Problems of control of insecticide resistant *Plutella xylostella*. Pestic Sci 23:177–188

Chisowa EH, Hall DR, Farman DI (1998) Chemical composition of the essential oil of Tagetes minuta L. from Zambia. J Essent Oil Res 10(2):183–184

Chowdhery HJ, Wadhwa BM (1984) Flora of Himachal Pradesh analysis, vol 1–3. Botanical Survey of India, Howrah

Dung NT, Bajpai VK, Yoon JI, Kang SC (2009) Anti-inflammatory effects of essential oil isolated from the buds of *Cleistocalyx operculatus* (Roxb) Merr and Perry. Food Chem Toxicol 47:449–453

Durden K, Sellars S, Cowell B, Brown JJ, Pszczolkowski MA (2011) *Artemisia annua* extracts, artemisinin and 1,8-cineole, prevent fruit infestation by a major, cosmopolitan pest of apples. Pharm Biol 49:563–568

Fradin MS, Day JF (2002) Comparative efficacy of insect repellents against mosquito bites. N Engl J Med 347:13–18

Gahloth D, Kumar PA, Dhaliwal HS, Sharma AK, Shukla U, Birah A, Gupta GP (2011) Bioinsecticidal activity of *Murraya koenigii* miraculin-like protein against *Helicoverpa armigera* and *Spodoptera litura*. Arch Insect Biochem Physiol 78:132–144

Grainge MS, Ahmed WC, Mitchell, Hylin JW (1984) Plant species reportedly possessing pest control properties – a database. Resource Systems East–West Center, Honolulu

Haghighian F, Sendi JJ, Aliakbar A, Javaherdashti M (2008) The growth regulatory, deterrency and ovicidal activity of worm wood (*Artemisia annua* L.) on *Tribolium confusum* Duv. and identification of its chemical constituents by GC-MS. Pestycydy 12:51–59

Hasan M, Sagheer M, Ullah E, Ahmad F, Wakil W (2006) Insecticidal activity of different doses of *Acorus calamus* oil against *Trogoderma granarium* (Everts). Pak J Agri Sci 43:1–2

Isman MB (2002) Insect antifeedants. Pestic outlook 13:152–157

Jacobson M (1982) Plants, insect and man-their interrelationship. Econ Bot 36:346–354

Jeyasankar A, Premalatha S, Elumalai K (2012) Biological activities of *Solanum pseudocapsicum* (Solanaceae) against cotton bollworm, *Helicoverpa armigera* Hübner and armyworm, *Spodoptera litura* Fabricius (Lepidoptera: Noctuidae). Asian Pac J Trop Biomed 2:981–986

Jiang ZL, Akhtar Y, Zhang X, Bradbury R, Isman MB (2012) Insecticidal and feeding deterrent activities of essential oils in the cabbage looper, *Trichoplusia ni* (Lepidoptera: Noctuidae). J Appl Entomol 136:191–202

Khosarvi R, Sendi JJ, Ghadamyari M, Yezdani E (2011) Effect of sweet wormwood *Artemisia annua* crude leaf extracts on some biological and physiological characteristics of the lesser mulberry pyralid, *Glyphodes pyloalis*. J Insect Sci 11:156

Kim Soon IP, Ohh MH, Cho HC, Ahn YJ (2003) Contact and fumigant activities of aromatic plant extracts and essential oils against *Lasioderma serricorne* (Coleoptera: Anobiidae). J Stored Prod Res 39:11–19

Koul O, Smirle MJ, Isman MB (1990) Asarones from *Acorus calamus* L. oil, their effect on feeding behavior and dietary utilization in *Peridroma saucia*. J Chem Ecol 16:1911–1920

Ladhari A, Omezzine F, Chaieb I, Laarif A, Haouala R (2013) Antifeeding and insecticidal effects of *Capparis spinosa* L. on *Spodoptera littoralis* (Boisduval) larvae. Afr J Agric Res 8:5232–5238

Leatemia JA, Isman MB (2004) Efficacy of crude seed extracts of *Annona squamosa* against diamondback moth, *Plutella xylostella* L. in the greenhouse. Int J Pest Manage 50:129–133

Lim GS, Sivapragasam A, Loke WH (1996) Crucifer insect pest problems: trends, issues and management strategies. In: The Management of Diamondback Moth and other Crucifer Pests: Proceedings of the third international workshop, Malaysian Agriculture Research and Development Institute, Kuala Lumpur, Oct 1996

Maggi ME, Mangeaud A, Carpinella MC, Ferrayoli CG, Valladares GR, Palacios SM (2005) Laboratory evaluation of *Artemisia annua* L. extract and artemisinin activity against *Epilachna paenulata* and *Spodoptera eridania*. J Chem Ecol 31:1527–1536

Moreira MD, Picanc MC, Barbosa LCA, Guedes RNC, Barros E, Campos MR (2007) Compounds from *Ageratum conyzoides*: isolation, structural elucidation and insecticidal activity. Pest Manage Sci 63:615–621

Padalia RC, Verma RS, Sundaresan V (2010) Volatile constituents of three invasive weeds of Himalayan region. Rec Nat Prod 4(2):109–114

Pathak CS, Tiwari SK (2012) Insecticidal action of neem seed (*Azadirachta indica* A. Juss) acetone extract against the life-cycle stages of rice-moth, *Corcyra cephalonica* Staint. (Lepidoptera: Pyralidae). World J Agric Sci 8:529–536

Pavela R (2011) Screening of Eurasian plants for insecticidal and growth inhibition activity against *Spodoptera littoralis* larvae. Afr J Agric Res 6:2895–2907

Prajapati V, Tripathi AK, Aggarwal KK, Khanuja SPS (2005) Insecticidal, repellent and oviposition-deterrent activity of selected essential oils against *Anopheles stephensi, Aedes aegypti* and *Culex quinquefasciatus*. Bioresour Technol 96:1749–1757

Rachokarn S, Piyasaengthong N, Bullangpoti V (2008) Impact of botanical extracts derived from leaf extracts *Melia azedarach* L. (Meliaceae) and *Amaranthus viridis* L. (Amaranthaceae) on populations of *Spodoptera exigua* (Hübner) (Lepidoptera: Noctuidae) and detoxification enzyme activities. Commun Agric Appl Biol Sci 73:451–457

Ramya S, Rajasekaran C, Sundararajan G, Alaguchamy N, Jayakumararaj R (2008) Antifeedant activity of leaf aqueous extracts of selected medicinal plants on VI instar larva of *Helicoverpa armigera* (Hübner). Ethnobot Leafl 12:938–943

Rao PJ, Kumar KM, Singh S, Subrahmanyam B (1999) Effect of *Artemisia annua* oil on development and reproduction of *Dysdercus koenigii* (Hem., Pyrrhocoridae). J Appl Entomol -Zeitschrift fur Angewandte Entomologie 123:315–318

Reichling J, Schnitzler P, Suschke U, Saller R (2009) Essential oils of aromatic plants with antibacterial, antifungal, antiviral, and cytotoxic properties – an overview. Forsch Komplementmed 16:9–90

Sacchetti G, Maietti S, Muzzoli M, Scaglianti M, Manfredini S, Radice M, Bruni R (2005) Comparative evaluation of 11 essential oils of different origin as functional antioxidants antiradicals and antimicrobials in foods. Food Chem 91:621–632

Sagar A, Kaur R (2010) Study on fungal associates of *Aesculus indica*. Biol Forum – Int J 2:49–52

Samant SS, Pant S, Singh M, Lal M, Singh A, Sharma A, Bhandari S (2007) Medicinal plants in Himachal Pradesh, north western Himalaya, India. Int J Biodivers Sci Manage 3:234–251

Senthilkumar A, Venkatesalu V (2012) Larvicidal potential of *Acorus calamus* L. essential oil against filarial vector mosquito *Culex quinquefasciatus* (Diptera: Culicidae). Asian Pac J Trop Dis 2:324–326

Shukla S, Tiwari SK (2011) Toxicological effects of *Dryopteris filix-mas* against the ontogeny of Rice-moth, *Corcyra cephalonica* (Staint). World Appl Sci J 12:6–20

Singh KM, Singh MP, Sureja AK, Bhardwaj R (2012) Insecticidal activity of certain plants of Zingiberaceae and araceae against *Spodoptera litura* F. and *Plutella xylostella* Saunders in cabbage. Indian J Entomol 74:62–68

Tewary DK, Bhardwaj A, Shanker A (2005) Pesticidal activities in five medicinal plants collected from mid hills of western Himalayas. Ind Crop Prod 22:241–247

Thakur M, Mahajan R, Sharma S, Mehta PK (2007) Bio efficacy of *Podophyllum hexandrum* and Podophyllotoxin against Cruciferous Pests. Indian J Plant Prot 35:43–46

Tripathi AK, Prajapati V, Aggarwal KK, Khanuja SPS, Kumar S (2000) Repellency and toxicity of oil from *Artemisia annua* to certain stored-product beetles. J Econ Entomol 93:43–47

Wang H, Jiang Z, Feng J, Zhang X (2007) Inhibitory effects of seven essential oils against *Botrytis cinerea*. Plant Prot 33:111–114

Wang GF, Ting ZY, Hua YM (2010) Antifeedant activity of three plant extracts including *Phytolacca acinosa* against *Pieris rapae*. J Board Plant Dis Pest 1:26–27

Weaver DK, Wells CD, Dunkel FV, Bertsch W, Sing SE, Sriharan S (1994) Insecticidal activity of floral, foliar, and root extracts of *Tagetes minuta* (Asterales: Asteraceae) against adult Mexican bean weevils (Coleoptera: Bruchidae). J Econ Entomol 87:1718–1725

Wheeler DA, Isman MB (2001) Antifeedant and toxic activity of *Trichilia Americana* extracts against the larvae of *Spodoptera litura*. Entomol Exp Appl 98:9–16

Wong KKY, Signal FA, Campion SH, Motion RL (2005) Citronella as an insect repellent in food packaging. J Agric Food Chem 53:4633–4636

Natural Insecticides from Actinomycetes and Other Microbes for Vector Mosquito Control

5

M. Gabriel Paulraj, P. Saravana Kumar, S. Ignacimuthu, and D. Sukumaran

Abstract

Mosquitoes are the most dreadful bloodsucking insects in the world, and though tiny in size, they inflict most human deaths worldwide. They transmit deadly pathogens like *Plasmodium*, chikungunya virus, yellow fever virus, dengue virus, Japanese encephalitis virus and West Nile virus. Worldwide, there are 3500 species of mosquitoes grouped into 41 genera, but only 100 species are reported as vectors of human and other vertebrate diseases. India contributes nearly 34 % of global dengue and 11 % of global malaria cases. During the year 2012, nearly 1.13 million people were infected with dengue, malaria and chikungunya in India, and 766 succumbed to these diseases. In India, three genera, namely, *Aedes*, *Anopheles* and *Culex*, are the most common groups of mosquitoes found almost in all regions. *Aedes* spp. transmit dengue, chikungunya and yellow fever, *Anopheles* spp. transmit malaria, and *Culex* spp. transmit filariasis and Japanese encephalitis. In recent years, a decrease in the malaria and filariasis cases has been reported, but the number of infected cases and mortality due to dengue is steadily increasing. The failure in mosquito control is mainly due to the inefficiency of synthetic pesticides and repellents. Mosquitoes have developed resistance to almost all types of chemical insecticides. The increasing number of mosquito breeding sites and the destruction of mosquitoes' natural enemies are also contributing to the sudden rise in mosquito population and mosquito-borne diseases. Application of synthetic chemicals in water bodies is unsafe to humans

M.G. Paulraj (✉) • P.S. Kumar • S. Ignacimuthu
Entomology Research Institute, Loyola College,
Chennai 600 034, Tamil Nadu, India
e-mail: gabriel_paulraj@yahoo.com; eriloyola@hotmail.com

D. Sukumaran
Defence Research and Development Establishment,
Jhansi Road, Gwalior 474 002, India

© Springer India 2016

Vijay Veer, R. Gopalakrishnan (eds.), *Herbal Insecticides, Repellents and Biomedicines: Effectiveness and Commercialization*, DOI 10.1007/978-81-322-2704-5_5

and nontarget organisms. Microbial pesticides and botanical pesticides are eco-friendly and target specific compared to synthetic pesticides. Microbial pesticides obtained from actinomycetes, *Bacillus thuringiensis* (*Bt*), *B. sphaericus* (*Bs*) and many other microorganisms are reported as eco-friendly alternatives for mosquito control. A large number of *Bt* strains have been reported to possess insecticidal properties against different groups of insects. *B. thuringiensis israelensis* (*Bti*) is an important pathogenic bacterium to mosquitoes. The secondary metabolites of some microorganisms are potential toxins against mosquito larvae at very low concentrations. Spinosad, a potent insecticide, has been isolated from the actinomycete bacterium *Saccharopolyspora spinosa*. In this review, potentially effective actinomycetes and other microorganisms against mosquito larvae and their effective bioactive compounds are described. The review also presents up-to-date information on the efficacy of microbial pesticides in mosquito control, their biosafety, field efficacy and commercial applications.

5.1 Introduction

Mosquitoes, the tiny dipteran insects, are known as the deadliest insects in the world, because they transmit lethal pathogens from one human to the other and kill millions of people every year. They have killed more people than all the wars in history. Malaria is the most dreadful mosquito-borne disease in the world, and in 2012, there was an estimated 627,000 malaria deaths and about 207 million malaria cases in the world (WHO 2013). Moreover, tens of millions of people are killed and harmed by other mosquito-borne diseases, namely, dengue, encephalitis, yellow fever, filariasis and chikungunya. Mosquitoes are highly adaptable to anthropogenic impacts on their habitats. Unlike other aquatic insects, mosquitoes can utilize a variety of aquatic habitats such as freshwater pools, ponds, brackish water, overhead tanks, sewage waters, rain water in small containers and tyres and drainage water from refrigerators and air conditioners for their development.

Several vector-borne diseases are emerging in new areas in the world, mainly due to the increasing anthropogenic activities and climate change (Patz et al. 2005; Pascual et al. 2006; Nerio et al. 2010). Outbreak of many vector-borne diseases like malaria and dengue is on the rise in the developing world. Man is fighting against mosquitoes for many centuries, but the war is not winnable. To escape from mosquito-borne diseases, we are following two main measures, namely, mosquito population eradication and personal protection. In mosquito eradication programmes, they are killed at their adult stage or immature stages. Adulticiding is mainly done in malaria control programmes, and larval control is done to eradicate filariasis, dengue and encephalitis (Mulla 1991).

Mosquito eradication and personal protection are largely relying upon synthetic chemicals. Controlling mosquito larvae at their breeding site depends on the application of chemical larvicides to water. The early larvicides such as DDT, BHC and methoxychlor were found to be ineffective after some years due to the development of pesticide resistance in mosquitoes. Synthetic chemicals such as chlorpyrifos, diflubenzuron, malathion, methoprene, pyriproxyfen, permethrin, petroleum oils, temephos and resmethrin are used to eradicate mosquitoes at the larval and adult stages (Brattsten et al. 2009).

Even though synthetic mosquitocides instantly kill mosquitoes, they leave behind many unwanted effects like environmental pollution and nontarget effects on humans and other organisms (Paulraj et al. 2011). Synthetic pesticides also cause the development of pesticide resistance in mosquitoes (Charles and LeRoux 2000). Due to these unwanted effects of synthetic chemicals, researchers are trying to formulate eco-friendly and target-specific pesticides especially from natural resources. Plants and microbes are promising sources of natural pesticides against agricultural pests and vector insects. Mosquito control properties of plant products (Zarroug et al. 1988; Ignacimuthu 2000; de Luna et al. 2005; Maheswaran et al. 2008; Mathew et al. 2009; Patil et al. 2010; Ramar et al. 2013a, b; Rajiv Gandhi et al. 2014; Reegan et al. 2013a, 2014a, b; Sivaraman et al. 2014) and microorganisms (Des Rochers and Garcia 1983; Lee and Zairi 2006; Rydzanicz et al. 2010; Rashad et al. 2012; Poopathi et al. 2014) have been extensively studied. In this review, the toxic principles reported in mosquitocidal bacteria, their mode of action, residual toxicity, nontarget effect and their importance in eco-friendly mosquito control at present and in the future are discussed.

5.2 Biorational Mosquitocides

Nature is providing abundant sources of beneficial molecules to be utilized by man for his welfare. Plants and microorganisms are important natural sources because they possess diverse groups of molecules and are easily available. Secondary metabolites of plants and microbes show many biological properties. In plants the secondary metabolites play an important role in plant defence against pathogens and herbivory. Researchers have found that the secondary metabolites of plants and microbes can be used as potential pesticides against vector mosquitoes and agricultural pests. Plant secondary metabolites such as alkaloids, phenolics and terpenoids have been extensively studied for their mosquito control properties (Lee 2000; Bilal et al. 2012; Liu et al. 2012; Gautam et al. 2013). Larvicidal effect of plant extracts against vector mosquitoes has been reported by many investigators (de Luna et al. 2005; Maheswaran et al. 2008; Mathew et al. 2009; Patil et al. 2010; Ramar et al. 2013a, b; Rajiv Gandhi et al. 2014; Reegan et al. 2013a, 2014a, b; Sivaraman et al. 2014). Ramar et al. (2014) have reported that essential oils of aniseed, calamus, cinnamon, clove, lemon, orange, thyme, tulsi and vetiver presented larvicidal activity against *Cx. quinquefasciatus*, and the toxicity was very high in clove and tulsi oil treatments. Reegan et al. (2014a) have isolated a protolimonoid compound, niloticin from *Limonia acidissima*. This compound showed 100 % larvicidal activity against *Ae. aegypti* at 2 ppm concentration. Niloticin also showed pupicidal, ovicidal, oviposition deterrent and growth-regulating activities at 2 ppm concentration.

Biological pest control is one of the eco-friendly methods, which involves the mass culture of biocontrol agents and release in infested areas. After the work of Bassi (1836), who identified *Beauveria bassiana* as a pathogen of silkworm, and the investigations of Louis Pasteur on different diseases of the silkworm, scientists concluded that microorganisms could be used to control insect pests (Johnson 1998). Biocontrol agents of mosquitoes include viruses, bacteria, fungal pathogens, nematodes, predatory invertebrates and vertebrates like fish. Some of the biocontrol agents like nematodes, predatory invertebrates and mosquito fish are less utilizable considering the difficulties in mass multiplication (Usta 2013). But bacteria and their toxins can be produced in large quantities at laboratories and industries. Innumerable bacterial species are present on earth, which provide chances of discovering new mosquitocidal agents.

Besides plants and microbes, some more natural sources also possess mosquito larvicidal and repellent principles. Reegan et al. (2013b, 2015) have studied the larvicidal effect of marine sponge *Cliona celata* against *An. stephensi*, *Cx. quinquefasciatus* and *Ae. aegypti* mosquitoes. Some synthetic derivatives of plant and microbial compounds are also reported as potential mosquito larvicides and adulticides. Paulraj et al. (2011) screened benzaldehyde and propionic acid for larvicidal, pupicidal and adult knock-down effects against *Ae. aegypti* and *Cx. quinquefasciatus*. Benzaldehyde killed 50 % populations of *Ae. aegypti* and *Cx. quinquefasciatus* at concentrations of 30.39 and 40.48 ppm, respectively. Benzaldehyde is a major compound in almond oil, and propionic acid is produced by *Propionibacterium* found in the sweat glands of humans.

5.3 Microbial Pesticides

Many bacterial strains with larvicidal activities have been identified, and some biopesticides have been formulated using their toxic principles for the eradication of mosquito larvae in their breeding places such as flood water, ponds, irrigation ditches, woodland pools, tidal water and fresh- or saltwater marshes. The most common bacterial strains that are reported as lethal to mosquitoes are *Bacillus thuringiensis (Bt)*, *B. thuringiensis* var. *israelensis* (*Bti*), *Lysinibacillus sphaericus* or *B. sphaericus*, some other strains in *B. thuringiensis* serotypes and *Clostridium bifermentans* serovar *malaysia* (Porter et al. 1993; WHO 1999; Foda et al. 2010). Among them, *B. thuringiensis israelensis* (*Bti*) and *B. sphaericus* (*Bs*) are widely exploited against different mosquito species. *Bt* was first isolated by Ishiwata from the mulberry silkworm *Bombyx mori* in 1901 (Ishiwata 1901). *Bt* was first scientifically described by Berliner in Germany in 1911. *Bti* was first discovered in 1976 in the Negev Desert of north-central Israel and was found to be useful to control mosquito and black fly (Margalit and Dean 1985). *Bti* is a spore-forming bacterium naturally found in soil and aquatic environments. *Bti* shows different levels of larval toxicity against different mosquito genera. *Culex* and *Aedes* were found to be highly susceptible to *Bti*, whereas *Anopheles* was less susceptible (WHO 1985; Charles et al. 1996). Furthermore, it shows species-specific activity within one genus of mosquito (Chui et al. 1995). *Bti* was found to be specifically toxic to larvae of 109 mosquito species Glare and O'Callaghan (1998)

B. sphaericus produces binary toxin during sporulation (Broadwell and Baumann 1986; Charles et al. 1988), and this binary toxin is composed of two polypeptides, namely, BinA (molecular weight, 41.9 kDa) and BinB (51.4 kDa) (Smith et al. 2004). The amino acid sequences of these two polypeptides are not similar to the amino acid sequence of crystal proteins of *B. thuringiensis*. The binary toxin, BinA and BinB, forms microcrystalline inclusions inside the mother cell and will be solubilized in the alkaline pH of the mosquito larval gut, if ingested (Smith et al. 2004). Rungrod et al. (2009) have stated that the mosquitocidal toxins, namely, Mtx1 and Mtx2, were species specific and very toxic against *Cx. quinquefasciatus* and *Ae. aegypti*, respectively. They cloned *mtx*1 and *mtx*2 genes into a single plasmid and expressed in *Escherichia coli*. The cells produced both Mtx1 and Mtx2 toxins and recorded high synergistic activity against *Ae. aegypti* larvae nearly 10 times more compared to the activity of a single toxin.

The toxic properties of these bacteria against mosquito larvae are due to the production of protein inclusion bodies during sporulation. These toxins are highly lethal to the larvae of mosquitoes, black flies some closely related dipteran flies when ingested (Gibbs et al. 1986). *Bt* produces 'Cry' (crystal) and 'Cyt' (cytolytic) toxins, and *Bs* produces 'Bin' (binary) and 'Mtx' (mosquitocidal) toxins (Charles et al. 1996; Charles and LeRoux 2000; Federici et al. 2003) (Table 5.1). It has been reported that *Bti* is producing different groups of toxic proteins, namely, Cry4Aa, Cry4Ba, Cry10Aa, Cry11Aa, Cyt1Aa and Cyt2Ba (Berry et al. 2002). The larvicidal effect of these bacterial strains depends mainly upon the mosquito species and the environmental conditions. One important advantage of microbial larvicides is that they can be used along with other mosquito control measures in integrated pest management (IPM) programmes.

Several *Bt* strains with mosquito larvicidal activity have been isolated after the discovery of *Bti*. The strains differ from each other by their mosquito larvicidal activity, serotype and polypeptide composition. Plenty of work has been done on isolation, larvicidal screening and residual efficacy of *Bti* and *Bs* against vector mosquitoes. Many reviews and research articles have been published on these two bacterial pesticides. The species-specific activities, nontarget effects and efficacy in integrated control strategies of these two microbes have been well documented.

In a review, Mulla (1991) has documented the larvicidal effects of *B. thuringiensis* and *B. sphaericus* against different mosquito species in laboratory and open field conditions. He also discussed the factors influencing the efficacy and

Table 5.1 Larvicidal activity of Bin, Cry and Cyt proteins against different mosquito species

Name of bacterial toxin	Bacterial strain	Target mosquito species	Reference
Bin	*Bacillus sphaericus* WBM 1-1-13	*Ochlerotatus taeniorhynchus*, *Culex quinquefasciatus*	Park et al. (2007)
BinAB (recombinant)	*B. sphaericus* 2362 SPH-28 (expressed in *Escherichia coli)*	*Culex quinquefasciatus*	Pinto da Silva et al. (2011)
Cry2Aa1	*Bacillus thuringiensis kurstaki* HD-1, HD-263	*Aedes aegypti*	Zeigler (1999)
Cry4Aa1	*Bt israelensis* 4Q2-72	*Anopheles stephensi*, *Aedes aegypti*, *Culex pipiens*	Zeigler (1999)
Cry4Ba1	*Bt israelensis* 4Q2-72	*Aedes aegypti* (Diptera: Culicidae)	Zeigler (1999)
Cry10Aa1	*Bt israelensis* ONR60A	*Aedes aegypti* (Diptera: Culicidae)	Zeigler (1999)
Cry11Aa1	*Bt israelensis* HD-567	*Anopheles stephensi*, *Aedes aegypti*, *Culex pipiens*	Zeigler (1999)
Cry11Ba1	*Bt jegathesan* 367	*Anopheles stephensi*, *Aedes aegypti*, *Culex pipiens* (Diptera: Culicidae)	Zeigler (1999)
Cry11Bb1	*Bt medellin*	*Anopheles albimanus*, *Aedes aegypti*, *Culex quinquefasciatus* (Diptera: Culicidae)	Zeigler (1999)
Cry16Aa1	*Clostridium bifermentans malaysia* CH18	*Anopheles stephensi*, *Aedes aegypti*, *Culex pipiens* (Diptera: Culicidae)	Zeigler (1999)
Cry19Aa1	*Bt jegathesan*	*Anopheles stephensi*, *Culex pipiens* (Diptera: Culicidae)	Zeigler (1999)
Cry20Aa1	*Bt fukuokaensis*	*Aedes aegypti* (Diptera: Culicidae)	Zeigler (1999)
Cry21Aa1	*Bt higo*	*Culex pipiens molestus* (Diptera: Culicidae)	Zeigler (1999)
Cry11, C*ry30*	*Bt* 147-8906	*Aedes aegypti*, *Culex quinquefasciatus*, *Anopheles albimanus*	Ibarra et al. (2003)
Cyt1Aa1	*Bt israelensis* IPS82	*Anopheles stephensi*, *Aedes aegypti*, *Culex pipiens* (Diptera: Culicidae)	Zeigler (1999)
Cyt1Ab1	*Bt medellin* 163-131	*Anopheles stephensi*, *Aedes aegypti*, *Culex pipiens* (Diptera: Culicidae)	Zeigler (1999)
Cyt2Aa1	*Bt kyushuensis*	*Anopheles stephensi*, *Aedes aegypti*, *Culex pipiens* (Diptera: Culicidae)	Zeigler (1999)
Cyt1 and C*yt2*	*Bt* 147-8906	*Aedes aegypti*, *Culex quinquefasciatus*, *Anopheles albimanus*	Ibarra et al. (2003)

nontarget effects of these two microorganisms. In Africa, *Bs* and *Bti* are reported as promising biocontrol agents of major vectors of malaria (Fillinger and Lindsay 2006; Fillinger et al. 2003; Majambere et al. 2007).

Studies on mosquito larvicidal potential of actinomycetes are scanty. A few studies indicate that secondary metabolites of actinomycetes are potent mosquito larvicides. Kumar et al. (2011) have isolated a compound, namely, 5-(2,4-dimethylbenzyl) pyrrolidin-2-one, from a marine *Streptomyces* sp. This compound recorded 100 % larvicidal activity against *An. stephensi* and *Cx. tritaeniorhynchus* in 24 h.

5.4 Commercial Larvicidal Products from Microbes

During the last three decades, different types of formulations were developed using different subspecies of *B. thuringiensis* against vector mosquitoes. Some of these biopesticides showed very high efficiency against target mosquitoes. The two species of *Bacillus*, namely, *B. thuringiensis israelensis* and *B. sphaericus*, are the main ingredients in the biolarvicides, which are commercially available to control mosquitoes. Table 5.2 shows some of the commercially available biolarvicides and their target mosquito species.

Bti was first registered in 1983 as an insecticide by the United States Environmental Protection Agency (US EPA). Nearly 25 *Bti* products have been registered in the USA. AquaBac, Teknar, VectoBac and LarvX are common trade names of some of the mosquito control products (US EPA 2000). *Bs* was first registered by US EPA in 1991 for eradicating different species of mosquitoes. VectoLex CG and WDG are registered *Bs* products, which are effective for nearly 1–4 weeks after application in the larval habitats (US EPA 2000).

Djènontin et al. (2014) have evaluated the larvicidal activity of VectoBac GR (potency 200 ITU/mg) prepared from *Bti* strain AM65-52 against *An. gambiae* and *Cx. quinquefasciatus* in simulated field and natural habitats in Benin. They found that VectoBac GR caused emergence inhibition of ≥80 % until 21 days for *An. gambiae* at 1.2 g/m^2 dose and 28 days for *Cx. quinquefasciatus* at 2 g/m^2 in simulated field habitats. They also reported that the efficacy of VectoBac GR in natural habitat was for 2–3 days against larvae and up to 10 days against pupae. Fillinger et al. (2003) have studied the larvicidal potential of VectoBac and VectoLex against *Anopheles gambiae*. They found that *An. gambiae* was more susceptible to VectoLex (*B. sphaericus* as ingredient) than VectoBac. Majambere et al. (2007) have reported that both VectoBac and VectoLex were effective in controlling *An. gambiae*.

Mousticide is a combination of TMOF (trypsin-modulating oostatic factor) and *B. thuringiensis israelensis* (Bti) serotype H-14. TMOF is a natural decapeptide hormone synthesized by the ovaries and the neuroendocrine system of mosquitoes. TMOF stops protein digestion in mosquito larvae and causes larval death. When TMOF is combined with Bti, it yields a potential product with synergistic effect of more than 200×.

Since our country has a rich source of plants and microbes, there is a scope for finding numerous active principles from plants and microbes for the purpose of mosquito eradication/management (Ignacimuthu and Paulraj 2009).

5.5 Actinomycetes: Promising Sources of Active Compounds for Mosquito Control

Actinomycetes are gram-positive soil bacteria. They contain high GC content in their DNA. Actinomycetes, particularly the genus *Streptomyces*, produce many economically important secondary metabolites (Subramani and Aalbersberg 2012). Very few actinomycetes like *Mycobacterium tuberculosis* are pathogenic to humans. But a large number are very useful to humans, because they produce useful compounds with antibiotic, antifungal, antitumor, immunosuppressive and pesticidal properties. The active compounds of actinomycetes are present in the extracellular metabolites secreted by them in the culture media (Bode et al. 2002). Actinomycetes synthesize the secondary metabolites when their growth is slowing or stopped (Doull and Vining 1990; Sanchez and Demain 2002).

The antimicrobial properties of secondary metabolites of actinomycetes are well studied (Chaudhary et al. 2013; Rana and Salam 2014; Phongsopitanun et al. 2014). In recent years, researchers are interested to examine the acute and chronic toxicities of actinomycetes on different vector mosquitoes. Many studies have proven that actinomycetes were toxic to different mosquito spp. Vijayakumar et al. (2010) screened 30 actinomycetes isolated from soil samples from Muthupet mangrove forest, Tiruvarur District, against *Anopheles* mosquito larvae. They used the culture filtrate for larvicidal screening and found that 23 isolates presented larvicidal activity, among which 2 isolates were significantly effective.

Table 5.2 Some of the *Bt*-based commercial microbial larvicides used against vector mosquitoes

Sl. no.	Trade name of microbial pesticide	Formulation type	Active ingredient	Name of the manufacturer	Target mosquito species
1	AquaBac® XT	AS	*Bacillus thuringiensis* subsp. *israelensis*, serotype H-14, strain BMP-144	Becker Microbial Products, Inc., USA	*Aedes* spp., *Culex* spp., *Psorophora columbiae*
2	AquaBac® (200G)	G	*Bacillus thuringiensis* subsp. *israelensis*, serotype H-14, strain BMP-144	Becker Microbial Products, Inc., USA	*Aedes* spp., *Culex* spp., *Psorophora columbiae*
3	Bacticide®	WP	*Bacillus thuringiensis* var. *israelensis*, serotype H-14, strain 164	Biotech International Ltd., India	*Aedes* spp., *Anopheles* spp., *Culex* spp., *Culiseta* spp., *Psorophora* spp., *Uranotaenia* spp., *Mansonia* spp.
4	Bactimos®	Pellets, tablets	*Bacillus thuringiensis* var. *israelensis*	Valent BioSciences Corporation	Mosquitoes and black flies
5	Bti (AS, WP)[a]	AS, WP	Bti	Kilpest India Ltd.	*Aedes*, *Culex*, *Anopheles*
6	Introban®	AS	Bti	Valent BioSciences Corporation, USA	Mosquitoes
7	Mousticide™	WP	*Bacillus thuringiensis israelensis* (Bti) serotype H-14	EntoGeneX, Malaysia	Mosquitoes
8	Skeetal®	FC	*Bacillus thuringiensis* var. *israelensis*	Renovita	Mosquitoes
9	Sphericide®	WP	*Bacillus sphaericus*; serotype H-5a, 5b; strain B-101	Biotech International Ltd., India	*Aedes* spp., *Anopheles* spp., *Culex* spp., *Culiseta* spp., *Psorophora* spp., *Uranotaenia* spp., *Mansonia* spp.
10	Teknar®	SC	*Bacillus thuringiensis* var. *israelensis*	Valent BioSciences Corporation	Mosquitoes
11	VectoBac®	WDG	*Bacillus thuringiensis* subsp. *israelensis*, serotype H-14, strain HD-14	Valent BioSciences Corporation	Mosquitoes and black flies
12	VectoBac®	AS	*Bacillus thuringiensis* subsp. *israelensis* strain AM65-52	Valent BioSciences Corporation	Mosquitoes and black flies
13	VectoLex®	WDG	*Bacillus sphaericus*	Valent BioSciences Corporation	*Culex* spp., *Psorophora* spp., *Culiseta* spp.
14	VectoMax®	WSP	*Bacillus thuringiensis* var. *israelensis*, *Bacillus sphaericus*	Valent BioSciences Corporation	Mosquitoes
15	VectoPrime®	FG	*Bti* strain AM65-52 with (S)-methoprene	Valent BioSciences Corporation	Mosquitoes

AS aqueous suspension, *FG* fine granules, *G* granules, *SC* suspension concentrate, *WDG* water-dispersible granules, *WP* wettable powder, *WSP* water-soluble pouch

[a]No trade name

In India, some investigators have explored the anti-insect properties of actinomycete metabolites against insects including mosquitoes. Mishra et al. (1987) have reported that metabolites of actinomycetes are potential alternatives to synthetic insecticides. Rao et al. (1990) have reported the isolation of active compounds from actinomycetes against mosquitoes. Vijayan and Balaraman (1991) have studied the ovicidal, larvicidal and adulticidal activities of the secondary metabolites of fungi and actinomycetes against *Cx. quinquefasciatus*, *An. stephensi* and *Ae. aegypti*. They reported that the metabolites of 34 fungi and 3 actinomycetes, 133 fungi and 35 actinomycetes and 17 fungi were found to kill the eggs, larvae and adults, respectively. Dhanasekaran et al. (2010) have isolated 30 actinobacteria from Muthupet mangrove environment. Four isolates belonging to the genera *Streptomyces*, *Streptosporangium* and *Micropolyspora* showed strong larvicidal activity against *Anopheles* larvae.

Gadelhak et al. (2005) have isolated three efficient chitinase enzyme producing actinomycetes from 38 different strains collected from the United Arab Emirates soil. They found that the application of two isolates *Streptomyces clavuligerus* and *Actinoplanes philippinensis* in combination gave higher effects as this treatment reduced the pupation of *Drosophila melanogaster*. The compounds, namely, tetranectin (Ando 1983), avermectins (Pampiglione et al. 1985), faeriefungin (Anonymous 1990) and macrotetrolides (Zizka et al. 1989), are produced by *Streptomyces aureus*, *Streptomyces avermitilis*, *Streptosporangium albidum* and *Streptomyces griseus*, respectively. These compounds were reported to be lethal to different mosquito species.

There is a big scope for isolating potential actinomycete strains with significant mosquito control property from forest, desert, mangrove and marine environments. Our recent studies have resulted in the identification of 8 potential actinomycete isolates from a total of 283 pure isolates obtained from soil samples collected from Nilgiris and Kalakkad Mundanthurai Tiger Reserve in Tirunelveli District. An important finding in this study was that the active isolates showed species-specific activity against different mosquito species. Among the eight active isolates, CFR-16 (collected from Coonoor forest soil, Nilgiris) was found to be the most effective isolate against *Ae. aegypti*, *An. stephensi* and *Cx. quinquefasciatus*. Based on 16S rRNA characterization studies, the most effective isolate (CFR-16) was identified as a *Streptomyces* sp. (Fig. 5.1) (unpublished data).

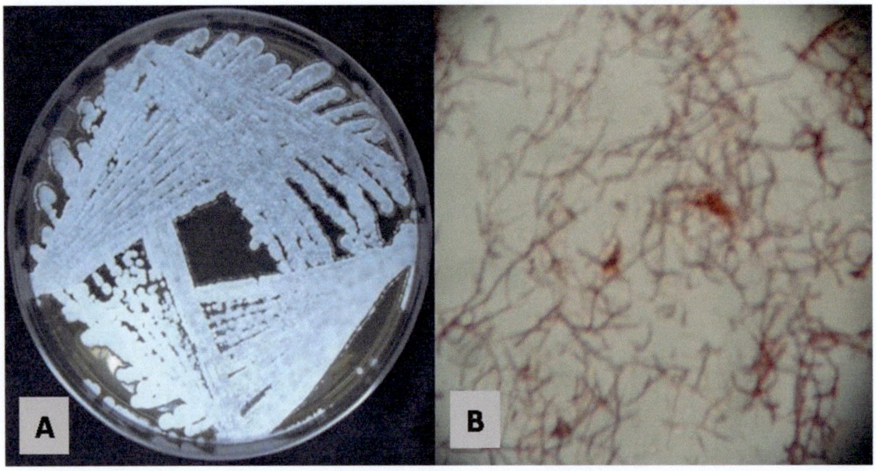

Fig. 5.1 The active isolate, *Streptomyces* sp. (CFR-16 Strain) collected from Coonoor Forest soil, Nilgiris, Tamil Nadu. (**a**) Surface colony morphology of *Streptomyces* sp. (CFR-16) AIA. (**b**) Gram stained photomicrograph of *Streptomyces* sp. (CFR-16) (100×)

5.6 Spinosad: A Promising Molecule from Actinomycetes for Mosquito Control

Spinosad is a biorational insecticide produced during the fermentation of the actinomycete *Saccharopolyspora spinosa*. Spinosad is a mixture of two tetracyclic macrolide neurotoxins, namely, spinosyns A and D. It targets the nicotinic acetylcholine and GABA receptors of the insect's nervous system, which leads to paralysis and death (Salgado 1997, 1998).

Spinosad is currently used to control coleopteran, dipteran, lepidopteran and thysanopteran pests of agricultural and forestry plants in different countries (Biondi et al. 2012). Spinosad has very little toxicity to vertebrates and has been approved for use against mosquito larvae in drinking water (WHO 2010). The United States Environmental Protection Agency (US EPA) has classified spinosad as a reduced-risk material due to its very low mammalian toxicity and favourable ecotoxicological profile (Thompson et al. 2000). Spinosad was registered in 1997 under the trade name Tracer®. It was found to be effective in reducing the development of immature stages of *Ae. aegypti*, *Ae. albopictus*, *An. gambiae*, *An. pseudopunctipennis*, *An. albimanus*, *Cx. pipiens* and *Cx. quinquefasciatus* (Hertlein et al. 2010). Spinosad is primarily a stomach poison with some contact activity and is particularly active against Lepidoptera, Diptera, some Coleoptera, termites, ants and thrips (Bret et al. 1997). Exposure resulted in cessation of feeding followed later by paralysis and death. Due to its selective toxicity and its favourable environmental profile, spinosad is considered by IPM practitioners as an important new-generation biorational pesticide (Schneider et al. 2004).

Many investigators have reported spinosad as a potentially valuable tool for the control of different vector mosquito species (Bond et al. 2004; Darriet et al. 2005; Romi et al. 2006). Marina et al. (2012) have studied the efficacy of spinosad against *Ae. aegypti*, *Ae. albopictus*, *Cx. quinquefasciatus* and *Cx. coronator* larval control in car tyres in southern Mexico. Much of the toxicity studies of spinosad on mosquitoes have been conducted under laboratory conditions; very few studies have been done in natural habitats of mosquitoes. Bond et al. (2004) have reported that spinosad at 1 ppm resulted in complete inhibition of reproduction of *Ae. aegypti* and *Culex* spp. for 8 and 15 weeks, respectively, in field trials. At 10 ppm concentration, spinosad completely eliminated reproduction of both mosquitoes during the entire period of 22 weeks.

5.7 Residual Toxicity of Bacterial Toxins on Mosquitoes

Jahan et al. (2013) have studied the residual toxicity of *B. thuringiensis* var. *israelensis* (technical powder and water-dispersible granules) and *B. sphaericus* against laboratory-reared *An. stephensi* and field-collected *Cx. quinquefasciatus* larvae. They reported that the residual toxicity decreased with decreasing concentrations. The residual activity of *B. thuringiensis israelensis* technical powder varied from 1 to 51 days against laboratory-reared *A. stephensi* larvae at 0.0001 and 100 ppm concentrations, respectively. *B. sphaericus* technical powder had a residual effect for 2–18 days at 0.0001 and 100 ppm concentrations, respectively, against the same species.

Lee and Zairi (2006) have studied the residual efficacy of *B. thuringiensis* H-14 against *Aedes* mosquitoes at field conditions with two different test designs. In one design, treated water was replenished daily with seasoned water, and in the other one, treated water was not replenished, but evaporated water was replenished. They reported that *Bt* showed a residual effect against *Aedes* mosquito larvae up to a period of 40 days with 80 % mortality, and the residual effect continued up to 60 days of study, but the larval mortality was reduced below 54 %. When the treated water was daily replenished, 100 % larval mortality was recorded for the first 3 days only. Without daily replenishment of treated water, 100 % larval mortality was recorded for the first 5 days.

Majambere et al. (2007) have tested the larval toxicity and residual effect of formulations of commercial *B. sphaericus* strain 2362 (*Bs*,

VectoLex®) and *B. thuringiensis* var. *israelensis* strain AM65-52 (*Bti*, VectoBac®) against *An. gambiae* in the Gambia. In their study, they found that *Bs* had no residual activity against anopheline larvae. But both microbes presented complete eradication of larvae when applied weekly and recorded 100 % larval mortality at 24–48 h post-application. There was 94 % reduction in pupa development at weekly retreatment intervals. Their results showed that the lethal concentration (LC) to kill 95 % of third instar larvae of *An. gambiae s.s.* after 24 h was 0.023 mg/l (14.9 BsITU/l) for *Bs* water-dispersible granules (WDG) and 0.132 mg/l (396 ITU/l) for *Bti* WDG.

5.8 Mode of Action of Microbial Toxins

The microbial toxins generally damage the gut epithelial cells of mosquito larvae. Singh and Gill (1988) and Poopathi et al. (1999a, b) have studied the cytopathological effects of microbial toxins. The Bin toxin affected the epithelial cells in the midgut of mosquito larvae by binding to Cpm1 (*Culex pipiens* maltase 1), a digestive enzyme, and causes severe intracellular damage, including a dramatic cytoplasmic vacuolation (Opota et al. 2011). Cyt toxins also affect insect midgut cells and are able to increase the insecticidal property of some Cry toxins. Moreover, the Cyt toxins are able to overcome resistance to Cry toxins in mosquitoes Soberón et al. (2013). It was found that Cyt1Aa was able to overcome the resistance to Cry4 or Cry11Aa toxins of the *Cx. quinquefasciatus* populations (Crickmore et al. 1995; Wirth et al. 1997).

5.9 Nontarget Effects of Microbial Larvicides

WHO (1999) has reported that biocontrol agents are better than chemical larvicides since they are very species specific and environmentally safe. Many studies have proven that microbial and botanical larvicides are non-toxic or less toxic to nontarget organisms like natural enemies (Theiling and Croft 1988). All microbial pesticides are thoroughly screened for their safety to nontarget organisms prior to registration. Extensive testing showed that microbial larvicides are safe to wildlife, to nontarget organisms and to the environment. The *Bti* or *B. sphaericus* products are non-toxic to humans when they are used according to the directions given in the label (Miura et al. 1980). An isolate of *B. thuringiensis* designated as PG-14 obtained from the Philippines was highly toxic to the mosquitoes *Ae. aegypti* and *Cx. molestus* but non-toxic to the silkworm, *Bombyx mori*, and adults of a daphnid. The degree of toxicity to mosquito larvae was the same as that of the reference strain of *B. thuringiensis* subsp. *israelensis* (serotype 14) (Padua et al. 1984).

5.10 Development of Actinomycete-Based Pesticides

Development of microbial pesticides, especially actinomycete-based pesticides, involves many steps. The sequence of the steps is sampling, isolation of actinomycetes, preliminary bioassay using optimized culture media, mass production of promising isolate, crude extraction, bioassay of crude extract, bioassay-guided fractionation and isolation of active compound, structural elucidation and identification of active compound, preparation of pesticidal formulation using the active compound, toxicological studies and registration.

The places of sampling of actinomycetes are generally chosen on the basis of certain evidences of the presence of beneficial microorganisms, such as dead arthropods, disease-suppressive soils or healthy plants in epidemic areas (Montesinos 2003). Extreme environments may contain useful actinomycetes. Pilot trials with pesticide formulation under real conditions of application are very important in which biosafety and nontarget effects of the microbial pesticide should be given priority.

5.11 Registration and Commercialization of Microbial Pesticides

Application of microbial pesticides in mosquito breeding sites is an eco-friendly and efficient way of prevention of mosquito-borne diseases. In India, the manufacture, commercial use, transport, import and distribution of microbial pesticides or any biopesticide fall under the Insecticide Act (1968) under which microbial pesticides should be registered with the Central Insecticides Board (CIB) of the Ministry of Agriculture (Anonymous 2013). Registration of microbial pesticides is mandatory for commercialization in India since 2006. As of October 2009, 14 primary microbial pesticide products and their formulations were registered in India, and nearly 150 companies were involved in the production of microbial pesticides (Devi et al. 2012).

Commercial production of microbial pesticides needs large-scale production of microbes, their preservation, storage at optimum conditions and formulation (Powell and Jutsum 1993). A pesticidal formulation is the process of converting an active compound into a product that can be applied by practical methods to permit its effective, safe and economic use (Taborsky 1992). Before registration of the formulated microbial pesticide, it should be studied for nontarget effects on fishes, birds, earthworms, honeybees and silkworm and for its ecotoxicity. After the completion of required studies, the microbial pesticide formulation should be patented for legal protection. Taborsky (1992) has given a detailed account on production techniques and commercialization of microbial pesticides at small scale.

Cost effectiveness is an important criterion for any pesticide. Economic feasibility is one of the important advantages of microbial pesticides compared to chemical pesticides. Very few investigators have studied the cost effectiveness of microbial pesticides for mosquito control. Fillinger and Lindsay (2006) have reported that the cost of providing protection to human population from *Anopheles* by using *B. thuringiensis* var. *israelensis* and *B. sphaericus* was less than US$0.09/person/year.

5.12 Limitations of Microbial Pesticides and Possible Solutions to Overcome

Some investigators have proposed that environmental factors may affect the effectiveness of microbial pesticides. According to Boisvert (2005), the activity of *Bti* or *Bs* against target organisms can be influenced by environmental factors such as organic pollution, water temperature and the presence of colloidal particles. Rydzanicz et al. (2010) found that sunlight decreased the activity of *Bti* and *Bs* against *Ochlerotatus caspius* mosquitoes.

Another important concern with microbial pesticides is that mosquitoes are developing resistance to certain bacterial toxins. But a study indicated that a combination of *B. sphaericus* 2362 in a 10:1 ratio with a strain of *B. thuringiensis* subsp. *israelensis* that produces Cyt1A reduced resistance by >30,000-fold. Resistance was suppressed completely when *B. sphaericus* was combined with purified Cyt1A crystals in a 10:1 ratio (Wirth et al. 2000).

5.13 Future Prospects of Microbial Control of Mosquitoes

Mulla (1994) has stated that microbial control agents will become important components in vector mosquito control during the first quarter of the twenty-first century. Due to their target-specific activities, non-toxicity to vertebrates and human beings and economic feasibility, microbial pesticides are considered as the most reliable mosquito control agents. The limitations of these excellent biopesticides should be succeeded in the future. The persistence of microbial pesticides in all types of aquatic habitats for longer duration and their UV stability should be improved. So research should be focused on these aspects in the future.

5.14 Conclusion

In conclusion, microbial pesticides are reliable control agents for mosquito population due to their target-specific effect. Microbial pesticides can be produced in large quantities without disturbing natural resources, and so it will ensure a continuous supply at low cost. Future research should focus on reducing the limitations of microbial pesticides particularly to avoid the pesticide resistance caused by some bacterial toxins by novel techniques. Government should give priority to such research activities to strengthen the mosquito control programme throughout the country.

Acknowledgement The authors acknowledge Entomology Research Institute for facilities. We acknowledge Dr. Vijay Veer, Director of Defence Research Laboratory, Tezpur, for his encouragement. We also thank DRDO for financial assistance through a project (No. ERIP/ER/1004554 M/01/1357).

References

Ando K (1983) How to discover new antibiotics for insecticidal use. In: Takahashi T, Yoshioka H, Misato T, Matusunaka S (eds) Pesticide chemistry: human welfare and the environment, vol 2, Natural products. Pergamon Press, New York, pp 253–259

Anonymous (1990) Biologically active KSB-1939L3 compound and its production-pesticide with insecticide and acaricide activity production by *Streptomyces* sp. culture. Biotechnol Abst 9(19):58. Japan Patent no 273961, 1988

Anonymous (2013) Biopesticides – quality assurance. Policy paper 62. National Academy of Agricultural Sciences, New Delhi

Bassi A (1836) Del mal del segno e di altre malattie dei bachi da seta. Parte seconda. Practica Tipografia Orcesi, Lodi: 58

Berry C, O'Neill S, Ben-Dov E et al (2002) Complete sequence and organization of pBtoxis, the toxin-coding plasmid of *Bacillus thuringiensis* subsp. *israelensis*. Appl Environ Microbiol 68:5082–5095

Bilal H, Akram W, Ali-Hassan S (2012) Larvicidal activity of *Citrus* limonoids against *Aedes albopictus* larvae. J Arthropod Borne Dis 6(2):104–111

Biondi A, Mommaerts V, Smagghe G et al (2012) The non-target impact of spinosyns on beneficial arthropods. Pest Man Sci. doi:10.1002/ps.3396

Bode HB, Bethe B, Höfs R, Zeeck A (2002) Big effects from small changes: possible ways to explore nature's chemical diversity. ChemBioChem 3:619–627

Boisvert M (2005) Utilization of *Bacillus thuringiensis* var. *israelensis* (*Bti*)-based formulations for the biological control of mosquitoes in Canada. In: Abstracts of the 6th Pacific rim conference on the biotechnology of *Bacillus thuringiensis* and its environmental impact, Victoria, pp 87–93

Bond JG, Marina CF, Williams T (2004) The naturally derived insecticides spinosad is highly toxic to *Aedes* and *Anopheles* mosquito larvae. Med Vet Entomol 18:50–56

Brattsten LB, Hamilton GC, Sutherland DJ (2009) Insecticides recommended for mosquito control in New Jersey. New Jersey Agricultural Experiment Station, Publication no P-08001-01-08. http://www.hudsonregional.org/mosquito/Files/bmpmcnj.pdf

Bret BL, Larson LL, Schoonover JR et al (1997) Biological properties of spinosad. Down to Earth 52:6–13

Broadwell AH, Baumann P (1986) Sporulation-associated activation of *Bacillus sphaericus* larvicide. Appl Environ Microbiol 52:758–764

Charles JF, LeRoux CN (2000) Mosquitocidal bacterial toxins: diversity, mode of action and resistance phenomena. Mem Inst Oswaldo Cruz 95:201–206

Charles JF, Kalfon A, Bourgouin C, de Barjac H (1988) *Bacillus sphaericus* asporogenous mutants: morphology, protein pattern and larvicidal activity. Ann Inst Pasteur Microbiol 139:243–259

Charles JF, LeRoux C, Delécluse A (1996) *Bacillus sphaericus* toxins: molecular biology and mode of action. Annu Rev Entomol 41:451–472

Chaudhary HS, Yadav J, Shrivastava AR et al (2013) Antibacterial activity of actinomycetes isolated from different soil samples of Sheopur (a city of central India). J Adv Pharm Technol Res 4:118–123

Chui VWD, Wong KW, Tori KW (1995) Control of mosquito larvae (Diptera: Culicidae) using *Bti* and teflubenzuron: laboratory evaluation and semi-field test. Environ Int 21:433–440

Crickmore N, Bone EJ, Williams JA, Ellar DJ (1995) Contribution of the individual components of the δ-endotoxin crystal to the mosquitocidal activity of *Bacillus thuringiensis* subsp. *israelensis*. FEMS Microbiol Lett 131:249–254

Darriet F, Duchon S, Hougard JM (2005) Spinosad: a new larvicide against insecticide-resistant mosquito larvae. J Am Mosq Control Assoc 21:495–496

de Luna JS, dos Santos AF, de Lima MR et al (2005) A study of the larvicidal and molluscicidal activities of some medicinal plants from northeast Brazil. J Ethnopharmacol 97:199–206

Des Rochers B, Garcia R (1983) The efficacy of *Bacillus sphaericus* in controlling mosquitoes breeding in sewer effluent. Proc and papers Calif Mosq Vect Cont Assoc 51:35–37

Devi PSV, Rao GVR, Gopalakrishnan S, Sivakumar G (2012) Environmental impact of microbial pesticides. In: Sharma HC, Dhillon MK, Sehrawat KL (eds) Environmental safety of biotech and conventional IPM technologies. Studium Press LLC, Houston, pp 261–272

Dhanasekaran D, Sakthi V, Thajuddin N, Panneerselvam A (2010) Preliminary evaluation of *Anopheles* mosquito larvicidal efficacy of mangrove Actinobacteria. Int J Appl Biol Pharm Technol 1:374–381

Djènontin A, Pennetier C, Zogo B et al (2014) Field efficacy of Vectobac GR as a mosquito larvicide for the control of Anopheline and Culicine mosquitoes in natural habitats in Benin, West Africa. PLoS One 9, e87934. doi:10.1371/journal.pone.0087934

Doull J, Vining L (1990) Nutritional control of actinorhodin production by *Streptomyces coelicolor* A (3) 2: suppressive effects of nitrogen and phosphate. Appl Microbiol Biotechnol 32:449–454

Federici BA, Park HW, Bideshi DK et al (2003) Recombinant bacteria for mosquito control. J Exp Biol 206:3877–3885

Fillinger U, Lindsay SW (2006) Suppression of exposure to malaria vectors by an order of magnitude using microbial larvicides in rural Kenya. Trop Med Int Health 11:1629–1642

Fillinger U, Knols BG, Becker N (2003) Efficacy and efficiency of new *Bacillus thuringiensis* var *israelensis* and *Bacillus sphaericus* formulations against Afrotropical anophelines in Western Kenya. Trop Med Int Health 8:37–47

Foda MS, Fawkia M, El-Beih FM et al (2010) Physiological formation of mosquitocidal toxin by a novel *Bacillus thuringiensis* isolate under solid state fermentation. Life Sci J 7:144–152

Gadelhak GG, EL-Tarabily K, AL-Kaabi FK (2005) Insect control using chitinolytic soil actinomycetes as biocontrol agents. Int J Agric Biol 7:627–633

Gautam K, Padma Kumar, Poonia S (2013) Larvicidal activity and GC-MS analysis of flavonoids of *Vitex negundo* and *Andrographis paniculata* against two vector mosquitoes *Anopheles stephensi* and *Aedes aegypti*. J Vector Borne Dis 50:171–178

Gibbs KE, Brautigam FC, Stubbs CS, Zibilske LM (1986) Experimental applications of Bti for larval black fly control: persistence and downstream carry, efficacy, impact on non-target invertebrates and fish feeding, Technical bulletin 123. Maine Agricultural Experiment Station, University of Maine, pp 1–25

Glare TR, O'Callaghan M (1998) Environmental and health impacts of *Bacillus thuringiensis israelensis*. Report for the Ministry of Health, Biocontrol and Biodiversity, Grasslands Division, AgResearch, Lincoln

Hertlein MB, Mavrotas C, Jousseaume C et al (2010) A review of spinosad as a natural mosquito product for larval mosquito control. J Am Mosq Control Assoc 26:67–87

Ibarra JE, del Rincón MC, Noriega SOD et al (2003) Diversity of *Bacillus thuringiensis* strains from Latin America with insecticidal activity against different mosquito species. Appl Environ Microbiol 69:95269–95274

Ignacimuthu S (2000) The role of botanicals in combating mosquitoes. In: John William S, Vincent S (eds) Recent trends in combating mosquitoes. School of Entomology and Centre for Natural Resources Management, PG and Research Department of Zoology, Loyola, Chennai, College pp 62–70

Ignacimuthu S, Paulraj MG (2009) Non-chemical insect pest management. Curr Sci 97:136

Ishiwata S (1901) On a kind of severe flacherie (sotto disease). Dainihon Sanshi Kaiho 114:1–5

Jahan N, Jamali EH, Qamar MF (2013) Residual activity of *Bacillus thuringiensis* var. *israelensis* and *Bacillus sphaericus* against mosquito larvae. J Anim Plant Sci 23(4):1052–1059

Johnson T (1998) CRC ethnobotany desk reference. CRC Press, Boca Raton

Kumar S, Govindasamy R, Krishnan K et al (2011) Larvicidal activity of isolated compound 5-(2,4-dimethylbenzyl) pyrrolidin-2-one from marine *Streptomyces* VITSVK5 sp. against *Rhipicephalus (Boophilus) microplus*, *Anopheles stephensi* and *Culex tritaeniorhynchus*. Parasitol Res 112:215–226

Lee SE (2000) Mosquito larvicidal activity of pipernonaline, a piperidine alkaloid derived from long pepper, *Piper longum*. J Am Mosq Control Assoc 16:245–247

Lee YW, Zairi J (2006) Field evaluation of *Bacillus thuringiensis* H-14 against *Aedes* mosquitoes. Trop Biomed 23:37–44

Liu ZL, Liu QZ, Du SS, Deng ZW (2012) Mosquito larvicidal activity of alkaloids and limonoids derived from *Evodia rutaecarpa* unripe fruits against *Aedes albopictus* (Diptera: Culicidae). Parasitol Res 111:991–996

Maheswaran R, Sathish S, Ignacimuthu S (2008) Larvicidal activity of *Leucas aspera* (Willd.) against the larvae of *Culex quinquefasciatus* Say. and *Aedes aegypti* L. Int J Integr Biol 2:214–217

Majambere S, Lindsay SW, Green C et al (2007) Microbial larvicides for malaria control in the Gambia. Malar J 6:76. doi:10.1186/1475-2875-6-76

Margalit J, Dean D (1985) The story of *Bacillus thuringiensis* var. *israelensis* (B.t.i.). J Am Mosq Cont Assoc 1:1–7

Marina CF, Bond JG, Muñoz J et al (2012) Spinosad: a biorational mosquito larvicide for use in car tires in southern Mexico. Parasit Vectors 5:95

Mathew N, Anitha MG, Bala TS et al (2009) Larvicidal activity of *Saraca indica*, *Nyctanthes arbor-tristis*, and *Clitoria ternatea* extracts against three mosquito vector species. Parasitol Res 104:1017–1025

Mishra SK, Keller JE, Miller JR et al (1987) Insecticidal and nematicidal properties of microbial metabolites. J Ind Microbiol 2:267–276

Miura T, Takahashi RM, Mulligan FS (1980) Effects of the bacterial mosquito larvicide, *Bacillus thuringiensis* serotype H-14 on selected aquatic organisms. Mosq News 40:619–622

Montesinos E (2003) Development, registration and commercialization of microbial pesticides for plant protection. Int Microbiol 6:245–252

Mulla SM (1991) Biological control of mosquitoes with entomopathogenic bacteria. Chinese Journal of Entomology special publication no 6, Proceedings of the IV national vector control symposium, Taichung, ROC, pp 93–104

Mulla SM (1994) Mosquito control then, now, and in the future. J Am Mosq Cont Assoc 10:574–584

Nerio LS, Olivero-Verbel J, Stashenko E (2010) Repellent activity of essential oils: a review. Bioresour Technol 101:372–378

Opota O, Gauthier NC, Doye A et al (2011) *Bacillus sphaericus* binary toxin elicits host cell autophagy as a response to intoxication. PLoS One 6, e14682. doi:10.1371/journal.pone.0014682

Padua LE, Ohba M, Aizawa K (1984) Isolation of a *Bacillus thuringiensis* strain (serotype 8a:8b) highly and selectively toxic against mosquito larvae. J Invertebr Pathol 44:12–17

Pampiglione S, Majori G, Petrangeli G, Romi R (1985) Avermectins, MK-933 and MK- 936, for mosquito control. Trans R Soc Trop Med Hyg 79:797–799

Park HW, Mangum CM, Zhong H, Hayes SR (2007) Isolation of *Bacillus sphaericus* with improved efficacy against *Culex quinquefasciatus*. J Am Mosq Control Assoc 23:478–480

Pascual M, Ahumada JA, Chaves LF et al (2006) Malaria resurgence in East African highlands: temperature trends revisited. Proc Natl Acad Sci U S A 103:5829–5834

Patil SV, Patil CD, Salunkhe RB, Salunke BK (2010) Larvicidal activities of six plants extracts against two mosquito species, *Aedes aegypti* and *Anopheles stephensi*. Trop Biomed 27:360–365

Patz JA, Campbell-Lendrum D, Holloway T, Foley JA (2005) Impact of regional climate change on human health. Nature 438:310–317

Paulraj MG, Reegan AD, Ignacimuthu S (2011) Toxicity of Benzaldehyde and Propionic acid against immature and adult stages of *Aedes aegypti* (Linn.) and *Culex quinquefasciatus* (Say) (Diptera: Culicidae). J Entomol 8:539–547

Phongsopitanun W, Suwanborirux K, Tanasupawat S (2014) Identification and antimicrobial activity of *Streptomyces* strains from Thai mangrove sediment. Thai J Pharm Sci 38:1–56

Pinto da Silva L, Gonçales RA, Conceição FR et al (2011) Stability, oviposition attraction, and larvicidal activity of binary toxin from *Bacillus sphaericus* expressed in *Escherichia coli*. Appl Microbiol Biotechnol. doi:10.1007/s00253-011-380

Poopathi S, Mani TR, Rao DR et al (1999a) Effect of *Bacillus sphaericus* and *Bacillus thuringiensis* var. *israelensis* on the ultrastructural changes in the midgut of *Culex quinquefasciatus* Say (Diptera: Culicidae). J Entomol Res 23:347–357

Poopathi S, Kabilan L, Mani TR et al (1999b) A comparative ultrastructural studies on the midgut of *Bacillus sphaericus* resistant and susceptible *Culex quinquefasciatus* Say. Insect Environ 5:129–130

Poopathi S, Thirugnanasambantham K, Mani C et al (2014) Isolation of mosquitocidal bacteria (*Bacillus thuringiensis*, *B. sphaericus* and *B. cereus*) from excreta of arid birds. Ind J Exp Biol 52:739–747

Porter AG, Davidson EW, Liu JW (1993) Mosquitocidal toxins of *Bacilli* and their genetic manipulation for effective biological control of mosquitoes. Microbiol Rev 57:838–861

Powell KA, Jutsum AR (1993) Technical and commercial aspects of biocontrol products. J Pestic Sci 37:315–321

Rajiv Gandhi M, Reegan AD, Sivaraman G et al (2014) Larvicidal and repellent activities of *Tylophora indica* (Burm. F.) Merr. (Asclepiadaceae) against *Culex quinquefasciatus* Say and *Aedes aegypti* L. (Diptera: Culicidae). Int J Pure App Zool 2:113–117

Ramar M, Paulraj MG, Ignacimuthu S (2013a) Screening of pupicidal activity of some essential oils against *Culex quinquefasciatus Say*. Peak J Med Plant Res 1:9–12

Ramar M, Paulraj MG, Ignacimuthu S (2013b) Preliminary screening of plant essential oils against larvae of *Culex quinquefasciatus* Say (Diptera: Culicidae). Afr J Biotechnol 12:6480–6483

Ramar M, Ignacimuthu S, Paulraj MG (2014) Biological activity of nine plant essential oils on the filarial vector mosquito, *Culex quinquefasciatus* Say (Insecta: Diptera: Culicidae). Int J Res Biol Sci 4:1–5

Rana S, Salam MD (2014) Antimicrobial potential of Actinomycetes isolated from soil samples of Punjab, India. J Microbiol Exp 1:00010

Rao KV, Chattopadhyay SK, Reddy GC (1990) Flavonoids with mosquito larval toxicity-tangeratin, daidzein and genistein crystal production, isolation and purification; *Streptomyces* spp. culture; Insecticide. J Agric Food Chem 38:1427–1430

Rashad FM, Saleh WD, Nasr M, Fathy HM (2012) Identification of mosquito larvicidal bacterial strains isolated from north Sinai in Egypt. AMB Express 2:9

Reegan AD, Paulraj MG, Ignacimuthu S (2013a) Larvicidal, ovicidal, repellent and histopathological effects of orange peel (*Citrus sinensis* Osbeck) extracts on *Anopheles stephensi* Liston mosquitoes (Diptera: Culicidae). Int J Appl Biol 1:24–29

Reegan AD, Kinsalin VA, Paulraj MG, Ignacimuthu S (2013b) Larvicidal, ovicidal, and repellent activities of marine sponge *Cliona celata* (Grant) extracts against *Culex quinquefasciatus* Say and *Aedes aegypti* L. (Diptera: Culicidae). ISRN Entomol. doi:10.1155/2013/315389

Reegan AD, Kinsalin VA, Paulraj MG, Ignacimuthu S (2015) Larvicidal, ovicidal and repellent activities of marine sponge *Cliona celata* (Grant) extracts against *Anopheles stephensi* Liston (Diptera: Culicidae). Asian Pac J Trop Med 8(1):29–34

Reegan AD, Paulraj MG, Ignacimuthu S (2014a) Effect of Niloticin, a protolimonoid isolated from *Limonia acidissima* L. (Rutaceae) on the immature stages of dengue vector *Aedes aegypti* L. (Diptera: Culicidae). Acta Trop 139:67–76

Reegan AD, Rajiv Gandhi M, Paulraj MG, Ignacimuthu S (2014b) Larvicidal activity of medicinal plant extracts against *Culex quinquefasciatus* Say and *Aedes aegypti* L. mosquitoes (Diptera: Culicidae). Int J Pure Appl Zool 2:205–210

Romi R, Proietti S, Di Luca M, Cristofaro M (2006) Laboratory evaluation of the bioinsecticide spinosad for mosquito control. J Am Mosq Control Assoc 22:93–96

Rungrod A, Tjahaja NK, Soonsanga S et al (2009) *Bacillus sphaericus* Mtx1 and Mtx2 toxins co-expressed in *Escherichia coli* are synergistic against *Aedes aegypti* larvae. Biotechnol Lett 31:551–555

Rydzanicz K, Sobczyński M, Guz-Regner K (2010) Comparison of the activity and persistence of microbial insecticides based on *Bacillus thuringiensis israelensis* and *Bacillus sphaericus* in organic polluted mosquito-breeding sites. Pol J Environ Stud 19:1317–1323

Salgado VL (1997) The modes of action of spinosad and other insect control products. Down to Earth 52:35–43

Salgado VL (1998) Studies on the mode of action of spinosad: insect symptoms and physiological correlates. Pestic Biochem Physiol 60:91–102

Sanchez S, Demain AL (2002) Metabolic regulation of fermentation processes. Enzyme Microb Technol 31:895–906

Schneider M, Smagghe C, Viñuela E (2004) Comparative effects of several insect growth regulators and spinosad on the different developmental stages of the endoparasitoid *Hyposoter didymator* (Thunberg). Pesticides and Beneficial Organisms. IOBC/WPRS Bull 27:13–19

Singh GJP, Gill SS (1988) An electron microscope study of the toxic action of *Bacillus sphaericus* in *Culex quinquefasciatus* larvae. J Invertebr Pathol 52:237–247

Sivaraman G, Paulraj MG, Rajiv Gandhi M et al (2014) Larvicidal potential of *Hydnocarpus pentandra* (Buch.-Ham.) Oken seed extracts against *Aedes aegypti* Linn. and *Culex quinquefasciatus* (Say) (Diptera: Culicidae). Int J Pure Appl Zool 2:109–112

Smith AW, Camara-Artigas A, Allen JP (2004) Crystallization of the mosquito-larvicidal binary toxin produced by *Bacillus sphaericus*. Acta Crystallogr D: Biol Crystallogr 60:952–953

Soberón M, López-Díaz JA, Bravo A (2013) Cyt toxins produced by *Bacillus thuringiensis*: a protein fold conserved in several pathogenic microorganisms. Peptides 41:87–93

Subramani R, Aalbersberg W (2012) Marine actinomycetes: an ongoing source of novel bioactive metabolites. Microbiol Res 167:571–580

Taborsky V (1992) Small-scale processing of microbial pesticides, FAO agricultural services bulletin no 96. Food and Agriculture Organization of the United Nations, Rome

Theiling KM, Croft BA (1988) Pesticide side-effects on arthropod natural enemies: a database summary. Agri Ecosyst Environ 21:191–218

Thompson GD, Dutton R, Sparks TC (2000) Spinosad-a case study: an example from a natural products discovery programme. Pest Manag Sci 56:696–702

United States Environmental Protection Agency (US EPA) (2000) For your information: larvicides for mosquito control. www.cmmcp.org/larvfs.pdf

Usta C (2013) Microorganisms in biological pest control – A review (bacterial toxin application and effect of environmental factors). In: Marina Silva-Opps (ed) Current progress in biological research. InTech. Croatia. doi:10.5772/55786. ISBN 978-953-51-1097-2

Vijayakumar R, Murugesan S, Cholarajan A, Sakthi V (2010) Larvicidal potentiality of marine actinomycetes isolated from Muthupet mangrove, Tamil Nadu, India. Int J Microbiol Res 1:179–183

Vijayan V, Balaraman K (1991) Metabolites of fungi & actinomycetes active against mosquito larvae. Indian J Med Res 93:115–117

WHO (World Health Organization) (1985) Information consultation on the development of *Bacillus sphaericus* as a Microbial larvicide, vol 85. World Health Organization, Geneva, pp 1–24

WHO (World Health Organization) (1999) International programme on chemical safety (IPCS): microbial pest control agent *Bacillus thuringiensis*. Environ Health Crit 217:1–105

WHO (World Health Organization) (2010) Spinosad DT in drinking-water: use for vector control in drinking-water sources and containers. WHO/HSE/WSH/10.01/12. WHO, Geneva

WHO (World Health Organization) (2013) Global malaria programme. World malaria report. WHO, Geneva

Wirth M, Georghiou GP, Federici BA (1997) CytA enables CryIV endotoxins of *Bacillus thuringiensis* to overcome high levels of CryIV resistance in the mosquito *Culex quinquefasciatus*. Proc Natl Acad Sci U S A 9:10536–10540

Wirth MC, Walton WE, Federici BA (2000) Cyt1A from *Bacillus thuringiensis* restores toxicity of *Bacillus sphaericus* against resistant *Culex quinquefasciatus* (Diptera: Culicidae). J Med Entomol 37:401–407

Zarroug IMA, Nugud AD, Bashir AK, Mageed AA (1988) Evaluation of Sudanese plant extracts as mosquito larvicides. Pharm Biol 26:77–80

Zeigler DR (1999) *Bacillus* genetic stock center catalog of strains, 7th edn, Part 2: *Bacillus thuringiensis* and *Bacillus cereus*. Department of Microbiology, The Ohio State University, Columbus

Zizka Z, Weiser J, Blumauerova M, Jizba J (1989) Ultrastructural effects of macroterrolides of *Streptomyces griseus* LKS-1 in tissues of *Culex pipiens* larvae- monactin, dinactin, triactin and nonactin preparation; insecticide activity. Cytobios 58:85–91

Extraction, Purification and Characterisation of Insecticidal Compounds from Plants

6

Shaswat Barua, Reji Gopalakrishnan, and Vijay Veer

Abstract

In recent times, a global concern has been raised against the utility of synthetic insecticides in households and fields. Many of such synthetic insecticides were banned in due course of time because of their toxicity to the nontarget flora and fauna. In the quest of developing sustainable and environment-friendly commercial products, different plant components have been studied extensively. Although there are many reports on the insecticidal potential of plant-based compounds, very few reports describe the actual active component responsible for such efficacy. Lack of adequate information on the bioactive compounds creates major challenges for the commercialisation of plant-derived products for pest control. The chapter describes the different classes of plant-derived insecticidal compounds such as alkaloids, flavonoids and terpenoids. The qualitative and quantitative studies of plant compounds need vast expertise in isolation and characterisation methods. The development of chromatographic and spectrometric tools eased the extraction of plant active components. The different techniques involved in the extraction, purification and characterisation of insecticidal compounds extracted from plants along with some salient examples available in the literature are discussed. This chapter may be helpful to construct a bridge between the biologists and chemists to work together in collaboration for elucidating the actual structure and activity of plant-based compounds, which have very significant insecticidal property.

6.1 Introduction

In the quest of developing sustainable and environment-friendly commercial products, different plant components have been studied extensively. Medicinal plants attained copious attention

S. Barua • R. Gopalakrishnan (✉) • Vijay Veer
Defence Research Laboratory (Defence Research and Development Organisation),
Tezpur 784 001, Assam, India
e-mail: rejidrl@gmail.com; vijayveer50@yahoo.com

© Springer India 2016

Vijay Veer, R. Gopalakrishnan (eds.), *Herbal Insecticides, Repellents and Biomedicines: Effectiveness and Commercialization*, DOI 10.1007/978-81-322-2704-5_6

for primary health care in many developing countries. Traditional knowledge-based findings proved that nature has answers to various human ailments and inconveniences. Thus, a range of plant-based products has been commercialised for their utility as medicines, cosmetics, food additives, pesticides, insecticides, etc.

In recent times, a global concern has been raised against the utility of synthetic insecticides in households and fields. Especially, the residual amount of the insecticide retained in the food or other close vicinities may create devastating loss to human life and wellbeing. Many of such synthetic insecticides were banned in due course of time because of their toxicity to the nontarget flora and fauna. Insecticides are categorised as of chemical and biological origin. The prime purpose of such materials is to repel or kill insects that are unfriendly to human. Natural insecticides are the chemical compounds or substances which can be obtained from living organisms. Such products utilises living organisms (both of zoological or botanical origin) to control pests (DeBach and Rosen 1991). A number of microorganisms such as *Bacillus thuringiensis* were found to be effective against mosquitoes, cabbage loopers, aphids, whitefly, caterpillars, ants, beetles, termites, etc.

Plant-based compounds have been extracted by people for controlling or repelling insects since ancient time. Chrysanthemum was reported to be used during 400 BC to control pests in paddy fields (Silva-Aguayo 2009). In the following times, a huge number of plants were studied for their insecticidal activity. They were usually extracted by soaking and separating plant products like leaves, roots, barks, etc. The main disadvantage associated with such product is that it is very difficult to ascertain the active components. Once, it is recognised, it further demands tedious efforts for isolation. Moreover, the isolated product may not be available in abundance, or it loses its activity during the isolation process. These facts stand as a barrier for commercialisation of such plant-based insecticides. Thus, qualitative and quantitative studies of plant compounds need vast expertise in isolation and characterisation methods (Smith 2002; Sasidharan et al. 2011). In broader sense, plant-based insecticides can be categorised into three classes, viz. alkaloid, flavonoid and terpenoid based, depending upon the active compound(s). Development of chromatographic and spectrometric tools eased the extraction of plant active components. In this chapter, we will concentrate on the extraction methods and characterisation tools which are utilised vastly for insecticidal plant compounds.

6.2 Plant-Derived Compounds with Insecticidal Activity

Plant-derived insecticides have shown tremendous potentiality for pest control. They are associated with a range of merits like rapid degradation, sustainability, environmentally benign nature, etc. Despite such advantages, these materials are not commercially popular because of some concerns like the lack of abundance and economic supply of the plant, lack of understanding about the active compound in an extract, limited scope for intellectual property right protection of the traditional ethnobotanical knowledge, etc. Thorough study of these aspects is an essential prerequisite to formulate plant-based commercial insecticides.

The history of natural insecticides dates back to the ancient times, when tobacco was used extensively for controlling pests in households. In broader sense, plant-based insecticides can be categorised into three classes, viz. alkaloid, flavonoid and terpenoid based, depending upon the active compound(s).

6.2.1 Alkaloid-Based Insecticides

Alkaloids are naturally occurring organic compounds, which mostly contain nitrogen. Although alkaloids are derived from a variety of organisms including bacteria, fungi, plants and animals, the most common source are the plants. Recent classification of alkaloids is based on the carbon skeleton present in the structures, such as indole, isoquinoline, pyridine, etc. Nicotine, the major compound of tobacco, contains a pyridine fragment. Tobacco has been utilised for a long time in insect control, where the alkaloid nicotine is the active ingredient. Some other pyridine-based alkaloids have been studied, which exhibited such activity against a range of organisms. Dihydronicotyrine, imidacloprid, anabaseine and anatabine are such alkaloids to name a few (Yamamoto 1965). Figure 6.1 demonstrates the structures of a few pyridine-based alkaloids.

Some piperidine alkaloids like pipernonaline and piperoctadecalidine, isolated from *Piper longum* L., showed profound insecticidal and acaricidal activity (Byeoung-Soo 2002; Porzel 2013). Further, alkaloids extracted from plants such as *Pergularia tomentosa*, *Annona squamosa* and *Corydalis bulbosa* were highly effective against the larva of *Locusta migratoria cinerascens*, *Anopheles stephensi* and *Drosophila melanogaster*, respectively (Acheuk and Doumandji-Mitiche 2013; Saxena et al. 1993; Miyazawa et al. 1998).

Alkaloids are extracted by treating the plant material with alkaline solutions, followed by extracting in organic solvents like chloroform, benzene, 1,2-dichloroethane, diethyl ether, etc.

6.2.2 Flavonoid-Based Insecticides

Flavonoids are water-soluble polyphenols, with the general structure of a 15-carbon skeleton. They consist of two phenyl rings and one heterocyclic ring. Flavonoids are mainly classified into three classes, viz. bioflavonoids, isoflavonoids and neoflavonoids. Figure 6.2 shows a flavone backbone structure.

Flavonoids are found abundantly in plants. They are plant pigments important for flower colouration. Important sources of flavonoids are tea, soybean, fruits, etc. Some flavonoids show inhibitory effect against plant pathogens such as *Fusarium oxysporum*. Flavonoid compounds extracted from various plant sources exhibit strong insecticidal activity. Flavonoids extracted from *Calotropis procera* were found to be highly toxic to the insect, *Callosobruchus chinensis* (Salunke et al. 2005). Again, *Tephrosia purpurea*-based flavonoids inhibited *Callosobruchus maculatus* grubs (Diwan and Saxena 2010). Flavonoids present in *Derris trifoliata* extract exhibited string larvicidal activity against the mosquito species, *Aedes aegypti* (Yenesew et al. 2009). Kaempferol, quercetin, lupinifolin (Fig. 6.3), etc. are some of the flavonoid compounds which are responsible for such activities.

Moreover, flavonoid compounds have high medicinal value as anticancer, anti-inflammatory and antimicrobial drugs.

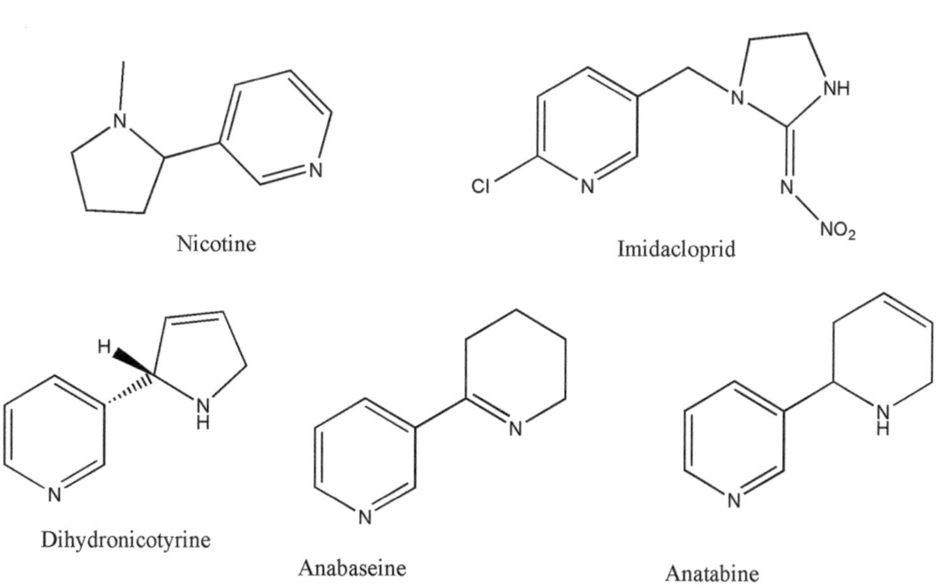

Fig. 6.1 Structure of some pyridine-based alkaloids

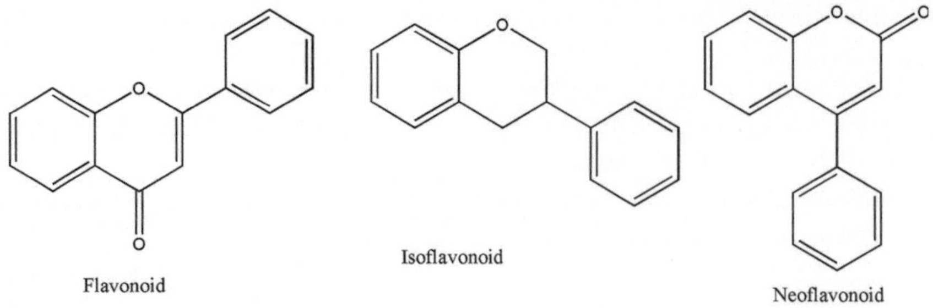

Fig. 6.2 Structure of flavone backbones

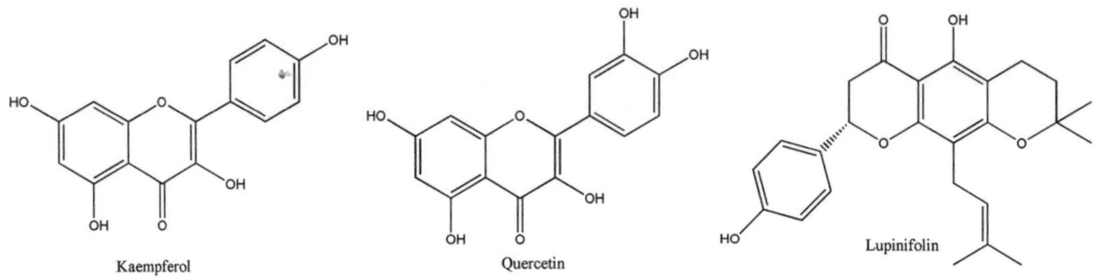

Fig. 6.3 Structure of some insecticidal flavonoids

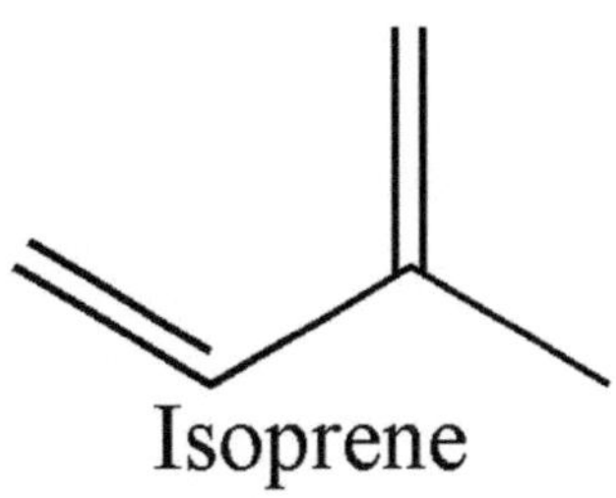

Fig. 6.4 Isoprene unit

6.2.3 Terpenoid-Based Insecticides

Terpenoids are naturally occurring organic compounds, derived from the five-carbon isoprene unit (Fig. 6.4).

Terpenoid compounds are generally found in plant essential oils. Such terpenoids are known for their distinct aroma. They have vast utility as antibacterial, larvicidal and pesticidal agents. Citral, menthol and camphor (Fig. 6.5) are some of the terpenoid compounds, which have strong insect repellent activity due to their characteristic pungent odour (Koul et al. 2008).

Terpenoids are categorised into different classes, such as hemiterpenoids, monoterpenoids, diterpenoids, polyterpenoid and sesquiterpenoids, as per the combination of the isoprene units. Plant essential oils exhibit broad spectrum efficacy against insects, pathogenic microorganisms and pests and thus help to formulate different insect repellents and insecticides. Terpenoids are primly responsible for such efficacy. Methyl chavicol or estragole found in *Ocimum basilicum* and linalool found in *Homalomena aromatica* (Fig. 6.6) are highly effective against mosquito larva, armyworm, *Spodoptera litura*, common housefly, etc. (Koul et al. 2008).

Despite such multifaceted activity, commercialisation of essential oil-based pesticides is problematic due to the tedious extraction and purification processes. Thus, a great expertise and ethnobotanical knowledge are required to bring such products to the market.

6.3 Extraction

Extraction is the primary step for isolating plant components. Separation, characterisation and identification are only possible after efficient extraction process. A number of extraction methods are known, which were based on the behav-

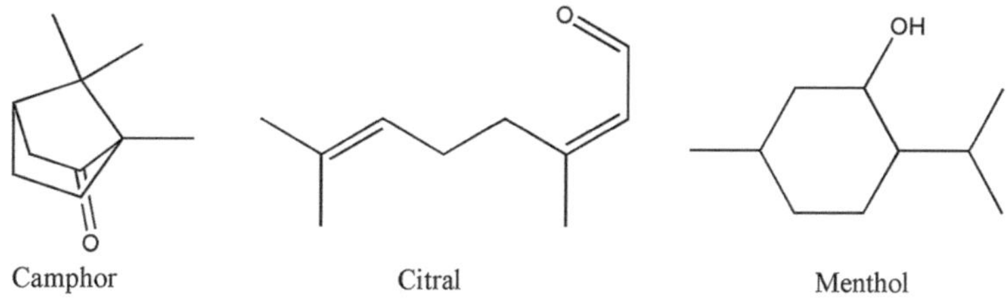

Fig. 6.5 Structures of citral, menthol and camphor

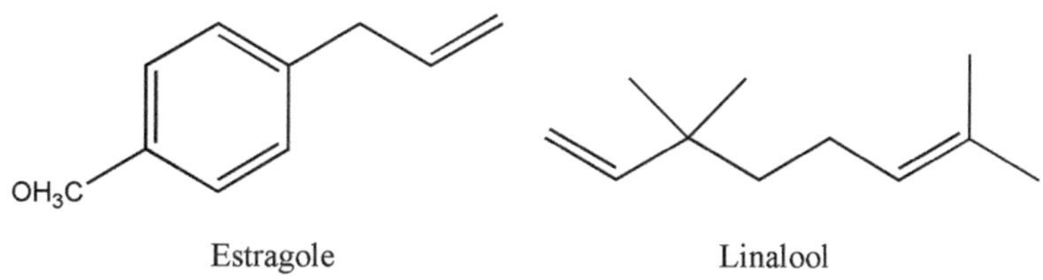

Fig. 6.6 Structures of estragole and linalool

iour of the targeted compounds. The conventional extraction techniques include Soxhlet extraction, maceration and hydrodistillation.

6.3.1 Soxhlet Extraction (SE)

German chemist Franz Ritter von Soxhlet in 1879 first used this technique for extraction of lipid. This proved to be the most widely used technique for extracting plant compounds for a long time. In this method, dry sample is placed in thimbles inside a distillation flask. Solvents are used as per the knowledge on the structure and solubility of the desired compounds.

This is then refluxed at specific temperature and the solution is aspirated. A siphon is attached to the apparatus, which pushes the solvent back to the distillation flask. The process continues till the completion of the extraction (Brusottia et al. 2013). A typical Soxhlet apparatus is shown in Fig. 6.7.

6.3.2 Maceration Extraction (ME)

This technique is used for a long time in home-made tonics. It is helpful to extract essential oils and active compounds from plants. Firstly, the samples are ground to fine mass which increases the surface area of the material. Then, it is mixed with an appropriate solvent in closed vessels. The liquid is then strained off, and the solid residue is pressed to obtain occluded solutions. This is filtered, generally, with muslin cloth to obtain the pure extract. As per requirement, the extract is concentrated by drying either by using heat or keeping under sunlight (Sasidharan et al. 2011).

6.3.3 Hydrodistillation Process (HP)

This method is very effective for extraction of essential oils and important bioactive components from plants. Here, the plant material is packed in a container and sufficient amount of water is added. It is boiled for a specific time, or direct steam can be used for extraction. The supplied heat energy helps to unbind the active components. The extract is then collected after condensing. If essential oils are present, a separator is used to collect the oil and water separately. This technique involves hydrodiffusion and hydrolysis mechanisms (Brusottia et al. 2013). Beyond these traditional extraction methods, some modern techniques are in use from the past few decades. Such approaches include supercritical fluid extraction (SFE), pulsed electric field (PEF) extraction, enzyme-assisted extraction (EAE), microwave-assisted extraction (MAE), ultrasound-assisted extraction (UAE), pressurised liquid extraction (PLE), etc.

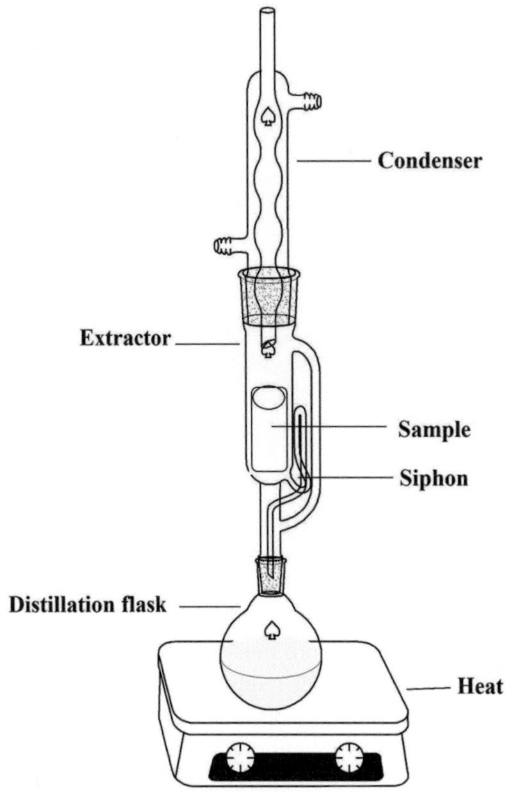

Fig. 6.7 A Soxhlet apparatus

6.3.4 Supercritical Fluid Extraction (SFE)

Hannay and Hogarth in 1879 revealed that supercritical fluids can be used for the extraction of plant components. However, after a long time, in 1964, Zosel utilised this approach for decaffeination of coffee using SFE. In supercritical state, a specific liquid or gas loses its properties like liquification or vaporisation by application of temperature and pressure. They are associated with both liquid- and gas-like attributes such as viscosity, diffusion, surface tension, density, solvation, etc. A typical SFE system is composed of a tank containing CO_2, the mobile phase, a pump, a co-solvent system, trapper and collecting vessel. High pressure is maintained in the system for efficient performance. Lang and Wai (2001) in 2001 described the advantages of using SFE for bioactive plant components, some of which are narrated below:

1. Extraction time is reduced in SFE as compared with the conventional methods.
2. Repeated reflux provides complete extraction.
3. Solvation capacity can be altered by varying temperature and pressure.
4. SFE can be performed at room temperature, which keeps the activity of the plant components intact.
5. The process is recyclable.
6. Scaling up is possible.

Many researchers have utilised this technique for extracting plant components. Flavonoids from *Strobilanthes crispus* leaves were extracted by this method (Liza et al. 2012). A wide range of essential oils were isolated, which proved potentiality as insect repellents or insecticides. SFE is a vital technique to extract such essential oils. Terpenes and their derivatives are mainly responsible for the characteristic aroma of essential oils, which repels insects. However, due to their thermolabile behaviour, steam or hydrodistillation processes are not favourable for extraction. SFE is advantageous in this regard as it can be operated at room temperature. A highly useful insecticidal essential oil from *Curcuma longa* has been extracted by SFE method (Cheng et al. 2012). Further, SFE was used to extract alkaloids like caffein, theophylline, theobromine, etc. from *Ilex paraguariensis* (Saldana et al. 1999). Supercritical CO_2 modified with alcohol was used to extract many important polyphenol compounds from plants.

6.3.5 Pulsed Electric Field Extraction (PEF)

Here the pulsed electric field helps in disrupting the cell membrane of plant components and enhances the extraction. The electric potential separates the components as per the charge on the cell membrane. The use of environmentally benign solvents, like alcohol and alkanes, beyond their boiling points speeds up the extraction process. This process is quite effective for extracting polyphenols, carotenoids, etc. A range of polyphenol compounds possess insecticidal efficacy. Thus, this technique may be a highly useful tool for their extraction, under the dictates of 'green' chemistry. The extraction of phytosterols from maize and isoflavonoids from soybeans increased by 32.4 % and 20–21 % when PEF technique was employed (Guderjan et al. 2005). Capsaicinoid was efficiently extracted from *Capsicum annuum* by this technique (Liu et al. 2013). Amongst the different solvents, water is regarded as the most preferred for its green credentials. Pressurised hot water extraction (PHWE) was employed to improve the extraction yield of isoxanthohumol, the anti-inflammatory flavonoid from *Humulus lupulus* (Gil-Ramírez et al. 2012). PEF is mainly used to improve the release of intracellular compounds bound within the plant tissues by the aid of increasing cell membrane permeability (Toepfl et al. 2006). A typical pulsed electric field extractor is shown schematically in Fig. 6.8.

6.3.6 Enzyme-Assisted Extraction (EAE)

The solvent extraction methods are not proficient in extracting the plant components present in the polysaccharide–lignin network due to extensive hydrogen bonding. Specific enzymes help in breaking the cell walls and to release the active compounds. A number of enzymes like cellulase, α-amylase, pectinase, etc. were used in extraction, which enhanced the process by hydrolysing the polysaccharides and lipid (Rosenthal et al 1996; Singh et al. 1999). Enzyme-assisted aqueous extraction (EAAE) method is used to hydrolyse the cell wall of seeds, which is governed by the composition and concentration of the enzyme and time of hydrolysis. Again, enzyme-assisted cold pressing (EACP) is useful for the extraction of bioactive components from oilseeds, owing to its non-inflammable nature (Latif and Anwar 2009). Improved extraction was evident for phenolic antioxidants, phenolic acids, non-anthocyanin flavonoids and anthocyanins by using EAE.

6.3.7 Microwave-Assisted Extraction (MAE)

MAE stimulates the penetration of solvents into the plant materials and fosters the solubility of bioactive components. Microwave irradiation directly impacts the polar components and induces heat by ionic conduction and dipolar rotation (Azmir et al. 2013). Polyphenols and caffeine were extracted from tea leaves within 4 min by employing MAE technique (Pan et al. 2003). Further, MAE was used for extracting polyphenols, flavolignin, guggolsterone, and tannin from different plants, within a very short time with efficient yields (Asghari et al. 2011).

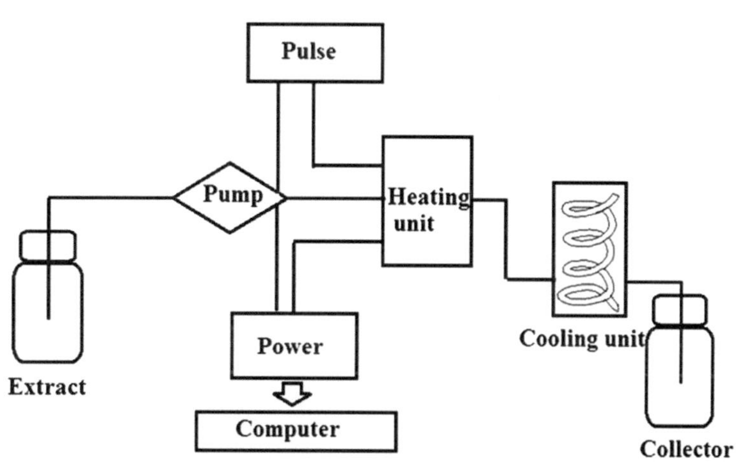

Fig. 6.8 Schematic diagram of a pulsed electric field extractor

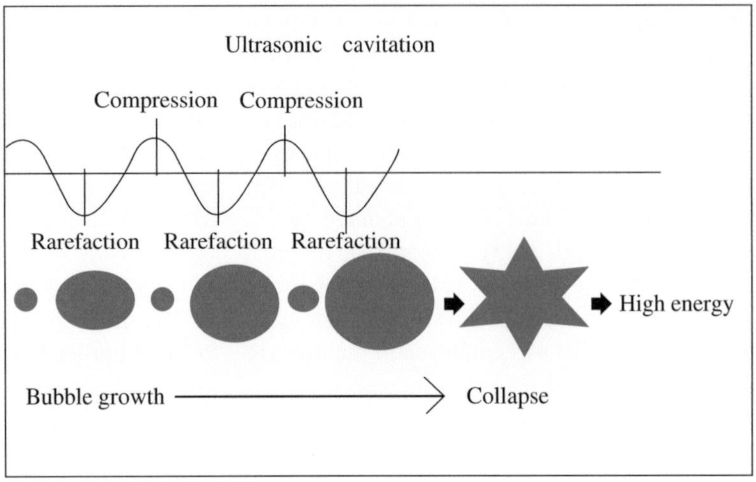

Fig. 6.9 Formation of ultrasonic cavitations

6.3.8 Ultrasound-Assisted Extraction (UAE)

Amongst the different extraction methods, UAE is the most widely used these days. Ultrasound produces a huge amount of energy by converting the kinetic energy to heat, which creates cavitations, that is, production, aggregation and collapse of bubbles. These bubbles possess temperatures beyond 5000 K according to Suslick and Doktycz (Suslick et al. 1990). Figure 6.9 demonstrates the process of formation of ultrasonic cavitations. UAE facilitates the access of the solvent to plant material, which is followed by diffusion of the solvent across the cell wall and rinsing the contents (Mason et al. 1996). Daidzin, glycitin, genistin and malonyl genistin were extracted from soybean by this technique (Rostagno et al. 2003). The extraction of curcumin and lycopene enhanced to many folds by the use of ultrasonic energy (Konwarh et al. 2010; Konwarh et al. 2012). Polyphenols like naringin, naringenin, rutin, ellagic acid, quercetin and kaempferol, polycarboxylic acids, alkaloids like catharanthine, vindoline, vinblastine, etc. have been extracted with high yield by using UAE method (Herrera and Luque de Castro 2004).

In addition to the described techniques, a number of other techniques such as pressurised fluid extraction (PFE), enhanced solvent extraction (ESE), high-pressure solvent extraction (HSPE), high-pressure MAE (HPMAE), vacuum MAE (VMAE), nitrogen-protected MAE (NPMAE), solvent-free MAE (SFMAE), dynamic MAE (DMAE), etc. are available for extracting insecticidal plant active components. The different extraction methods are schematically categorised in Fig. 6.10.

6.4 Purification of Plant Extracts

Plant extracts contain a myriad of bioactive compounds with differential activities. Thus, the identification and separation of the desired compound is quite a tedious task. In most of the cases, chromatographic techniques are useful to separate such compounds.

6.4.1 Chromatographic Techniques

Chromatography is a tool for separation or identification of different components present in a liquid or gaseous mixture, where the components are distributed between two phases. One is stationary phase, while the other is the mobile phase. The principle of these techniques is the differences in the distribution constant of each component.

According to the mobile phase used, chromatographic techniques can be divided into two broad classes, viz. liquid and gas chromatography.

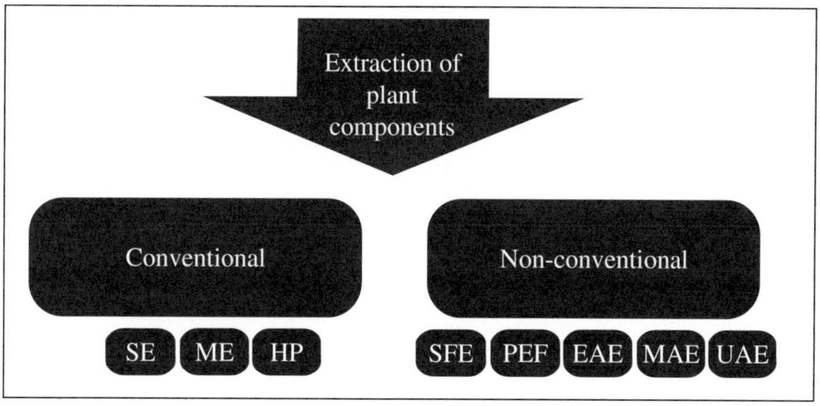

Fig. 6.10 Different extraction methods

6.4.1.1 Liquid Chromatography

Here, the mobile phase is a liquid. Different kinds of solvent are generally used as mobile phases, which can dissolve the components to be separated. Some liquid chromatographic tools used for the separation and purification of insecticides are briefly explained below.

Thin-Layer Chromatography (TLC)

This is the simplest, quick and inexpensive chromatographic technique, which can depict the number of components present in a plant extract. This method is based on the polarity of the compounds and their mobility in the used solvent system. Here, silica gel, alumina, cellulose, etc. are used as stationary phases bound on glass, plastic or aluminium foils. TLC includes three steps, where the first step is the spotting of the plant extract (very dilute solution) on the stationary phase by the use of a capillary tube. Next phase is the development, where the plate is eluted with an appropriate solvent system. The solvent travels up by capillary action, carrying the components of the mixture, depending upon the polarity of each component. The final step is visualisation, which can be done by naked eye for coloured spots. However, for colourless compounds, UV light is used to detect the spots. Besides, plates are also placed in iodine chambers, because most of the organic compounds form dark colour complexes in contact of iodine vapour.

Different compounds in the plant extract travel at different rates owing to their interaction with the stationary phase. Mobility of the spot is quantified by Rf value ($0<Rf<1$), which is equal to the distance travelled by the test compound divided by the distance travelled by the solvent system. Rf value of the compounds can be compared with known compounds, which helps in the preliminary identification process. A typical TLC plate is shown in Fig. 6.11, which represents the separate spots (C1, C2 and C3) for different components present in a plant extract. This technique is greatly helpful for preliminary study of insecticidal plant extracts that contain a mixture of polyphenols, alkaloids, terpenoids, etc.

Column Chromatography

Column chromatography is generally used for the separation and purification of different compounds present in a plant extract. This is a preparative technique, which can be scaled up to kilogramme level. A solvent or mixture of solvents is used as the mobile phase, and a fine solid surface is taken as the stationary phase. The latter adsorbs different components present in the mixture as per their binding affinity, and the solvent carries the components as it passes through the stationary phase. The solvent and compounds compete for binding onto the adsorbent surface, and thus weakly adsorbed species will elute first.

The most widely used adsorbents are silica gel and alumina. The particles of the adsorbent should have a large surface area for effective adsorption. As the stationary phase is highly polar, they have affinity towards the polar com-

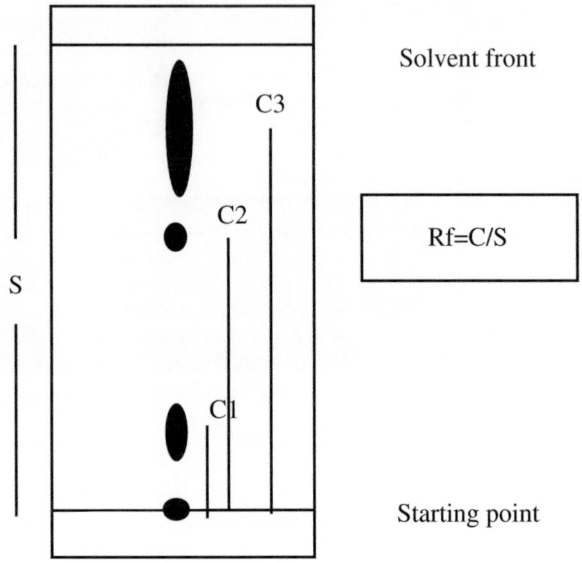

Fig. 6.11 A typical TLC plate

pounds of the extract. Consequently, non-polar compounds elute first. The choice of the solvent also depends on the polarity of the compounds desired to be extracted. However, in most of the separations, the solvents, which are less polar than the compounds, are selected. An elutropic series, i.e. the order of polarity of solvents for silica gel and alumina, is available as follows: hexane < carbon tetrachloride < toluene < dichloromethane < chloroform < diethyl ether < ethyl acetate < acetone < propanol < ethanol < methanol < acetic acid < water. The appropriate solvent system is generally selected by TLC experiments before running a column.

Each component, separated by this technique, is again analysed by TLC and collected in separation collection tubes. These eluents are concentrated to obtain the desired compounds. In the case of plant extracts, different compounds are recognised from the spots noted in the TLC experiments. As per the polarity of the desired compound, solvent system is considered for column. Researchers have used TLC and column chromatography for detection and purification of insecticides, acaricides and fungicides (Neicheva et al. 1990).

High-Performance Liquid Chromatography (HPLC)

HPLC is used for the separation, identification and quantification of different components present in a plant extract. Here, a pressurised liquid solvent is passed, which contains the sample through a stationary phase packed inside a column. Every component interacts with the adsorbent in differential manner. Such interactions are generally of physical type, including dipole–dipole, hydrophobic and ionic interactions. Consequently, the flow rates differ for each component, which helps to collect them separately.

HPLC is helpful for quantitative analysis. An automatic injector is used here, which provides reproducible injection volumes. HPLC is one kind of liquid chromatography, which uses a liquid mobile phase. Further, reversed phase HPLC is available where the mobile phase is polar relatively than the stationary phase. The stationary phase is commonly solid such as silica, polymers, etc. The mixture to be separated is introduced to the mobile phase through the column, in very small volume (microlitres). The time taken by a specific compound to elute is termed as its retention time, which is determined under particular conditions for identifying a compound in reference to a known compound. Various insecticidal compounds were isolated from different plant resources like *Chrysanthemum cinerariifolium* (pyrethrum), *Azadirachta indica* (neem), *Piper longum* (long pepper), etc. by this technique.

Revathy et al. isolated three curcuminoid compounds, curcumin, desmethoxycurcumin and bisdemethoxycurcumin from *Curcuma longa* L. by HPLC technique. Each compound showed single peak on HPLC at the respective retention times of 10, 12 and 13 min (Revathy et al. 2011). Again, HPLC was used to isolate the active component from a butanol fraction of *Citrullus colocynthis* which showed abundance at retention time 4.8 min at a wavelength of 275 nm (Torkey et al. 2009). Finally the compound was identified to be 2-O-β-D-glucopyranosylcucurbitacin E by ^{13}C NMR spectroscopy.

6.4.1.2 Gas Chromatography

Gas chromatography (GC) is used for the separation and identification of compounds present in a plant extract, which can be vaporised without decomposition. This is a preparative chromatography, which can be used to isolate the desired pure compounds from a mixture. Here, the mobile phase is an inert carrier gas like helium or argon. The stationary phase is packed inside a column. The vapour of the mixture under analysis interacts with the sidewalls of the column, containing the stationary phase. Each compound elutes at different retention times. The compounds, which have greater interaction with the column sidewalls are retarded greater and get separated from the ones with smaller interaction. The eluent containing the components is quantified by a detector, followed by collection.

Generally, there are two types of GC, viz. gas–solid chromatography (GSC) and gas–liquid chromatography (GLC). GSC uses a solid stationary phase and separates the analyte by physical adsorption. Contrarily, GLC is used to separate ions or molecules. This technique is highly useful for detecting residual amount of pesticides or insecticides retained in the food items. Acetylenic 2-phenylethylamides, the larvicidal compound, were isolated from the plant, *Acmella oleracea*, by GC technique (Simas et al. 2013). However, this technique is rarely used without the combination of mass spectrometry. Below, some of the chromatographic techniques are briefed, which are used in combination with mass spectrometry.

6.4.1.3 Combined Techniques

Liquid Chromatography–Mass Spectrometry (LC-MS)

This technique uses the combined merits of liquid chromatography and mass spectrometry. LC-MS is a very high sensitive and selective technique and selectivity and so is useful in many applications. It is greatly useful in the separation, purification and identification of various compounds present in natural extract. Preparative LC-MS is used for mass-directed purification of different compounds from a plant extract. High-pressure LC-MS (HPLC-MS) is an efficient tool for the purification of natural extracts.

Ingelse et al. detected polar pesticides by using LC-MS (Ingelse et al. 2001). Barrek and his group (Barrek et al. 2004) determined the content of azadirachtin and chemical composition in neem seed oil by LC-MS.

Gas Chromatography–Mass Spectrometry (GC-MS)

GC-MS is an analytical tool, which is a combination of the features of gas chromatography and mass spectrometry to separate and detect various components in a test sample. It is being used in drug or pesticide detection, environmental analysis and most importantly in the identification of unknown compounds in natural extracts. This technique is very useful for the separation and identification of insecticides in a crude plant sample. Biochemical components present in the rhizomes of *Nervilia aragoana* was analysed by Thomas et al. using GC-MS (Thomas et al. 2013). Insecticidal components were isolated from *Psoralea corylifolia*-based essential oil with this method (Gupta et al. 2013). Larvicidal compounds were isolated by GC-MS analysis from the extract of *Euphorbia lactea* (Samidurai and Mathew 2014). The insecticidal activity of *Tanacetum argenteum*-based essential oil and the individual components isolated by GC-MS technique against the mosquito species, *Aedes aegypti*, were studied. It was revealed that α-pinene is the highest constituent (54–68 %) of the essential oil, which showed LC_{50} at <40 ppm (Ali et al. 2014).

Abou-Elnaga (2014) isolated active components from sage oil and nutmeg oil by GC-MC, and their structures were ascertained from ^{1}H NMR spectroscopy. The compounds showed larvicidal and adulticidal activities against *Culex pipiens* and *Aedes aegypti*. In the quest of formulating low-cost mosquito repellent from natural resources, researchers have analysed the *Vitex negundo* (Verbenaceae) extract by GC-MS. The study revealed that hexane extract contains the following chemical compounds, viz. viridiflorol (>19.55 %), β-caryophyllene (>16.59 %) and sabinene (>12.07 %), in abundance. Further, the volatile fractions of *Aframomum elliotii*, *A. strobilaceum*, *A. geocarpum*, *A. longiscarpum* and *A. sceptrum* were found to possess 52 chemical entities (Diomandé et al. 2012) such as β-pinene, β-caryophyllene, eucalyptol, linalool, caryophyllene oxide, etc. which was confirmed from GC and GC-MS analyses. Again, the major component of *Mentha pulegium* leaf (a known insect repellent) was ascertained to be piperitone (35.56 %), by GC-MS (Derwich et al. 2010). A thorough literature survey establishes GC-MS to be the most widely used chromatographic technique for isolation of insecticidal compounds present in a plant extract.

6.4.2 Non-chromatographic Techniques

The non-chromatographic techniques for structural characterisation of plant-based insecticides are described below.

6.4.2.1 Fourier Transform Infrared Spectroscopy (FTIR)

FTIR is a very useful spectroscopic tool in chemistry, which reveals important information associated with the chemical functionalities of a compound or substance. This is used to record the infrared spectrum of absorption or emission of solid, liquid or gas. Fourier transformation converts a signal (a function of time) into frequencies (in terms of wave number), which helps in providing the actual spectrum.

After extraction and purification of an insecticidal compound from a plant, it is very important to identify the functional groups in it. Chemical functionalities are primly responsible for the activity of the compound. In this quest, GC-IR (gas chromatography–infrared spectrometry) is used where the separated components of a plant extract are directly analysed by a FTIR spectrometer. Further, LC-IR is also used for similar purpose.

To understand the aforesaid importance, the FTIR spectrum of a widely used plant-derived insecticide, nicotine, is elaborated briefly here. FTIR spectrum of nicotine provides the information about the chemical linkages present in the structure. For each bond, there is a corresponding wave number (in terms of cm^{-1}), which represents the stretching or bending of a particular chemical bond. For nicotine, a sharp band is observed near 1677 cm^{-1}, which is due to the C=N double bond stretching. Aromatic C=C double bond stretching is evident from the band at around 1691 cm^{-1}. Bands near 717 and 904 cm^{-1} are because of the out-of-plane C–H bond bending for the pyridine moiety. C–N stretching bands are found at 1200–1350 cm^{-1}. The chemical linkages present in the structure of nicotine and their corresponding FTIR bands are shown in Fig. 6.12.

Thus, we can visualise that this can be a very useful tool to identify the chemical entities present in an isolated and purified plant-based insecticide.

6.4.2.2 Nuclear Magnetic Resonance Spectroscopy (NMR)

NMR technique is the mostly used technique for elucidating the structure of any chemical compound. Nucleus present in an atom absorbs and re-emits electromagnetic radiation under the influence of an external magnetic field, and this principle is used in NMR spectroscopy to identify the chemical environment of the atom and its surroundings. NMR is also used in biomedical imaging such as MRI (magnetic resonance imaging). A wide range of nuclei are studied by NMR

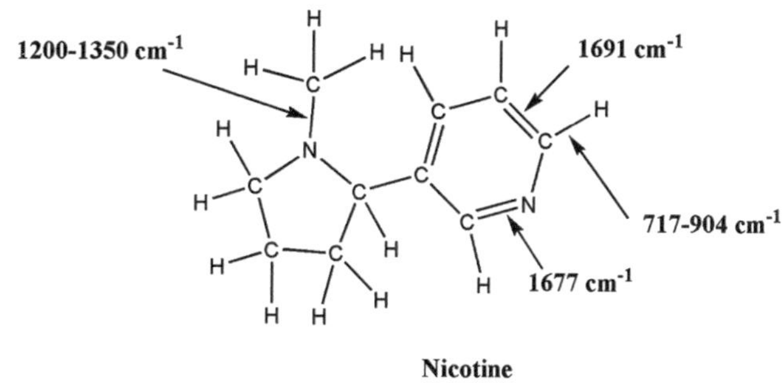

Fig. 6.12 FTIR bands of nicotine

spectroscopy, which includes ^{1}H, ^{13}C, ^{6}Li, ^{10}B, ^{11}B, ^{14}N, ^{29}Si, ^{31}P, etc.

However, ^{1}H and ^{13}C NMR are popularly used for the structural elucidation of a chemical compound. In case of plant-based insecticides, though this is not abundantly used, some researchers tried to explain the chemical environment of particular insecticides. Chemical shift is determined by NMR in terms of parts per million (ppm). Again, here we cite the example of the insecticide, nicotine. The chemical shift values (in ppm) for the different protons present in the structure of nicotine are depicted in Fig. 6.13.

Again, the carbons with different environments in nicotine are shown in Fig. 6.14, in terms of their chemical shift values. Simas and co-workers analysed the insecticidal compound, acetylenic 2-phenylethylamides, isolated from *Acmella oleracea* extract, by the use of both ^{1}H and ^{13}C NMR spectroscopy (Simas). Thus, NMR spectroscopy is a strong tool to characterise the chemical structure of the active compounds, isolated from plants. However, this technique has not been explored extensively in case of plant-based insecticidal compounds.

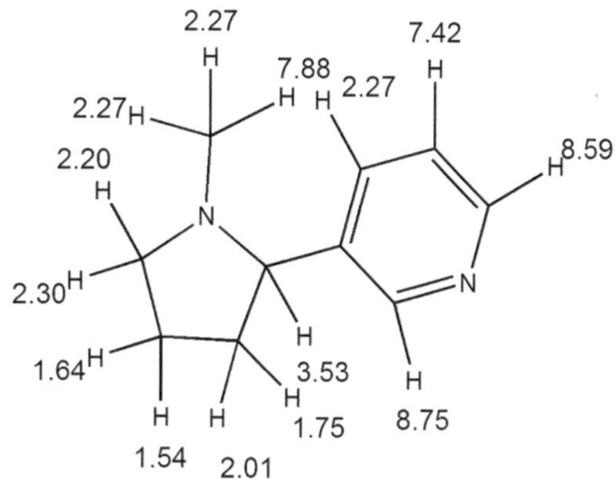

Fig. 6.13 1H NMR chemical shift value for nicotine

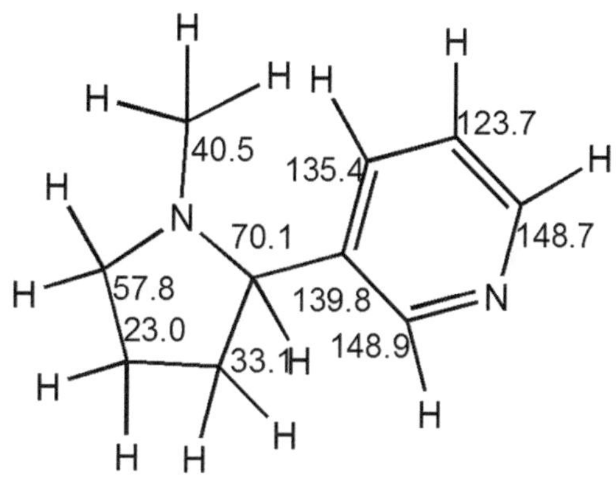

Fig. 6.14 ^{13}C NMR chemical shift value for nicotine

6.5 Conclusion

This chapter encompasses the different techniques involved in the isolation, purification and characterisation of insecticidal compounds extracted from plants. The conventionally used methods and the modern methods are discussed briefly herein. The discussion also included some salient examples available in the literature. The literature reports a huge number of papers where plant-based compounds are analysed for their insecticidal potential. However, very few reports describe the actual active component responsible for such efficacy. This creates major challenges for commercialisation of those products. Thus, this chapter may be helpful to construct a bridge between the biologists and chemists to work together in collaboration for elucidating the actual structure and activity of plant-based compounds, which have very significant insecticidal property.

References

Abou-Elnaga ZS (2014) Insecticidal bioactivity of eco-friendly plant origin chemicals against *Culex pipiens* and *Aedes aegypti* (Diptera: Culicidae). J Entomol Zool Stud 2:340–347

Acheuk F, Doumandji-Mitiche B (2013) Insecticidal activity of alkaloids extract of Pergularia tomentosa (Asclepiadaceae) against fifth instar larvae of Locusta migratoria cinerascens (Fabricius 1781) (Orthoptera: Acrididae). Int J Sci Adv Tech 3:8–13

Ali A, Tabanca N, Kurkcuoglu M et al (2014) Chemical composition, larvicidal, and biting deterrent activity of essential oils of two subspecies of Tanacetum argenteum (Asterales: Asteraceae) and individual constituents against Aedes aegypti (Diptera: Culicidae). J Med Entomol 51:824–830

Asghari J, Ondruschka B, Mazaheritehrani M (2011) Extraction of bioactive chemical compounds from the medicinal Asian plants by microwave irradiation. J Med Plants Res 5:495–506

Azmir J, Zaidul ISM, Rahman MM et al (2013) Techniques for extraction of bioactive compounds from plant materials: a review. J Food Eng 117:426–436

Barrek S, Paisse O, Grenier-Loustalot MF (2004) Analysis of neem oils by LC-MS and degradation kinetics of azadirachtin-A in a controlled environment. Characterization of degradation products by HPLC-MS-MS. Anal Bioanal Chem 378:753–763

Brusottia I, Cesaria A, Dentamaroa G et al (2013) Review: isolation and characterization of bioactive compounds from plant resources: the role of analysis in the ethnopharmacological approach. J Pharm Biomed Anal 87:218–28

Byeoung-Soo P, Sung-Eun L, Won-Sik C et al (2002) Insecticidal and acaricidal activity of pipernonaline and piperoctadecalidine derived from dried fruits of *Piper longum* L. Crop Protect 21:249–251

Cheng SB, Wu LC, Hsieh YC et al (2012) Supercritical carbon dioxide extraction of aromatic turmerone from *Curcuma longa* L. induces apoptosis through reactive oxygen species-triggered intrinsic and extrinsic pathways in human hepatocellular carcinoma HepG2 cells. J Agric Food Chem 60:9620–9630

DeBach P, Rosen D (1991) Biological control by natural enemies, 2nd edn. Cambridge University Press, New York

Derwich E, Benziane Z, Taouil R (2010) GC/MS analysis of volatile compounds of the essential oil of the leaves of *Mentha pulegium* growing in Morocco. Chem Bull "POLITEHNICA" Univ (Timisoara) 55:103–106

Diomandé GD, Koffi AM, Tonzibo ZF (2012) GC and GC/MS analysis of essential oil of five *Aframomum* species from Côte D'ivoire. Middle-East J Sci Res 11:808–813

Diwan RK, Saxena RC (2010) Insecticidal property of flavonoid isolated from *Tephrosia purpuria*. Int J Chem Sci 8:777–782

Gil-Ramírez A, Mendiola JA, Arranz E et al (2012) Highly isoxanthohumol enriched hop extract obtained by pressurized hot water extraction (PHWE). Chemical and functional characterization. Innov Food Sci Emerg 16:54–60

Guderjan M, Töpfl S, Angersbach A et al (2005) Impact of pulsed electric field treatment on the recovery and quality of plant oils. J Food Eng 67:281–287

Gupta M, Gupta A, Gupta S (2013) Chemical composition and insecticidal activity of essential oil obtained from DCM extracts of *Psoralea corylifolia* against agricultural pest. Curr World Environ 8:309–312

Herrera MC, Luque de Castro MD (2004) Ultrasound-assisted extraction for the analysis of phenolic compounds in strawberries. Anal Bioanal Chem 379:1106–1112

Ingelse BA, van Dam RC, Vreeken RJ et al (2001) Determination of polar organophosphorus pesticides in aqueous samples by direct injection using liquid chromatography-tandem mass spectrometry. J Chromatogr A 918:67

Konwarh R, Saikia JP, Karak N et al (2010) Poly(ethylene glycol)-magnetic nanoparticles-curcumin' trio: directed morphogenesis and synergistic free-radical scavenging. Colloids Surf B Biointerfaces 81:578–586

Konwarh R, Pramanik S, Kalita D et al (2012) Ultrasonication – a complementary 'green chemistry' tool to biocatalysis: a laboratory-scale study of lycopene extraction. Ultrason Sonochem 19:292–299

Koul O, Walia S, Dhaliwal GS (2008) Essential oils as green pesticides: potential and constraints. Biopestic Int 4:63–84

Lang Q, Wai CM (2001) Supercritical fluid extraction in herbal and natural product studies—a practical review. Talanta 53:771–782

Latif S, Anwar F (2009) Physicochemical studies of hemp (*Cannabis sativa*) seed oil using enzyme-assisted cold-pressing. Eur J Lipid Sci Technol 111:1042–1048

Liu A, Han C, Zhou X et al (2013) Determination of three capsaicinoids in *Capsicum annuum* by pressurized liquid extraction combined with LC–MS/MS. J Sep Sci 3:1–6

Liza MS, Rahman RA, Mandana B et al (2012) Supercritical fluid extraction of bioactive flavonoid from *Strobilanthes crispus* (pecah kaca) and its comparison with solvent extraction. Int Food Res J 19:503–508

Mason TJ, Paniwnyk L, Lorimer JP (1996) The uses of ultrasound in food technology. Ultrason Sonochem 3:253–260

Miyazawa M, Yoshio K, Ishikawa Y et al (1998) Insecticidal alkaloids from Corydalis bulbosa against *Drosophila melanogaster*. J Agric Food Chem 46:1914–1919

Neicheva A, Kovacheva E, Karageorgiev D (1990) Simultaneous determination of insecticides, acaricides and fungicides by thin-layer chromatography. J Chromatogr 15:263–269

Pan X, Niu G, Liu H (2003) Microwave-assisted extraction of tea polyphenols and tea caffeine from green tea leaves. Chem Eng Process 42:129–133

Porzel A, Kuster PM (2013) Acetylenic 2-phenylethylamides and new isobutylamides from *Acmella oleracea* (L.) R. K. Jansen, a Brazilian spice with larvicidal activity on *Aedes aegypti*. Phytochem Lett 6:67–72

Revathy S, Elumalai S, Benny M et al (2011) Isolation, purification and identification of curcuminoids from turmeric (*Curcuma longa* L.) by column chromatography. J Exp Sci 2:21–25

Rosenthal A, Pyle DL, Niranjan K (1996) Aqueous and enzymatic processes for edible oil extraction. Enzyme Microb Technol 19:402–420

Rostagno MA, Palma M, Barroso CG (2003) Ultrasound-assisted extraction of soy isoflavones. J Chromatogr A 1012:119–128

Saldana MDA, Mohamed RS, Baer MG et al (1999) Extraction of purine alkaloids from mate (*Ilex paraguariensis*) using supercritical CO_2. J Agric Food Chem 47(9):3804–3808

Salunke BK, Klotkar HM, Mendki PS et al (2005) Efficacy of flavonoids in controlling *Callosobruchus chinensis* (L.) (Coleoptera: Bruchidae), a post-harvest pest of grain legumes. Crop Prot 24:888–893

Samidurai K, Mathew N (2014) Bioassay guided fractionation and gc-ms analysis of *Euphorbia lactea* extract for mosquito larvicidal activity. Int J Pharm Pharm Sci 6:344–347

Sasidharan S, Chen Y, Saravanan D et al (2011) Extraction, isolation and characterization of bioactive compounds from plants' extracts. Afr J Tradit Complement Altern Med 8:1–10

Saxena RC, Harshan V, Saxena A et al (1993) Larvicidal and chemosterilant activity of annona squamosa alkaloids against *Anopheles stephensi*. J Am Mosq Control Assoc 9:84–87

Silva-Aguayo, G (2009) Botanical insecticides. In Radcliffe EB, Hutchison WD, Cancelado RE (eds) Radcliffe's IPM world textbook. University of Minnesota, St. Paul, http://ipmworld.umn.edu/chapters/SilviaAguayo.htm

Simas NK, Dellamora ECL, Schripsema J et al (2013) Acetylenic 2-phenylethylamides and new isobutylamides from *Acmella oleracea* (L.) R. K. Jansen, a Brazilian spice with larvicidal activity on *Aedes aegypti*. Phytochem Lett 6:67–72

Singh RK, Sarker BC, Kumbhar BK et al (1999) Response surface analysis of enzyme-assisted oil extraction factors for sesame, groundnut, and sunflower seeds. J Food Sci Technol 36:511–514

Smith RM (2002) Extractions with superheated water. J Chromatogr A 975:31–46

Suslick KS, Doktycz SJ (1990) The effects of ultrasound on solids. In: Mason, TJ (ed) Advances in Sonochemistry, vol. 1. JAI Press, New York, pp. 197–230.

Thomas E, Aneesh T, Thomas DG et al (2013) GC-MS Analysis of phytochemical compounds present in the rhizomes of *Nervilia aragoana* gaud. Asian J Pharm Clin Res 6:68–74

Toepfl S, Mathys A, Heinz V et al (2006) Review: potential of high hydrostatic pressure and pulsed electric fields for energy efficiency and environmentally friendly food processing. Food Rev Int 22:405–423

Torkey HM, Abou-Yousef HM, Abdel-Azeiz AZ et al (2009) Insecticidal effect of Cucurbitacin E glycoside isolated from *Citrullus colocynthis* against *Aphis craccivora*. Aust J Basic Appl Sci 3:4060–4066

Yamamoto I (1965) Nicotinoids as insecticides. In: Metcalf RL (ed) Advances in pest control research, vol VI. Wiley, New York, pp 231–260

Yenesew A, Twinomuhwez H, Kabaru JM et al (2009) Antiplasmodial and larvicidal flavonoids from *Derris trifoliata*. Bull Chem Soc Ethiop 23:409–414

Recent Advancements in Bio-botanical Pesticide Formulation Technology Development

7

Megha Pant, Saurabh Dubey, and P.K. Patanjali

Abstract

The misuse and overuse of synthetic pesticides have led to the vast destruction of beneficial organisms along with detrimental effects on environment. The use of biopesticides including microbial agents, biochemical pesticides, and botanicals is considered as a more convincing strategy for the management and control of various insect pests. The biochemical pesticides and the botanicals possess pesticidal as well as repellent properties and have come up as better substitutes to synthetics as they are eco-friendly with target specificity and biodegradability and are economically feasible. The chapter provides information on the different types of the biological product-based pesticide formulations for the control of mosquitoes and other disease vectors.

7.1 Introduction

The escalating problems of pest resistance and pesticide residues along with the contamination of the ecosphere due to the irrational use of synthetic chemical pesticides have necessitated the replacement of the same with the pesticides ensuring biodegradability and target specificity. Also, their excessive use for controlling stored grain pests and agricultural pests makes the edibles completely unpalatable for human consumption. In the case of household pests, viz., mosquitoes and cockroaches, there is every possibility of insecticide coming in contact with the human beings. An awareness was circulated worldwide to develop newer class of pesticides (Rajashekhar et al. 2012). The new category of insecticides must have the characteristics, viz., non-phytotoxicity, target specificity, no or low mammalian toxicity, availability, cost-effectiveness, etc. (Hermawan et al.. 1997).

Biopesticides refer to all biologically derived pesticides which help in reducing the pest population (Koul and Walia 2009) such as plants (e.g., pyrethrum – *Chrysanthemum* sp., *Azadirachta* sp., etc.) microbes, nematodes, bac-

M. Pant • P.K. Patanjali (✉)
Institute of Pesticide Formulation Technology,
Sector-20, Udyog Vihar, Gurgaon, 122 016, India
e-mail: pkpatanjali@yahoo.com

S. Dubey
Centre for Rural Development and Technology,
Indian Institute of Technology Delhi,
New Delhi 110 016, India

© Springer India 2016

Vijay Veer, R. Gopalakrishnan (eds.), *Herbal Insecticides, Repellents and Biomedicines: Effectiveness and Commercialization*, DOI 10.1007/978-81-322-2704-5_7

teria (e.g., *Bacillus thuringiensis*), viruses (e.g., nucleopolyhedrosis virus), fungi (e.g., *Beauveria* sp., *Metarhizium* sp.), and the transgenic plants containing a pest-combating gene (e.g., Bt cotton). The biochemical pesticides include pheromones, plant extracts and oils, plant growth regulators, and insect growth regulators. Biopesticides work in harmony with the integrated pest management programs. A number of plant derivatives are known to function as insecticides, insect repellents, antifeedants, and insect growth and development regulators. The effects of plant secondary metabolites responsible for insecticidal activity are described as suppression of calling behavior, growth retardation, toxicity, oviposition deterrence, feeding inhibition, and reduction of fecundity and fertility.

7.2 Categorization of Biopesticides

7.2.1 Microbial Insecticides

7.2.1.1 Bacteria

The most potent and effective microbial insecticide is *Bacillus thuringiensis*. More than 90 % of the bacteria species with different strains that are manufactured and sold throughout the world are for the control of caterpillar pests, mosquitoes, and black flies. The specific orders against which these organisms are useful are Lepidoptera, Diptera, and Coleoptera. Basically the technique involves the utilization of pathogenic microorganisms isolated from diseased insects (Dubey et al. 2008).

7.2.1.2 Viruses

Baculoviruses are rod-shaped DNA viruses, which initiate their life cycle by reproducing inside the cells. Generally, caterpillar cells are reservoirs for baculoviruses, which multiply and get incorporated into protective polyhedron-shaped protein structures called occlusion bodies. The caterpillars infected with the baculoviruses die and contaminate the leaf surfaces with the occlusion bodies. Then the healthy caterpillars ingest the occlusion bodies and release the virus while feeding on contaminated leaves, thus continuing the life cycle of infection and replication. They eliminate some of the caterpillar populations like tobacco budworm, cotton bollworm, etc. (Ben 2002).

7.2.1.3 Fungi

Epizootic outburst in some areas with insect pest populations is a result of inoculating the populations with entomopathogenic fungi, especially with *Entomophthorales* that often successfully regulate the insect pest populations. It instigates an efficient biological control of pests, most notably against the gypsy moth. The most common method of employing fungi for insect control is through inundatory means. The conidia of most of the entomophthoralean fungi species are relatively difficult to produce, and their primary conidia are short lived, making the timing of the inundator applications difficult (Isman and Akhtar 2007). The utility of these fungi has been increased by developing effective methods for production of resting spores and competent mycelia of entomophthoralean species. A broad range of insect pests, including whiteflies, aphids, thrips, termites, grasshoppers and locusts, and beetles, were chosen to investigate the effect of entomopathogenic hyphomycetes.

7.2.2 Biochemical Pesticides

7.2.2.1 Essential Oils

Steam distillation of aromatic plants yields the volatile organic compounds, essential oils produced as secondary metabolites in plants. Approximately 3,000 essential oils are known, out of which 300 are commercially important for cosmetics, perfume, and pharmaceutical industries besides their pesticidal potential (Chang and Cheng 2002, Dubey et al. 2011). Various plant families, for example, Myrtaceae, Lauraceae, Rutaceae, Lamiaceae, Asteraceae, Apiaceae, Cupressaceae, Poaceae, Zingiberaceae, and Piperaceae, are known to produce essential oils. Since the Middle Ages, essential oils have been widely used for bactericidal, virucidal, fungicidal, parasiticidal, and insecticidal applications.

Also, the essential oils can be exploited as insect repellents and insect-feeding behavior alterants, along with those affecting ecdysis (moulting) and behavior during mating and oviposition.

7.2.2.2 Insect Growth Regulators

Insect growth regulators (IGRs) have a unique mode of action different from most chemical insecticides. Generally, these products avert insects from reaching a reproductive stage, thereby plummeting the spreading out of pest populations. The direct impact of IGRs on target pests combined with the preservation of beneficial insects and pollinators aids the growers in maximizing yield and product quality. IGRs can be divided into two broad categories, i.e., those that disrupt the hormonal regulation of insect metamorphosis and those that disrupt the synthesis of chitin, a principal component of insect exoskeletons. As far as the agricultural applications are concerned, the first category of compounds is mainly focused, known as "hormone mimics." The most widely used botanical insect growth regulators is azadirachtin, which interrupts moulting due to structural resemblance to the natural insect moulting hormone, ecdysone.

7.2.3 Botanicals

7.2.3.1 Pyrethrum

Pyrethrin is one of the most widely and heavily used botanical insecticide worldwide, used in household aerosols for fast knockdown of pests. 20–25 % of pyrethrins are present in the technical grade pyrethrum (resin), which is used in formulating commercial insecticides. Piperonyl butoxide (PBO), derived from sassafras or N-octyl bicycloheptene dicarboximide, is added to it to increase insect mortality and shelf life of the product.

7.2.3.2 Rotenone

Rotenone, an isoflavonoid, is a traditional botanical insecticide mainly used for organic food production. It is obtained from the roots or rhizomes of tropical legumes in the genera *Derris*, *Lonchocarpus*, and *Tephrosia*. The dust formulation of rotenone contains 1–5 % active ingredients and is most commonly marketed for household and garden use. The liquid formulations used in organic agriculture can contain as much as 8–15 % of total rotenoids (Isman et al. 2011).

7.2.3.3 Nicotine

The aqueous extracts of tobacco (*Nicotiana* spp.: Solanaceae) and *Anabasis aphylla* (Chenopodiaceae) yield the alkaloid, nicotine. Nicotine is a synaptic poison and therefore induces high level of insecticidal effects as it mimics the neurotransmitter, acetylcholine. The symptoms of poisoning caused by the neonicotinoids are similar to those seen with organophosphate and carbamate insecticides. Most commonly, nicotine is used as a fumigant in greenhouses against soft-bodied pests.

7.2.3.4 Neem

Azadirachta indica (neem) is widely grown in other Asian countries and in tropical and subtropical areas of Africa, America, and Australia and is considered a traditional botanical insecticide (Allan et al. 2002). Azadirachtin was isolated in 1968 and was used as a most potent locust antifeedant. In addition to that, neem possesses fungicidal, nematicidal, bactericidal, and molluscidial properties. Also it exhibits immunomodulatory, anti-inflammatory, antimalarial, antiviral, antioxidant, antimutagenic, and anticarcinogenic effects. Azadirachtin is a persistent insecticide against the crop pests as it has systemic effects (Al-Quraishy et al. 2012).

7.3 Active Ingredients in Botanicals Accountable for Biopesticidal Properties

7.3.1 Terpenes and Terpenoids

Terpenes are a group of molecules whose structure is based on a various but definite number of isoprene units (methylbuta-1,3-diene, named hemiterpene, with 5 carbon atoms). Terpenes are classified on the basis of the number of isoprene units integrated in the basic molecular skeleton as monoterpenes with two isoprene units, sesquiterpenes with three, diterpenes with four, sesterterpenes with five, and triterpenes with six isoprene units.

Modified terpenes, where methyl groups are moved or removed or oxygen atoms are added, are known as terpenoids. Terpenoids are represented as the most widespread group of natural products. Camphor, menthol, eucalyptus oil, lemongrass oil, orange oil, etc. are some of the commercially available terpenoids (Kokete et al. 2012). The essential oil of the basil shrub, *Ocimum kilimandscharicum*, contains camphor as a marker compound which was evaluated against the stored product beetles, *Sitophilus granarius*, *S. zeamais*, *Tribolium castaneum*, etc., in terms of toxicity, grain protectant potential, and repellency (Obeng-Ofori et al. 1998).

7.3.2 Flavonoids

Flavonoids are a group of polyphenolic phytochemicals that include flavones, isoflavones, (iso) flavonones, catechins, and chalcones. Fruits, vegetables, nuts and grains, and herbs and spices contain a high concentration of flavonoids. *P. trifoliata* yields flavonoids that can be employed in the development of commercial mosquitocidal products (Rajkumar and Jebanesan 2008).

7.3.3 Glycosides

Glycosides are colorless, crystalline, or amorphous solid substances generally poisonous in nature. They are optically active (L), usually soluble in water and alcohol, but insoluble in ether and chloroform. Glycosides are the condensation product of hydroxyl group or other groups of aglycon and hemiacetal hydroxyl group of sugar. The aglycon should possess at least one hydroxyl group to which glycosidal hydroxyl group of sugar joints. Glycosides are termed as O-, N-, S-, and C-glycosides on the basis of glycosidic linkage (Shah 2009). Cyanogenic glycosides, which liberate cyanide, are plant-based pesticides which block cytochrome c oxidase and NIS, hence toxic for a large group of parasites and herbivores (Kokete et al. 2012).

7.3.4 Alkaloids

Alkaloids refer to the nitrogen heterocycles extracted from plants with the help of organic solvent, e.g., methanol. They are found in plants as their salts of carboxylic acids as citric, lactic, oxalic, acetic, malic, and tartaric acids. Angiosperms are known to be the major source of alkaloids. The plant families identified with the highest alkaloid levels are Papaveraceae, Berberidaceae, Leguminosae, Liliaceae, etc.(Kokete et al. 2012; Shah 2009). Nicotine (*N. tabacum*, *N. glauca*) is considered as the most potent alkaloid derivative used for the control of piercing and sucking insects, e.g., aphids, thrips, etc. It mimics the neurotransmitter, acetylcholine, in the insect's central nervous system (Dayan et al. 2009).

7.4 Formulations of Bio-botanicals

Very few pesticidal materials, either synthetic or botanical, can be used as they exist in technical forms. Most of these, besides being highly hazardous, are sticky, oily, gummy, greasy, or solid materials which are insoluble in water and difficult to handle. For example, essential oils are highly volatile in nature due to the presence of oxygenated monoterpenes. These have, therefore, to be brought to forms in which they could be used as such (Koul et al. 2008). The process of mixing a pesticide with other materials like wetters, spreaders, stickers, deflocculators, stabilizers, etc. to give it certain desirable properties constitutes its "formulation." It also includes combining with inerts like silica and synergists like PBO to increase the effectiveness.

7.5 Advantages of Biological Pesticides over Synthetic Chemical Pesticides

1. As compared with the persistent and highly toxic synthetic chemical pesticides, the biological pesticides are biodegradable in nature and do not leave any harmful residue.

2. The biopesticides are eco-friendly in contrast to the synthetic ones, which are hazardous to the environment and users.
3. They are non-phytotoxic with greater selectivity toward the targeted pest unlike the synthetic pesticides having phytotoxicity and mammalian toxicity.
4. The biopesticides have an array of constituents responsible for the insecticidal action, so the chances of pests developing resistance are relatively less, whereas in case of synthetic chemical pesticides, the phenomenon of pest resurgence is quite common.
5. Availability of biological components (derived from naturally occurring plants) is trouble-free as compared to the synthetic ones which are always not easily procured.
6. The biopesticidal products have a great economic feasibility as low-cost infrastructure (equipments, raw materials, etc.) is used in their production in contrast to the synthetic-based products which use high-cost machines and other ingredients.

7.6 Types of Bio-botanical Formulations

7.6.1 Emulsions

7.6.1.1 Microemulsions

Microemulsions are isotropically clear dispersion of two immiscible liquids, having a droplet size less than 0.1 μm, and are thermodynamically stable and transparent. They form a single phase of three components, i.e., oily liquid or solid dissolved in organic solvent, water, and surfactant/co-surfactant containing micelles (Singla and Patanjali 2013) in which the nonaqueous phase of the active ingredient and solvent is dissolved by the surfactant aggregate system. The preparation of microemulsions involves the addition of two different types of surfactants, one water soluble (anionic with high HLB) and one oil soluble (ionic and low HLB), and a co-surfactant, like butanol, hexanol, etc., with low HLB value. As compared to the o/w emulsion, the microemulsion can have a total concentration of surfactants as high as 10–30 %, which is 5 % for o/w emulsions (Tadros 1995).

For example, microemulsion formulation of eucalyptus oil, citronella oil, etc. possesses certain desirable properties worth exploiting for pest management but has the drawback of being volatile in nature. However, this can be overcome by encapsulating the oils at the surfactant aggregate system, thereby reducing its volatility and increasing the effectiveness.

7.6.1.2 Nanoemulsions

Nanoemulsions are oil-in-water dispersions of an oil and water phase in combination with a surfactant with the droplet size ranging from 100 to 600 nm (Solans et al. 2003). They are thermodynamically and kinetically stable (Bouchernal et al. 2004). For example, nanoemulsions of citronella oil, hairy basil oil, and vetiver oil with mean droplet size ranging from 150 to 220 nm showed long-lasting mosquito repellent activity against *Aedes aegypti* (Nuchuchua et al. 2009).

Nanoemulsions utilizing eucalyptus oil (Fig. 7.1a) and aqueous extracts of botanicals (*Pongamia glabra* and *Jatropha curcas*) (Bar et al. 2009) have been developed at IPFT, Gurgaon, with long-term stability and increased efficacy against the stored grain pest, *Tribolium castaneum*. They also provide a better surface coverage property of the targeted pest due to nanoparticle size and therefore increased surface area.

Advantages

(a) Micro- and nanoemulsions are aqueous-based formulations.
(b) The use of solvents is eliminated and the pesticide dosage is also reduced to enhance their biological activity.
(c) They have an easy application.
(d) Very low particle size of nanoemulsions makes them a promising alternative for pest control.

Disadvantages

(a) The emulsifier level used for microemulsion is high.

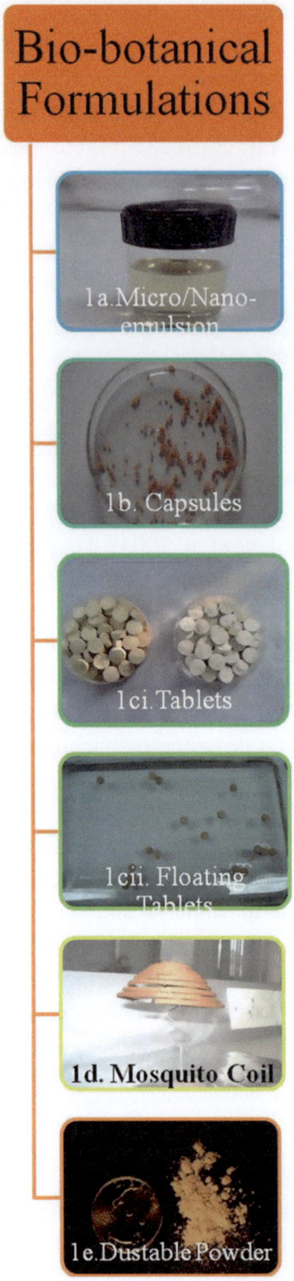

Fig. 7.1 Types of bio-botanical formulations developed at IPFT, Gurgaon

7.6.2 Controlled Release Formulations (CRF)

The technology refers to the triggered/controlled release of an active ingredient to the targeted site to achieve an intended effect on the target pest. Four major categories of controlled release formulations are as follows: (1) coated pesticide granules, (2) matrix systems containing physically trapped particles, (3) polymer systems containing covalently bound particles, and (4) polymer membrane-pesticide reservoir system, e.g., microencapsulation. More popularization of microencapsulation technique has come about in last some years (Beestman 2003). The technique uses the principle of interfacial polymerization (Knowles 2008). In this process the active ingredient, usually a liquid or low-melting waxy solid, is dissolved in an aqueous solution of monomer, e.g., sodium alginate along with slow addition of surfactant to get a homogeneous mixture of all the contents. This mixture is allowed to react with an aqueous solution of a cross-linking agent, e.g., calcium chloride. Interfacial polymerization occurs at the interface where the active component gets encapsulated. The rate of release of the pesticide is, therefore, a diffusion-controlled process (Fernández-Pérez 2007). Examples of CRF formulations are microencapsulations, capsulated suspensions, nanoencapsulated gels, etc. The efficacy of nanoparticle-encapsulated garlic essential oil as compared to free garlic oil was evaluated against *Tribolium castaneum*, the red flour beetle, where the nanoparticle loaded with essential oil gave good mortality over 5 months time indicating a slow and persistent release of active components (Yang et al. 2009). Microencapsulation (Fig. 7.1b) of neem seed oil and karanja oil has been successfully done at IPFT, Gurgaon, for the control of larvae of *Aedes aegypti* (Pant et al. 2012).

Advantages

(a) This technology allows the triggered release of pesticide.
(b) It doesn't affect the nontarget organisms and is safe for the humans.
(c) It also reduces the level of pesticides in the environment.

Disadvantages

The technology requires the use of expensive process/inert ingredients.

7.6.3 Botanical Tablet Formulations

Botanical active ingredients are compressed into a solid mass, i.e., tablet (Fig. 7.1ci), with the help of a tablet machine. The basic components of a pesticide tablet are the active material (e.g., powders of various active components derived from naturally occurring plants), diluents or filler, binder, lubricants, and wetting/dispersing agent (Patanjali 2012). Floating tablets are (Fig. 7.1cii) used for aquatic insects and pests like mosquito larvae. Botanical tablets have been successfully developed by IPFT, Gurgaon (patent filed, application No.:2705/DEL/2012), by using different botanical powders having insecticidal property and botanical wastes (deoiled cakes of *Karanja* and *Jatropha*) alone and in combination. These tablets were tested against the household pest, *Periplaneta americana* (American cockroach).

Advantages

(a) It is an eco-friendly innovation which is completely biodegradable and safe to the user.
(b) The formulation has no mammalian toxicity, phytotoxicity, and dispersion of insect populations like the synthetic sprays.
(c) All ingredients are available throughout India.
(d) Tablets are easy to formulate, store, and use as compared to commercially available synthetic products with an accurate dosage.

Disadvantages

Fungal infestation may occur when high concentration of botanicals is used.

7.6.4 Botanical Coil Formulations

Botanical coil formulations (Fig. 7.1d) comprise of botanical active ingredients, inerts, burning materials and binders (derived from naturally occurring plants), and preservatives (Patanjali 2010). The ingredients are properly dried, powdered, mixed well, and finally extruded through a coil machine to get the product, i.e., coil. Preservatives are used to preserve the repellant properties for a longer period and to increase their shelf life. Coil formulations have been successfully developed by IPFT, Gurgaon (patent filed, Application No.: 365/DEL/2010), by using different botanical powders having insecticidal property along with burning materials, binders, and preservatives in various combinations. The coils prepared were tested against adults of *Aedes aegypti*.

Advantages

(a) The coil formulations are non-hazardous to the environment, less toxic, and safe to the user.
(b) They are economical as they do not consume electricity.
(c) They possess mosquito repellant and knockdown properties.

Disadvantages

Burning of coil produces smoke, which may cause irritation when inhaled.

7.6.5 Dustable Powder

It consists of active ingredients (botanical powders) along with inerts and carriers like china clay. The active ingredient, either solid or liquid, is gradually added in china clay. After complete addition of the active ingredient, it is ground in a mixer to get a uniform composition. It is easy to formulate and use. Dustable powder formulations (Fig. 7.1e) have been developed by IPFT, Gurgaon, for the control of stored grain pests, and a patent has been filed.

Advantages

(a) It is a low-cost formulation and has a good stability aspect.

Disadvantages

(a) It has a low-technology image due to its dustiness and inconvenient application, which create toxic hazards on handling.
(b) It is not target specific, hence has poor bioefficacy.

7.7 Development and Testing of Botanical-Based Pesticide Formulations at IPFT

The bioefficacy of any formulation evaluated against a certain pest is calculated by determining the LC_{50} value using the probit analysis. The LC_{50} value measures the concentration of any pesticide which is able to kill 50 % of the population of the test organisms to which it is exposed. The lower the LC_{50} value, the higher the toxicity of pesticide and vice versa (Mikhaiel 2011).

To conclude the result of any bioefficacy trial, probit analysis is being performed. It relates with the transformation of the observed mortalities to log probit. The doses/concentrations are expressed as logarithmic values. The doses versus observed mortalities give a linear regression parameter. The LC_{50} value is calculated from the linear regression parameter.

Considering the intensive nature of calculations for the estimated LC_{50} and associated 95 % confidence interval using the probit method, the data analysis by a computer program is highly recommended (Rath et al. 2011). The bioefficacy results along with LC_{50} of two botanical pesticide formulations developed at IPFT, Gurgaon, are given below.

7.7.1 Microencapsulation

Encapsulation of neem and karanja oil was done using interfacial polymerization technique where calcium alginate beads were prepared incorporating the oils individually and in combination (Pant et al. 2012). The beads were oven-dried at 45 °C for 2 days and then used for the bioefficacy trials against the *Aedes aegypti* larvae.

Table 7.1 shows the synergistic larvicidal effect of the combination of neem and karanja oil as compared to individual oils. Better mortality was observed with the combination of neem and karanja oils. The LC_{50} value was also found to be less in this combination as compared to LC_{50} individual oils. FTIR analysis of the calcium alginate beads was done to confirm the stability of active ingredient. This formulation has shelf life of more than 6 months and can be exploited for commercial purpose.

7.7.2 Botanical-Based Mosquito Coils

Mosquito coils were developed using neem kernel powder (NKP) and keekar powder (KP) as active ingredients along with binders, preservatives, and burning materials. The efficacy of the coils was evaluated in terms of knockdown and repellency against the *Aedes aegypti* adults (Patent, Abstract published, Application No.: 365/DEL/2010).

The bioefficacy of the coils was evaluated against the mosquitoes, and the percent protection was calculated as % protection = [(No. of mosquitoes landing in control-No. of mosquitoes landing in tested)/(No. of mosquitoes landing in

Table 7.1 Bioefficacy of *Aedes aegypti* with encapsulated neem oil, karanja oil, and neem + karanja oil calcium alginate beads

Serial no.	Amount of calcium alginate beads (mg)	% Mortality in 48 h (Neem oil encapsulation) N	(Karanja oil encapsulation) K	(Neem + karanja oil encapsulation) N + K
1.	100	10.00	20.00	43.33
2.	150	13.33	43.33	53.33
3.	200	16.66	46.67	66.67
4.	250	20.00	53.33	73.33
5.	300	23.33	73.33	86.67
6.	Control	0	0	0

LC_{50} (N) = 5.1 mg/l, LC_{50} (K) = 4.0 mg/l, LC_{50} (N + K) = 3.1 mg/l

control)]×100. The percent protection was 83.49 % whereas the knockdown effect was 39 %.

The results clearly show that mosquito repellent and knockdown activity of the neem (*Azadirachta indica*) kernel powder was enhanced due to the presence of kabuli keekar (*Prosopis juliflora*) pods powder. It is also shown in the results that the maximum mosquito repellent activities and mosquito knockdown activities were shown by the composition, whereas the neem (*Azadirachta indica*) kernel powder and kabuli keekar (*Prosopis juliflora*) pods powder were taken in a specific ratio. The results indicate that kabuli keekar (*Prosopis juliflora*) pods powder has a synergistic effect over the neem (*Azadirachta indica*) kernel powder resulting in enhanced mosquito repellent properties and knockdown activities.

7.8 Conclusion

The current chapter elaborates the importance of botanical-based products/formulations with reference to the formulations developed by IPFT, Gurgaon. The work is in progress with screening and identification of more botanicals as a source of natural insecticides to compensate the loss/damage caused by the excessive use of synthetic chemical-based pesticides. Burning of mosquito coils is a traditional but significant means to repel or kill adult mosquitoes. Many synthetic chemicals such as octachlorodipropyl ether in coils are found to be one of the important genotoxic agents (Pauluhn and Mohr 2000). On the contrary, plant-derived coils are mostly safe and nontoxic (Singh et al. 2011). Increasing attention is being paid to develop safer, effective, and more eco-friendly pesticide formulations. Formulation technology improves operator safety and reduces dose rate and wastage of pesticides applied to crops along with reducing environmental impact and increasing food safety. This has led to the development of water-based liquid formulations regarded as a new technology. The current chapter elaborated the new possible technological developments in pesticide formulations and how the Institute of Pesticide Formulation Technology is progressing in the development of user and eco-friendly pesticide formulations.

References

Allan EJ, Eeswaran JP, Jarvis AP, Mordue-Luntz AJ et al (2002) Induction of hairy root cultures of Azadirachta indica A. Juss. and their production of azadirachtin and other important insect bioactive metabolites. Plant Cell Rep 21:374–379

Al-Quraishy S, Abdel-Ghaffar F, Al-Rasheid KA et al (2012) Observations on effects of a neem seed extract (MiteStop®) on biting lice (mallophages) and blood-sucking insects parasitizing horses. Parasitol Res 110(1):335–339

Bar H, Bhui K, Sahoo D et al (2009) Green synthesis of silver nanoparticles using seed extract of Jatropha curcas. Colloid Surf A: Physicochem Eng Asp 348:212–216

Beestman GB (2003) Controlled release in crop protection: past experience and future potential. Chem crop Prot prog & prosp Sci Reg 272-279. Wiley-VCH Verlag GmbH & Co, Weinheim

Ben H (2002) Giving baculoviruses a better edge. Agric Res 50(1):14

Bouchernal K, Brianeon S, Perrier E et al (2004) Nano-emulsion formulation using spontaneous emulsification, solvent oil and surfactant optimization. Int J Pharmacol 280:41–25

Chang ST, Cheng SS (2002) Antitermite activity of leaf essential oils and their constituents from Cinnamomum osmophloeum. J Agric Food Chem 50:1389–1392

Dayan FE, Cantrell CL, Duke SO (2009) Natural products in crop protection. Bioorg Med Chem 17:4022–4034

Dubey NK, Srivastava B, Kumar A (2008) Current status of plant products as botanical pesticides in storage pest management. J Biopestic 1:182–186

Dubey, NK, Shukla R, Kumar A (2011) Global scenario on the application of natural products in integrated pest management programmes. In: NK Dubey (Ed.) Nat prod plant pest manage 1:1–20. CAB International. http://livedna.org/91.6368

Fernández-Pérez M (2007) Controlled release systems to prevent the agro-environmental pollution derived from pesticide use. J Environ Sci Health B 42:857–862

Hermawan W, Nakajima S, Tsukuda R et al (1997) Isolation of an antifeedant compound from Andrographis paniculata (Anthaceae) against the diamond back, Plutella xylostella (Lepidoptera: Yponomeutidae). App Entomol Zool 32(4):551–559

Isman MB (2006) Botanical insecticides, deterrents, and repellents in modern agriculture and an increasingly regulated world. Ann Rev Entomol 51:45–66

Isman MB, Akhtar Y (2007) Plant natural products as a source for developing environmentally acceptable insecticides. In: Shaaya I, Nauen R, Horowitz AR (eds) Insecticide designing using advanced technologies. Springer, Berlin/Heidelberg, pp 235–248

Isman MB, Miresmailli S, Machial C (2011) Commercial opportunities for pesticides based on plant essential oils in agriculture, industry and consumer products. Phytochem Rev 10:197–204

Knowles A (2008) Recent developments of safer formulations of agrochemicals. Environmentalist 28(1):35–44, 0251-1088

Kokete CK, Purohit AP, Gokhale SP (2012) Pharmacognosy. Moksha Publication House, Rampur

Koul O, Walia S (2009) Comparing impacts of plant extracts and pure allelochemicals and implications for pest control. CAB Rev: Perspect Agric Vet Sci Nutr Nat Res 4:1–30

Koul O, Walia S, Dhaliwal GS (2008) Essential oil as green pesticides: potential and constraints. Biopestic Int 4(1):63–84

Mikhaiel AA (2011) Potential of some volatile oils in protecting packages of irradiated wheat flour against Ephestia kuehniella and Tribolium castaneum. J Stored Prod Res 47:357–364

Nuchuchua O, Sakulku U, Uawongyart N et al (2009) In vitro characterization and mosquito (*Aedes aegypti*) repellent activity of essential-oils-loaded nanoemulsions. AAPS Pharm Sci Tech 10(4): 1234–1242

Obeng-Ofori D, Reichmuth CH, Bekele AJ et al (1998) Toxicity and protectant potential of camphor, a major component of essential oil of Ocimum kilimandscharicum, against four stored product beetles. Int J Pest Manag 44(4):203–209

Pant M, Dubey S, Raza SK et al (2012) Encapsulation of neem and karanja oil mixture for synergistic as well as larvicidal activity for mosquito control. J Sci Ind Res 71:348–352

Patanjali PK, Aggarawl A, Dubey S et al (2010) A novel synergistic mosquito repellent composition for preparation of mosquito coils. Patent, Abstract Published, Application no: 365/DEL/2010

Patanjali PK, Dubey S, Pant M et al (2012) Insecticidal compositions for controlling household pests. Patent, Abstract published, Application no: 2705/DEL/2012

Pauluhn J, Mohr U (2000) Inhalation studies in laboratory animals current concepts and alternatives. Toxicol Pathol 28:734–753

Rajashekhar Y, Bakthavatsalam N, Shivanandappa T (2012) Botanicals as grain protectants. Psyche 2012:13

Rajkumar S, Jebanesan A (2008) Bioactivity of flavonoid compounds from Poncirus trifoliata L. (family: Rutaceae) against the dengue vector, Aedes aegypti L. (Diptera: Culicidae). Parasitol Res 104(1):19–25

Rath S, Sahu MC, Dubey D et al (2011) Which value should be used as the lethal concentration 50 (LC_{50}) with bacteria? J Interdiscip Sci: Comput Life Sci 3:138–143

Shah BN (2009) Text book of pharmacognosy and phytochemistry. Elsevier Publications. http://whqlibdoc.who.int/publications/1998/9241545100.pdf

Singh S, Adhikari U, Chandra G (2011) Smoke repellency and mosquito larvicidal potentiality of Mesua ferra L. Leaf extract against filarial vector Culex quinquefasciatus. Asian Pac J Trop Biomed 1:S119–S123

Singla M, Patanjali PK (2013) Phase behaviour of neem oil based microemulsion formulations. Ind Crop Prod 44:421–426

Solans S, Esquena J, Forigianini A et al (2003) Absorption and aggregation of surfactants in solution. In: Mittal KL, Dinesh OS (eds) Nano-emulsion: form prop appl. Marcel Dekker, New York, pp 525–554

Tadros TF (1995) Surfactants in agrochemicals. Marcel Dekker, New York

Yang FL, Li XG, Zhu F, Lei CL (2009) Structural characterization of nanoparticles loaded with garlic essential oil and their insecticidal activity against *Tribolium castaneum* (Herbst) (Coleoptera: Tenebrionidae). J Agri Food Chem 57(21):10156–10162

Medicinal Plants and Phytomedicines

8

Brahma Singh

Abstract

The use of herbs for disease treatment, primary health and prophylaxis; as health promoters; and in other forms is practised by 60 % of the population globally. Some countries use them heavily, whereas the others, mostly advanced countries, use them sparingly to the extent of 40 % or less. While the traditional medicines are derived from medicinal plants, minerals and organic matter, the herbal drugs and phytomedicines are prepared from medicinal plants only. The use of plants as a source of medicine has been inherited and is an important component of the health-care system in India. Public, academic and government interest in traditional medicines is growing exponentially due to the increased incidence of the adverse drug reactions and economic burden of the modern system of medicine earlier considered safe without much of side effects. In rural India, 70 % of the population is dependent on the traditional system of medicine, the Ayurveda. The interest in plant medicines is growing, and they are taking considerable market share in most of the developed countries. Medicinal plants are being used as single plant extract or as synthetic analogue of natural molecules. These are known as phytomedicines or phytopharmaceuticals, and it has been estimated that the world market for these product is more than Rupees 2000 crores. The present chapter reviews the medicinal activity of plants along with the major bioactive compounds in them. The cultivation and conservation of medicinal plants and their potential use as phytomedicines are also discussed.

8.1 Introduction

Plants support human life in multiple manners by providing food, fuel, fodder, fibre, drugs, medicine, etc. Initially plants were found to be useful in combating human ailments by trial and error. The knowledge on plants graduated based on repeated

Brahma Singh (✉)
World Noni Research Foundation,
12, Rajiv Gandhi Road, Perungudi,
Chennai, 600 096, India

Defence Research and Development Organisation,
Rajaji Marg, New Delhi, 110011, India
e-mail: brahma88@gmail.com

© Springer India 2016

Vijay Veer, R. Gopalakrishnan (eds.), *Herbal Insecticides, Repellents and Biomedicines: Effectiveness and Commercialization*, DOI 10.1007/978-81-322-2704-5_8

use on human subjects for different ailments and distinct category of plants emerged as medicinal plants. Further studies of these plants accumulated knowledge on their therapeutic and prophylactic potential. Such plants were called as herbs and system of using them as herbal treatment under different names in different countries of the world. With the development of allopathy or modern system of medicine, herbs or herbal treatment is now known as traditional system of medicine. The use of herbs for disease treatment, primary health and prophylaxis; as health promoters; and in other forms is practised by 60 % of the population globally. Some countries use them heavily, whereas the others, mostly advanced countries, use them sparingly to the extent of 40 % or less.

India has vast knowledge about the traditional medicines and herbal medicines from the inherited sources. Most of the traditional healers or practitioners use herbal medicines or traditional medicines according to their own method of preparations, recipes, etc. However, it is true that rationality or claims are perhaps missing when scientifically judged. The traditional medicines are composed of the plants, minerals and organic matter, whereas the herbal drugs are prepared from medicinal plants and are having phytoconstitution only. We have witnessed that for the last few decades, the herbal drugs are playing a major role in the western world particularly for the treatment of metabolic disorders and terminal illnesses. Public, academic and government interest in traditional medicines is fast growing due to the increased incidence of the adverse drug reactions and economical burden of the modern system of medicine. Herbal drugs are considered safe without much of side effects.

India has one of the 12 M biodiversity centres with three major hotspots existing in the Eastern Himalayas, Western Ghats and Andaman and Nicobar Islands. About 45,000 reported plant species exist in India, out of which 3000 plants have medicinal importance as per the official documents. Interestingly the traditional practitioners are still using more than 6000 plants. India is the second largest producer of medicinal herbs after China and is appropriately called botanical garden of the world.

There are currently about 250,000 registered medical practitioners of the Ayurvedic system (total for all traditional systems: approximately 291,000), as compared to about 700,000 of the modern medicine system. In rural India, 70 % of the population is dependent on the traditional system of medicine, the Ayurveda. In Ayurveda texts plants have been classified in terms of their therapeutic actions. They are grouped into two categories. The first includes the promoters of vigour in healthy individuals (positive health), and the second includes destructors of disease in ailing individuals (curative). Plants of the second category are generally called as therapeutic plants.

The preparation made using therapeutic plants is called phytomedicine or phytopharmaceutical. Earlier these plant medicines were inducted in allopathic system of medicine as mixtures or tinctures in the beginning of last decades of the last century. In recent years they have come up in more sophisticated and refined form. In the last four decades, the developed countries are seeing an ever-growing interest in these plant medicines. These have come up as single plant extract or as synthetic analogue of natural molecules. These are known as phytomedicines or phytopharmaceuticals. It has been estimated that the world market for these products is more than Rupees 2000 crores. The growth rate in this sector is estimated to be around 15 %.

Phytochemicals are quite distinct from the traditional herbal medicinal preparations of Ayurveda, Chinese system of medicines and other traditional systems of medicine. Mostly, these classical medicinal formulations consist of several plants, and little is known about the active principles and it is difficult if not impossible to standardise them. On the other hand, plant-derived chemicals/pharmaceuticals are produced scientifically, standardised and clinically evaluated like the other conventional pharmaceuticals. The ESCOP (European Scientific Cooperative Phytotherapy) has listed 150 herbal drugs as beneficial in different ailments. In Germany, the Federal Ministry of Health has set up a special commission, which looks after various aspects of herbal drugs. It has evaluated and published monographs on more than 300 individual herbs. The medicines are available as prescription drugs. In France, herbal medicine is governed by similar regulations which cover over 200 medicinal plant products.

In the United States of America, herbal products are popular but are sold as dietary supplements as no therapeutic claim is unfortunately permitted. However, development of new herbal products continues as evidenced by increase in filing more investigational new drug applications for herbal or botanical medicines by pharmaceutical industry and accepted by the US Food and Drug Administration.

8.2 Scientific Validation of Phytopharmaceuticals

There are several relevant questions in popularity of plant products as medicines or drug. Is there scientific rationale for developing and using phytomedicines? Standardisation of herbal therapeutic agent is more problematic and expensive. On the other hand, question is raised why to load a patient or even a healthy individual with lot of inactive plant material by administering the plant extract instead of pure active principle or molecule(s). It looks like recommending compressed nutrients for human and animals in place of bulky food, fodder and feed comprising of so much bulk which is passed out as faecal matter. The necessity and importance of this bulk can hardly be overemphasised. The modern system of medicine knows it well but would not talk or make mention of it adequately.

In a mixture of compounds such as a herbal abstract, the biological or, in fact, any other activity of a certain individual compound is subject to influences of the other constituents. If the other constituents do not possess the same activity, they just act as diluents or may potentiate (synergism) or reduce (antagonism) the said activity. If other compounds also possess the same activity, the effect could be synergistic or additive. The same question arises when two or more extracts are mixed together. There are cases where the compounds isolated from plants and administered individually are inactive but effective as a mixture. The synergetic action of compounds is well known.

Synergism is a well-known phenomenon in biology. For example, in antimicrobials efflux pump inhibitors have been demonstrated to potentiate the activity. In treatment of cancer patients, different combinations of chemotherapy regiments are routinely employed. Even in the use of insecticides, piperonyl butoxide is a well-established synergist for natural and synthetic pyrethroids and rotenone is being used as adjunct.

Synergism, additive and adjunct phenomenon associated with plant medicines is worth researching and establishing the science and rationale behind it to further exploit the use of phytomedicines. This will help in promoting the use of phytomedicines or to get preference for phytomedicine.

Moreover, several medicinal plants like *Curcuma longa*, *Glycyrrhiza glabra*, *Tinospora cordifolia*, *Zingiber officinale* and others have several active compounds and several closely related secondary metabolites. These may exhibit synergism or additiveness in action. There are enough examples of herbs with synergistic interaction of constituents. For example, *Artemisia annua* has artemisinin, an antimalarial along with two co-occurring flavonoids, chrysoplenol-D and chrysoplenetin having no antimalarial activity, but they potentiate the antimalarial action of artemisinin.

8.3 Poly-herbs as Medicine

There are mounting evidence of herbal-herbal interactions and adjunct therapy. Certain pure compounds or extracts of some plants potentiate the pharmacological action of another plant extract or an active compound. This potentiation by an auxiliary is being exploited in some situations in conventional practice to curtail the therapeutic dose, thus limiting any side reactions. The adjunct compounds operate in several ways such as bioavailability enhancer, disabling infection/disease resistance, fortifying immune defence, etc.

8.4 Medicinal Activity of Plants

The medicinal activity of plants has been reviewed by Dev (2006). Plants having particular medicinal activity have been listed below for authentication and useful leads by researchers of medical science. The information provided hereunder is compiled from Dev (2006) and other sources listed in the references section, which were supplemented and confirmed by the author.

Abortifacient *Achyranthes aspera*, *Andrographis paniculata*, *Curcuma zedoaria*, *Momordica charantia*, *Moringa oleifera*.

Acetylcholinesterase inhibitor *Adhatoda vasica*, *Desmodium gangeticum*, *Fumaria indica*, *Inula racemosa*, *Nardostachys jatamansi*.

Adaptogenic *Asparagus racemosus*, *Bacopa monnieri*, *Crocus sativus*, *Curculigo orchioides*, *Phyllanthus emblica*, *Hippophae rhamnoides*, *Ocimum sanctum*, *Tinospora cordifolia*.

α-Adrenergic blocker *Rauvolfia serpentina*.

β-Adrenergic blocker *Inula racemosa*, *Litsea glutinosa*.

Aldose reductase inhibitor *Azadirachta indica*, *Butea monosperma*, *Curcuma longa* (cataract), *Phyllanthus fraternus*, *Swertia chirayita*.

Anabolic *Achyranthes aspera*, *Alpinia galangal*, *Centella asiatica*, *Tribulus terrestris*, *Withania somnifera*.

Analgesic *Aegle marmelos*, *Azadirachta indica*, *Boerhavia diffusa*, *Calotropis procera*, *Cannabis sativa*, *Cassia fistula*, *Cassia tora*, *Cedrus deodara*, *Celastrus paniculatus*, *Crocus sativus*, *Curcuma zedoaria*, *Cyperus rotundus*, *Desmodium gangeticum*, *Eclipta alba*, *Embelia ribes*, *Fumaria indica*, *Lawsonia inermis*, *Moringa oleifera*, *Nelumbo nucifera*, *Nigella sativa*, *Ocimum sanctum*, *Phyllanthus fraternus*, *Terminalia bellirica*, *Tinospora cordifolia*, *Vitex negundo*, *Zingiber officinale*.

Androgen *Curculigo orchioides*, *Zingiber officinale*.

Angiogenic *Aloe vera*.

Angiotensin-converting enzyme inhibitor *Phyllanthus fraternus*, *Terminalia chebula*, *Tribulus terrestris*.

Anthelmintic *Acorus calamus* (nematodes), *Alpinia galangal* (ascariasis), *Andrographis paniculata* (ascariasis, filariasis), *Asparagus adscendens* (filariasis), *Azadirachta indica* (ascariasis), *Bacopa monnieri* (nematodes), *Butea monosperma* (ascariasis), *Caesalpinia bonduc* (ascariasis), *Commiphora wightii* (tapeworms and hookworms), *Cyperus rotundus* (filariasis), *Embelia ribes* (ascariasis, tapeworm), *Fumaria indica* (strongyloidiasis), *Hedychium spicatum* (tapeworms), *Mallotus philippensis* (tapeworms), *Momordica charantia* (ascariasis), *Mucuna pruriens* (tapeworms), *Nigella sativa* (cestodes), *Ocimum sanctum* (nematodes), *Psoralea corylifolia* (cestodes, trematodes, nematodes), *Punica granatum* (cestodes, filariasis), *Tribulus terrestris* (nematodes), *Zingiber zerumbet* (nematodes).

Anti-abortion *Asparagus racemosus*.

Anti-ache	*Aloe vera*, *Commiphora wightii*, *Curcuma longa*, *Hemidesmus indicus*.
Anti-ageing	*Acacia catechu*, *Asparagus racemosus*, *Bacopa monnieri*, *Phyllanthus emblica*, *Litsea glutinosa*.
Antialcoholic	(toxic effects of alcohol): *Curcuma longa*, *Phyllanthus emblica*.
Anti-allergic	*Adhatoda vasica*, *Aloe vera*, *Alpinia galangal*, *Curcuma zedoaria*, *Inula racemosa*, *Nigella sativa*, *Plumbago zeylanica*, *Terminalia chebula*, *Tinospora cordifolia*.
Anti-Alzheimer	*Centella asiatica*, *Curcuma longa*.
Antiamoebic	*Berberis aristata*, *Caesalpinia bonduc*, *Coccinia grandis*, *Curcuma longa*, *Glycyrrhiza glabra*, *Holarrhena antidysenterica*, *Nigella sativa*, *Piper longum*.
Anti-androgen	*Azadirachta indica*, *Embelia ribes*.
Antianginal	*Berberis aristata*, *Inula racemosa*, *Pueraria tuberosa*, *Terminalia arjuna*, *Tribulus terrestris*, *Zingiber officinale*.
Antiangiogenesis	*Saussurea lappa*.
Antiarrhythmic	*Aegle marmelos*, *Azadirachta indica*, *Berberis aristata*, *Fumaria indica*, *Nelumbo nucifera*, *Rauvolfia serpentina*.
Antiarthritic	*Alpinia galangal* (osteoarthritis), *Azadirachta indica*, *Boswellia serrata*, *Coccinia grandis*, *Commiphora wightii*, *Crataeva nurvala*, *Curcuma longa*, *Cyperus rotundus*, *Picrorhiza kurroa*, *Saussurea lappa*, *Semecarpus anacardium*, *Swertia chirayita*, *Tinospora cordifolia*, *Zingiber officinale*.
Antiasthmatic	*Acorus calamus*, *Adhatoda vasica*, *Boswellia serrata*, *Calotropis procera*, *Cannabis sativa*, *Curcuma longa*, *Glycyrrhiza glabra* (bronchial), *Hedychium spicatum* (bronchial asthma), *Inula racemosa* (bronchial), *Nardostachys jatamansi*, *Nigella sativa* (bronchodilatory), *Opium sanctum*, *Picrorhiza kurroa*, *Piper longum*, *Saussurea lappa*, *Tinospora cordifolia*.
Antiatherosclerotic	*Acacia catechu*, *Andrographis paniculata*, *Celastrus paniculatus*, *Commiphora wightii*, *Crocus sativus*, *Curcuma longa*, *Picrorhiza kurroa*, *Plumbago zeylanica*, *Punica granatum*, *Rauvolfia serpentine*, *Semecarpus anacardium*, *Terminalia arjuna*, *Terminalia chebula*, *Zingiber officinale*, *Zingiber zerumbet*.
Antibacterial	*Acacia catechu*, *Acorus calamus*, *Allium sativum*, *Aloe vera*, *Alpinia galangal*, *Alstonia scholaris*, *Andrographis paniculata*, *Asparagus racemosus*, *Azadirachta indica*, *Bacopa monnieri*, *Berberis aristata*, *Caesalpinia bonduc*, *Calotropis procera*, *Cassia absus*, *Cassia fistula*, *Cassia tora*, *Cedrus deodara*, *Curcuma longa*, *Embelia ribes*, *Phyllanthus emblica*, *Ficus benghalensis*, *Garcinia morella*, *Glycyrrhiza glabra*, *Hedychium spicatum*, *Holarrhena antidysenterica*, *Inula racemosa*, *Lawsonia inermis*, *Leptadenia reticulata*, *Leucas aspera*, *Litsea glutinosa*, *Mallotus philippensis*, *Moringa oleifera*, *Nardostachys jatamansi*, *Nelumbo nucifera*, *Nigella sativa*, *Ocimum sanctum*, *Plumbago zeylanica*, *Psoralea corylifolia*, *Pterocarpus marsupium*, *Punica granatum*, *Rubia cordifolia*, *Swertia chirayita*, *Terminalia arjuna*, *Terminalia bellirica*, *Terminalia chebula*, *Vitex negundo*, *Zingiber officinale*.

Anticancer	*Aegle marmelos* (antineoplastic), *Aloe vera*, *Alpinia galangal* (tumours), *Alstonia scholaris*, *Asparagus adscendens*, *Azadirachta indica*, *Bacopa monnieri*, *Berberis aristata*, *Boerhavia diffusa* (antiproliferative), *Boswellia serrata* (leukaemia), *Butea monosperma* (tumour), *Caesalpinia bonduc* (tumours), *Calotropis procera* (cytotoxic), *Cannabis sativa*, *Cassia fistula* (tumours), *Cassia angustifolia* (tumours), *Catharanthus roseus*, *Cedrus deodara*, *Celastrus paniculatus*, *Centella asiatica* (tumour), *Coccinia grandis* (tumour), *Commiphora wightii*, *Convolvulus microphyllus*, *Crataeva nurvala*, *Crocus sativus* (leukaemia, other malignancies), *Curcuma longa* (leukaemia, tumours), *Curcuma zedoaria*, *Cyperus rotundus*, *Desmodium gengeticum*, *Eclipta alba*, *Embelia ribes* (tumour), *Phyllanthus emblica*, *Ficus benghalensis* (tumour), *Garcinia morella*, *Glycyrrhiza glabra*, *Hedychium spicatum*, *Hemidesmus indicus*, *Inula racemosa*, *Lawsonia inermis*, *Leptadenia reticulate*, *Mallotus philippensis*, *Momordica charantia*, *Moringa oleifera*, *Micuna puriens*, *Nardostachys jatamansi*, *Nelumbo nucifera*, *Nigella sativa*, *Ocimum sanctum*, *Piper longum*, *Plumbago zeylanica*, *Psoralea corylifolia*, *Rauvolfia serpentina*, *Rubia cordifolia*, *Santalum album*, *Saussurea lappa*, *Semecarpus anacardium*, *Swertia chirata*, *Terminalia arjuna*, *Terminalia chebula*, *Tinospora cordifolia*, *Tribulus terrestris*, *Trigonella foenum-graecum*, *Valeriana jatamansi*, *Vitex negundo*, *Withania somnifera*, *Zingiber officinale*, *Zingiber zerumbet*.
Anticataract	*Ocimum sanctum*, *Trigonella foenum-graecum*.
Anticolitis	*Curcuma longa*, *Litsea glutinosa*, *Picrorhiza kurroa*.
Anticomplement	*Tinospora cordifolia*, *Crataeva nurvala*.
Anticonjunctivitis	*Cyperus rotundus*, *Glycyrrhiza glabra*.
Anticonvulsant	*Aegle marmelos*, *Butea monosperma*, *Curculigo orchioides*, *Rauvolfia serpentina*.
Anticystic fibrosis	*Curcuma longa*.
Antidepressant	*Curcuma longa*, *Rhodiola rosea*, *Withania somnifera*.
Antidiabetic	*Achyranthes aspera*, *Adhatoda vasica*, *Aegle marmelos*, *Aloe vera*, *Andrographis paniculata*, *Azadirachta indica*, *Butea monosperma*, *Caesalpinia bonduc*, *Cannabis sativa*, *Cassia fistula*, *Catharanthus roseus*, *Cinnamomum tamala*, *Coccinia indica*, *Curcuma longa*, *Daucus carota*, *Embelia ribes*, *Phyllanthus emblica*, *Syzygium cumini*, *Ficus benghalensis*, *Gymnema sylvestre*, *Hedychium spicatum*, *Inula racemosa*, *Kyllinga monocephala*, *Mallotus philippensis*, *Momordica charantia*, *Moringa oleifera*, *Nelumbo nucifera*, *Ocimum sanctum*, *Phyllanthus fraternus*, *Picrorhiza kurroa*, *Piper longum*, *Psoralea corylifolia*, *Pterocarpus marsupium*, *Pueraria tuberosa*, *Punica granatum*, *Syzygium cumini*, *Swertia chirayita*, *Tecoma stans*, *Terminalia bellirica*, *Terminalia chebula*, *Tinospora cordifolia*, *Trigonella foenum-graecum*, *Withania somnifera*, *Zingiber officinale*.
Antidiarrhoeal	*Acacia catechu*, *Acorus calamus*, *Aegle marmelos*, *Berberis aristata*, *Caesalpinia bonduc*, *Calotropis procera*, *Syzygium*

	cumini, Ficus benghalensis, Nelumbo nucifera, Ocimum sanctum, Piper longum.
Antidrug misuse	*Benincasa hispida.*
Antidysentery	*Andrographis paniculata, Nelumbo nucifera, Saussurea lappa.*
Antidyspepsia	*Curcuma longa, Curcuma zedoaria.*
Antieczematic	*Andrographis paniculata, Azadirachta indica, Centella asiatica.*
Antiemetic	*Cannabis sativa, Cyperus rotundus, Zingiber officinale.*
Antiepileptic	*Centella asiatica, Nardostachys jatamansi.*
Antifatigue	*Bacopa monnieri, Celastrus paniculatus.*
Antifertility	*Andrographis paniculata, Azadirachta indica, Butea monosperma, Calotropis procera, Cassia fistula, Curcuma zedoaria, Embelia ribes, Moringa oleifera, Nelumbo nucifera, Pueraria tuberosa.*
Antifibrinolytic	*Boerhavia diffusa, Curcuma longa.*
Antifungal	*Acacia catechu, Aloe vera, Alpinia galangal, Azadirachta indica, Berberis aristata, Calotropis procera, Cassia fistula, Cassia tora, Cedrus deodara, Curcuma longa, Curcuma zedoaria, Fumaria indica, Glycyrrhiza glabra, Inula racemosa, Lawsonia inermis, Leucas aspera, Litsea glutinosa, Moringa oleifera, Nardostachys jatamansi, Nelumbo nucifera, Nigella sativa, Ocimum sanctum, Plumbago zeylanica, Pterocarpus marsupium, Punica granatum, Terminalia bellirica, Terminalia chebula, Tribulus terrestris, Zingiber officinale.*
Antigiardiasis	*Berberis aristata, Embelia ribes.*
Antigastritis	*Curcuma longa, Glycyrrhiza glabra.*
Antiglaucoma	*Cannabis sativa, Punica granatum.*
Antihaemorrhagic	*Terminalia arjuna, Terminalia bellirica.*
Antihaemorrhoides	*Centella asiatica.*
Antihistaminic	*Aloe vera, Azadirachta indica, Benincasa hispida.*
Antihyperthyroid	*Aegle marmelos, Convolvulus microphyllus, Phyllanthus emblica, Piper longum.*
Antihypertensive	*Allium sativum, Andrographis paniculata, Azadirachta indica, Berberis aristata, Cassia absus, Cassia tora, Cedrus deodara, Celastrus paniculatus, Cissus quadrangularis, Clerodendrum colebrookianum, Convolvulus microphyllus, Crocus sativus, Cyperus rotundus, Holarrhena antidysenterica, Leptadenia reticulata, Litsea glutinosa, Moringa oleifera, Nardostachys jatamansi, Nelumbo nucifera, Nigella sativa, Phyllanthus fraternus, Pterocarpus marsupium, Rauvolfia serpentina, Terminalia arjuna, Terminalia bellirica, Terminalia chebula, Tribulus terrestris.*
Antihypothyroid	*Commiphora wightii.*
Anti-incontinence	*Crataeva nurvala.*
Anti-infective	*Andrographis paniculata, Asparagus racemosus, Crataeva nurvala.*
Anti-inflammatory	*Achyranthes aspera, Adhatoda vasica, Aloe vera, Alpinia galangal, Andrographis paniculata, Azadirachta indica, Berberis aristata, Boerhavia diffusa, Boswellia serrata, Butea monosperma, Caesalpinia bonduc, Calotropis procera,*

	Cannabis sativa, Cassia absus, Cassia angustifolia, Cedrus deodara, Celastrus paniculatus, Centella asiatica, Coccinia grandis, Commiphora wightii, Crataeva nurvala, Crocus sativus, Curculigo orchioides, Curcuma longa, Curcuma zedoaria, Cyperus rotundus, Desmodium gangeticum, Eclipta alba, Embelia ribes, Phyllanthus emblica, Syzygium cumini, Fumaria indica, Glycyrrhiza glabra, Hedychium spicatum, Inula racemosa, Lawsonia inermis, Leucas aspera, Litsea glutinosa, Moringa oleifera, Nelumbo nucifera, Nigella sativa, Ocimum sanctum, Phyllanthus fraternus, Piper longum, Psoralea corylifolia, Rubia cordifolia, Saussurea lappa, Semecarpus anacardium, Swertia chirayita, Terminalia bellirica, Terminalia chebula, Tinospora cordifolia, Tribulus terrestris, Vitex negundo, Withania somnifera, Zingiber officinale, Zingiber zerumbet.
Antileishmanial	*Andrographis paniculata, Bacopa monnieri, Curcuma longa, Desmodium gangeticum, Swertia chirata.*
Antilepric	*Centella asiatica.*
Antileucodermic	*Psoralea corylifolia, Tribulus terrestris.*
Antilichen	*Aloe vera.*
Antimalarial	*Alstonia scholaris, Andrographis paniculata, Artemisia annua, Azadirachta indica, Calotropis procera, Coptis teeta, Cassia tora, Celastrus paniculatus, Crotalaria occulta, Curcuma longa, Cyperus rotundus, Hedychium spicatum, Momordica charantia, Ocimum sanctum, Phyllanthus fraternus, Plumbago zeylanica, Polygala persicariaefolia, Swertia chirayita, Terminalia bellirica, Vitex peduncularis.*
Antimenorrhagia	*Rubia cordifolia.*
Antimigraine	*Calotropis procera, Zingiber officinale.*
Anti-motion sickness	*Zingiber officinale.*
Anti-multiple sclerosis	*Curcuma longa, Cannabis sativa.*
Anti-muscle spasticity	*Cannabis sativus.*
Antinephritis	*Andrographis paniculata, Butea monosperma, Boerhavia diffusa, Tribulus terrestris.*
Antiobesity	*Acacia catechu, Boswellia serrata, Momordica charantia, Rauvolfia serpentina, Trigonella foenum-graecum.*
Antioedemic	*Boerhavia diffusa, Centella asiatica.*
Antiosteoporotic	*Cissus quadrangularis, Psoralea corylifolia.*
Antioxidant	*Acacia catechu, Aloe vera, Andrographis paniculata, Asparagus racemosus, Azadirachta indica, Bacopa monnieri, Butea monosperma, Caesalpinia bonduc, Cannabis sativa, Cassia fistula, Cassia tora, Cedrus deodara, Celastrus paniculatus, Centella asiatica, Cissus quadrangularis, Coccinia grandis, Commiphora wightii, Curcuma longa, Desmodium gangetum, Embelia ribes, Phyllanthus emblica, Syzygium cumini, Glycyrrhiza glabra, Hippophae rhamnoides, Hedychium spicatum, Hemidesmus indicus, Leucas aspera, Litsea glutinosa, Moringa oleifera, Nelumbo nucifera, Nigella sativa, Ocimum sanctum, Picrorhiza kurroa, Psoralea corylifolia, Punica granatum, Rubia cordifo-*

	lia, Semecarpus anacardium, Swertia chirata, Terminalia arjuna, Terminalia bellirica, Terminalia chebula, Tinospora cordifolia, Trigonella foenum-graecum, Withania somnifera, Zingiber officinale.
Antioxytocic	*Asparagus racemosus.*
Antipancreatitis	*Phyllanthus emblica.*
Antiparkinsonian	*Mucuna pruriens.*
Antiplatelet	*Andrographis paniculata, Cassia tora, Commiphora wightii, Curcuma longa, Fumaria indica, Litsea glutinosa, Nelumbo nucifera, Nigella sativa, Picrorhiza kurroa, Psoralea corylifolia, Pueraria tuberosa, Rauvolfia serpentine, Rubia cordifolia, Zingiber officinale.*
Antiprotozoal	*Garcinia morella.*
Antipsoriatic	*Aloe vera, Centella asiatica, Psoralea corylifolia.*
Antipsychotic	*Acorus calamus, Bacopa monnieri, Celastrus paniculatus, Centella asiatica, Cyperus rotundus, Nardostachys jatamansi, Nelumbo nucifera, Rauvolfia serpentine, Santalum album, Swertia chirayita, Valeriana jatamansi.*
Antipyretic	*Aegle marmelos, Azadirachta indica, Caesalpinia bonduc, Calotropis procera, Cassia fistula, Celastrus paniculatus, Cyperus rotundus, Phyllanthus emblica, Fumaria indica. Glycyrrhiza glabra, Hedychium spicatum, Lawsonia inermis, Nelumbo nucifera, Nigella sativa, Psoralea corylifolia, Tinospora cordifolia, Valeriana jatamansi, Zingiber officinale.*
Antiscabies	*Azadirachta indica.*
Antispasmodic	*Acorus calamus, Cedrus deodara, Curcuma longa, Fumaria indica, Glycyrrhiza glabra, Hedychium spicatum, Holarrhena antidysenterica, Litsea glutinosa, Moringa oleifera, Nardostachys jatamansi, Nigella sativa, Picrorhiza kurroa, Plumbago zeylanica, Pueraria tuberosa.*
Antistroke	*Withania somnifera, Zingiber officinale.*
Antithrombotic	*Andrographis paniculata, Curcuma longa, Glycyrrhiza glabra, Hemidesmus indicus.*
Antitubercular	*Andrographis paniculata, Inula racemosa, Lawsonia inermis, Psoralea corylifolia.*
Antitussive	*Acorus calamus, Adhatoda vasica, Asparagus racemosus, Calotropis procera, Phyllanthus emblica, Glycyrrhiza glabra, Nelumbo nucifera, Plumbago zeylanica, Rubia cordifolia, Zingiber officinale.*
Antityphoid	*Andrographis paniculata.*
Antiulcerative	*Aegle marmelos, Aloe vera, Asparagus racemosus, Alpinia galangal, Azadirachta indica, Bacopa monnieri, Benincasa hispida, Calotropis procera, Centella asiatica, Convolvulus microphyllus, Curcuma longa, Curcuma zedoaria, Phyllanthus emblica, Glycyrrhiza glabra, Hemidesmus indicus, Momordica charantia, Nardostachys jatamansi, Ocimum sanctum, Piper longum, Saussurea lappa, Swertia chirayita, Terminalia chebula, Trigonella foenum-graecum, Zingiber officinale.*

Antiurolithiasis	*Achyranthes aspera, Crataeva nurvala, Phyllanthus fraternus, Tribulus terrestris, Trigonella foenum-graecum.*
Antivenom	*Curcuma longa, Eclipta alba, Glycyrrhiza glabra, Hemidesmus indicus, Leucas aspera, Vitex negundo.*
Antivertigo	*Zingiber officinale.*
Antiviral	*Aegle marmelos, Allium sativum, Aloe vera, Andrographis paniculata, Azadirachta indica, Berberis aristata, Butea frondosa, Cassia angustifolia, Centella asiatica, Curcuma longa, Glycyrrhiza glabra, Momordica charantia, Moringa oleifera, Nigella sativa, Phyllanthus fraternus, Santalum album, Saussurea lappa, Swertia chirata, Terminalia arjuna, Terminalia chebula, Zingiber officinale.*
Antiwrinkle	*Phyllanthus emblica.*
Anxiolytic	*Azadirachta indica, Centella asiatica, Nardostachys jatamansi, Rauvolfia serpentina, Withania somnifera.*
Aphrodisiac	*Mucuna pruriens, Tribulus terrestris.*
Appetite stimulant	*Trigonella foenum-graecum, Cannabis sativa.*
Bioavailability enhancer	*Piper longum, Plumbago zeylanica.*
Bone fracture repair	*Cissus quadrangularis, Psoralea corylifolia.*
Bronchorelaxant	*Adhatoda vasica, Nigella sativa, Terminalia bellirica, Vitex negundo.*
Calcium channel antagonistic	*Bacopa monnieri, Boerhavia diffusa, Nelumbo nucifera, Nigella sativa.*
Calcium mobilisation	*Boswellia serrata, Caesalpinia bonduc.*
Carbonic anhydrase inhibitor	*Punica granatum.*
Cancer preventive	*Acacia catechu, Achyranthes aspera, Aloe vera, Crocus sativus, Cassia tora, Curcuma longa, Phyllanthus emblica, Glycyrrhiza glabra, Momordica charantia, Nigella sativa, Ocimum sanctum, Psoralea corylifolia, Punica granatum, Santalum album, Terminalia bellirica, Trigonella foenum-graecum, Withania somnifera, Zingiber officinale.*
Capillary protectant	*Asparagus racemosus.*
Cardioprotective	*Aegle marmelos, Andrographis paniculata, Azadirachta indica, Boerhavia diffusa, Phyllanthus emblica, Ocimum sanctum, Psoralea corylifolia, Punica granatum, Terminalia arjuna, Terminalia chebula, Withania somnifera, Zingiber officinale.*
Cathartic	*Aloe vera, Cassia fistula, Cassia angustifolia, Cassia tora, Garcinia morella, Mallotus philippensis, Picrorhiza kurroa, Terminalia chebula.*
Choleretic	*Andrographis paniculata, Boerhavia diffusa, Berberis aristata, Curcuma longa, Picrorhiza kurroa.*
Cholinergic	*Withania somnifera.*
Congestive heart failure	*Terminalia arjuna.*
Cyclooxygenase inhibitor	*Aloe vera, Nigella sativa, Ocimum sanctum, Punica granatum, Zingiber officinale, Zingiber zerumbet.*
Decongestant	*Coccinia grandis.*
Detoxicant	*Aegle marmelos, Aloe vera, Andrographis paniculata, Azadirachta indica, Crocus sativus, Curcuma longa, Phyllanthus emblica, Glycyrrhiza glabra, Hemidesmus indicus, Lawsonia*

	inermis, Litsea glutinosa, Momordica charantia, Moringa oleifera, Ocimum sanctum, Picrorhiza kurroa, Piper longum, Santalum album, Semecarpus anacardium, Swertia chirayita, Trigonella foenum-graecum, Withania somnifera, Zingiber officinale, Zingiber zerumbet.
Digestive stimulant	*Acorus calamus, Andrographis paniculata, Emblica ribes, Phyllanthus emblica, Piper longum, Valeriana jatamansi, Zingiber officinale.*
Diuretic	*Achyranthes aspera, Azadirachta indica, Boerhavia diffusa, Leucas aspera, Nigella sativa, Rauvolfia serpentina, Tribulus terrestris, Withania somnifera.*
Estrogenic	*Butea monosperma, Cyperus rotundus, Psoralea corylifolia, Pueraria tuberosa.*
Expectorant	*Acorus calamus, Adhatoda vasica, Glycyrrhiza glabra, Rubia cordifolia.*
Fibrinolytic	*Curcuma longa.*
Gastroprotective	*Nigella sativa.*
High-altitude sickness	*Rhodiola rosea, Hippophae rhamnoides.*
Haemostatic	*Acacia catechu, Asparagus racemosus, Nelumbo nucifera, Psoralea corylifolia, Rubia cordifolia.*
Hepatoprotective	*Alstonia scholaris, Andrographis paniculata, Azadirachta indica, Bacopa monnieri, Berberis aristata, Boerhavia diffusa, Butea monosperma, Cassia fistula, Cassia tora, Cyperus rotundus, Coccinia grandis, Curcuma longa, Eclipta alba, Embelia ribes, Phyllanthus emblica, Fumaria indica, Glycyrrhiza glabra, Hemidesmus indicus, Lawsonia inermis, Leucas aspera, Litsea glutinosa, Nardostachys jatamansi, Nelumbo nucifera, Nigella sativa, Ocimum sanctum, Phyllanthus fraternus, Picrorhiza kurroa, Piper longum, Psoralea corylifolia, Pueraria tuberosa, Rubia cordifolia, Swertia chirata, Tinospora cordifolia, Tribulus terrestris.*
Hypnotic	*Valeriana jatamansi.*
Hypocholesterolaemic	*Acorus calamus, Aloe vera, Caesalpinia bonduc, Cassia fistula, Cassia tora, Celastrus paniculatus, Commiphora wightii, Crocus sativus, Curcuma longa, Embelia ribes, Phyllanthus emblica, Syzygium cumini, Glycyrrhiza glabra, Hemidesmus indicus, Momordica charantia, Moringa oleifera, Ocimum sanctum, Phyllanthus fraternus, Plumbago zeylanica, Pterocarpus marsupium, Semecarpus anacardium, Terminalia arjuna, Terminalia chebula, Trigonella foenum-graecum, Withania somnifera, Zingiber officinale.*
Hypoglyceridaemic	*Aloe vera, Caesalpinia bonduc, Phyllanthus emblica, Momordica charantia.*
Immunomodulator	*Acorus calamus, Aloe vera, Azadirachta indica, Phyllanthus emblica, Holarrhena antidysenterica, Ocimum sanctum, Morinda citrifolia, Piper longum, Plumbago zeylanica, Tinospora cordifolia, Withania somnifera.*
Immunostimulant	*Acacia catechu, Alpinia galangal, Alstonia scholaris, Andrographis paniculata, Asparagus adscendens, Asparagus*

	racemosus, Azadirachta indica, Benincasa hispida, Berberis aristata, Cassia fistula, Curculigo orchioides, Curcuma longa, Inula racemosa, Glycyrrhiza glabra, Nigella sativa, Ocimum sanctum, Punica granatum, Terminalia bellirica, Trigonella foenum-graecum, Withania somnifera.
Immunosuppressant	*Boerhavia diffusa, Glycyrrhiza glabra, Hemidesmus indicus, Swertia chirayita, Terminalia chebula, Withania somnifera.*
Irritable bowel syndrome inhibitor	*Aegle marmelos, Boswellia serrata.*
Labour inducing	*Achyranthes aspera, Andrographis paniculata, Curculigo orchioides.*
Lactogogue	*Asparagus racemosus, Leptadenia reticulata, Nigella sativa.*
Leprosy	*Hydnocarpus kurzii.*
Lipoxygenase inhibitor	*Azadirachta indica, Boswellia serrata, Curcuma longa, Nigella sativa, Ocimum sanctum, Punica granatum, Semecarpus anacardium, Zingiber officinale.*
Male sexual dysfunction	*Tribulus terrestris.*
Memory enhancer	*Bacopa monnieri, Benincasa hispida, Celastrus paniculatus, Centella asiatica, Convolvulus microphyllus, Crocus sativus, Desmodium gangetum, Inula racemosa, Nardostachys jatamansi, Withania somnifera.*
Muscle contractant	*Cassia absus.*
Muscle relaxant	*Terminalia bellirica.*
Neuroprotective	*Acorus calamus, Benincasa hispida, Commiphora wightii, Crocus sativus, Curcuma longa, Psoralea corylifolia, Semecarpus anacardium, Swertia chirayita, Withania somnifera.*
Nitric oxide inhibitor	*Nigella sativa, Psoralea corylifolia, Semecarpus anacardium, Terminalia arjuna, Zingiber officinale, Zingiber zerumbet.*
Photoprotective	*Tinospora cordifolia.*
Radiation protective	*Aegle marmelos, Centella asiatica, Curcuma longa, Syzygium cumini, Glycyrrhiza glabra, Moringa oleifera, Ocimum sanctum, Tinospora cordifolia, Zingiber officinale.*
Radio sensitising	*Alstonia scholaris, Withania somnifera.*
Sedative	*Acorus calamus, Aegle marmelos, Centella asiatica, Crocus sativus, Curculigo orchioides, Nardostachys jatamansi, Withania somnifera.*
Spermicidal	*Azadirachta indica.*
Thermogenic	*Zingiber officinale.*
Thyroid modulating	*Moringa oleifera, Ocimum sanctum.*
Trypsin	*Trigonella foenum-graecum.*
Tuberculostatic inhibitor	*Azadirachta indica.*
Tyrosine kinase inhibitor	*Swertia chirayita.*
Urinary disorder	*Crataeva nurvala.*
Uterostimulant	*Adhatoda vasica, Crocus sativus, Rubia cordifolia.*
Vasodilator	*Centella asiatica, Piper longum, Pueraria tuberosa.*
Wound healing	*Aloe vera, Berberis aristata, Calotropis procera, Centella asiatica, Curcuma longa, Garcinia morella, Terminalia arjuna, Terminalia chebula.*

A thorough study of the above list indicates that the following plants have many medicinal values. Active molecules of these plants have been studied and documented by several laboratories world over. Genetic improvement of these would help their commercial cultivation to provide quality material for drug production.

Acacia catechu, *Acorus calamus*, *Aegle marmelos*, *Aloe vera*, *Asparagus racemosus*, *Azadirachta indica*, *Bacopa monnieri*, *Benincasa hispida*, *Caesalpinia bonduc*, *Cannabis sativa*, *Centella asiatica*, *Coccinia grandis*, *Commiphora wightii*, *Convolvulus microphyllus*, *Crocus sativus*, *Curcuma longa*, *Cyperus rotundus*, *Embelia ribes*, *Phyllanthus emblica*, *Garcinia morella*, *Glycyrrhiza glabra*, *Hemidesmus indicus*, *Hippophae rhamnoides*, *Inula racemosa*, *Lawsonia inermis*, *Litsea glutinosa*, *Momordica charantia*, *Morinda citrifolia*, *Moringa oleifera*, *Nigella sativa*, *Ocimum sanctum*, *Phyllanthus fraternus*, *Rauvolfia serpentina*, *Santalum album*, *Saussurea lappa*, *Swertia chirayita*, *Terminalia* sp., *Trigonella foenum-graecum and Withania somnifera.*

Hippophae rhamnoides and *Morinda citrifolia* have recently been scientifically studied for their pharmaceutical and nutraceutical properties. Several products of both are in demand in world market of nutraceuticals and pharmaceuticals.

8.5 Cultivation of Medicinal Plants

Cultivation of medicinal plants is different than that of the other agricultural crops. These plants are grown for secondary metabolites unlike the other crops and are expected to have zero quantity of pesticides or heavy metals. Therefore, World Health Organization has emphasised to develop Good Agricultural Practices (GAP) for medicinal plants to guarantee quality of raw drug. This would facilitate the standardisation of quality of resultant drugs for acceptance to the medical practitioners world over. In India, a large number of State Agricultural Universities, the Indian Council of Agricultural Research (ICAR) and the Council of Scientific and Industrial Research (CSIR) institutes are engaged in developing GAP for medicinal plants. However, the cultivation is not gaining due momentum in the country due to the prevailing unorganised marketing arrangements. Agro-technologies for cultivation of several medicinal plants have been developed by CSIR and ICAR institutions. Organic cultivation of medicinal plants is preferred.

8.6 Conservation and Sustainable Utilisation of Medicinal Plants

Medicinal plants support and improve livelihood and well-being. The conservation of plant diversity including medicinal plants has become an important aspect of sustainable development world over. In general, the conservation of plant diversity refers to the protection of wild species in addition to safeguarding the genetic diversity of cultivated and domesticated species and their relatives having food, nutrition and medicinal values. Conservation of flora is essential for improvement in cultivated crops to ensure food, feed, fodder and nutritional security to a nation in addition to providing primary health care. Plant diversity is the base for their commercial utilisation. The following measures are suggested for the conservation of medicinal plants:

(a) In situ conservation by the establishment of nature or biosphere reserves.
(b) Ex situ conservation through the establishment of herbal/botanical gardens in different parts of the country.
(c) Establishment of medicinal and aromatic gene bank in different parts of the country.
(d) Awareness on the importance of local medicinal plants and active role of public and private agencies including defence forces to preserve the local medicinal plant wealth.
(e) Propagation and cultivation study of important and useful medicinal plant genetic resources.
(f) Creation of ecological awareness among inhabitants of the area.

(g) Judicious and planned exploitation of medicinal plant resources.
(h) Regular training of stakeholders on cultivation, harvesting and sustainable utilisation of medicinal plant resources of the area.
(i) Grazing by migratory livestock in the pastures and alpine meadows should be watched and regulated and, if necessitated, be banned completely to save rare medicinal plants.
(j) Vulnerable or endangered plant species should be protected and multiplied in situ and ex situ.
(k) Promote herbal gardens in the country particularly in schools and colleges.

Multiplication through propagation, micropropagation and cultivation study is one of the important aspects for the conservation of medicinal floral diversity.

8.7 Phytomedicines

Shikimic acid from anise seeds, Taxol or paclitaxel from *Taxus brevifolia*, vinblastine and vincristine from *Catharanthus roseus*, topotecan from *Camptotheca acuminata*, etoposide and teniposide from *Podophyllum peltatum*, morphine from poppy, quinone from cinchona bark, salicin or aspirin from willow bark, atropine from belladonna, digoxin from foxglove, artemisinin from *Artemisia annua* and strychnine from *Strychnos nux-vomica* are some of the phytomedicines being used by medical practitioners world over.

The following table (Table 8.1) has the indicative chemicals present in important medicinal plants and their potential uses (Taylor 2000), which can provide research leads but cannot be taken for granted due to a number of factors.

8.8 Important Medicinal Plants Which Need to Be Thoroughly Researched upon for Medicinal Properties and Their Use Are as Follows

Turmeric (*Curcuma longa*), garlic (*Allium sativum*), bitter gourd (*Momordica charantia*), tulsi (*Ocimum sanctum*), methi (*Trigonella foenum-graecum*), aloe (*Aloe vera*), isabgol (*Plantago ovata*), periwinkle (*Catharanthus roseus*), senna (*Senna alexandrina*), brahmi (*Bacopa monnieri*), sea-buckthorn (*Hippophae rhamnoides*), noni (*Morinda citrifolia*), neem (*Azadirachta indica*), ashwagandha (*Withania somnifera*), mulethi (*Glycyrrhiza glabra*), amla (*Phyllanthus emblica*), bael (*Aegle marmelos*) and others mainly serving Indian communities for their primary health care.

Medicinal plants are used in preparation of medicines which are known as herbal medicines. Another way to use them is to know the active molecule(s) and synthesise them for use. Generally direct use of plants for preparation of medicine is preferred. This approach has following challenges:

1. Reproducibility of biological activity of herbal extracts
2. Suspected toxicity and adverse effects of herbal extracts
3. Adulteration and contamination
4. Standardisation of raw material
5. Possible herb-drug interaction
6. Postharvest processing of medicinal plants

Above aspects are being gradually attended by Ministry of Health and Family Welfare, Government of India.

8.9 Conclusion

The much-known importance of medicinal plants and phytomedicines has been re-emphasised in light of increased available evidences. An exhaustive list of plants indicating their different anti-ailment traits has been given for the benefit of students of medicines, researchers and all other stakeholders. Problems associated with preparation of medicine out of medicinal plants or their direct use in therapy have been listed to be taken care of by the promoters of medicinal plants in public and private sector in the country. The adequately documented treasure of traditional knowledge available on medicinal plants in the world is awaiting its exploitation for development of phytomedicines and nutraceuticals for

Table 8.1 Active chemicals/drug intermediates present in plants and their potential uses (Source: Taylor 2000; supplemented and confirmed by the author)

Active chemical	Potential use	Plant species
Acetyldigoxin	Cardiotonic	*Digitalis lanata*
Achyranthine	Diuretic	*Achyranthes aspera*
Adoniside	Cardiotonic	*Adonis vernalis*
Aegeline, marmelosin	Bowel diseases	*Aegle marmelos*
Aescin	Anti-inflammatory	*Aesculus hippocastanum*
Aesculetin	Antidysentery	*Fraxinus rhynchophylla*
Allicin	Atherosclerosis, antibacterial, antiviral	*Allium sativum*
Agrimophol	Anthelmintic	*Agrimonia eupatoria*
Ajmalicine	Circulatory disorders, antiarrhythmic agent	*Rauvolfia serpentina*, *Catharanthus roseus*
Allantoin	Vulnerary	Several plants
Allyl isothiocyanate	Rubefacient	*Brassica nigra*
Anabasine	Skeletal muscle relaxant	*Anabasis aphylla*
Andrographolide	Bacillary dysentery	*Andrographis paniculata*
Anisodamine	Anticholinergic/anticancer	*Anisodus tanguticus*
Anisodine	Anticholinergic/anticancer	*Anisodus tanguticus*
Anthocyanins	Antidiabetic	*Syzygium cumini*
Asparanin A, asparanin B, sarsasapogenin	Fertility enhancer	*Asparagus adscendens*
Arboreal	Tonic, stomachic	*Gmelina arborea*
Arecoline	Anthelmintic	*Areca catechu*
Ascorbic acid	Antioxidant activity	*Phyllanthus emblica*, *Hippophae rhamnoides*
Asiaticoside	Memory enhancer	*Centella asiatica*
Atropine	Anticholinergic	*Atropa acuminata*
Benzyl benzoate	Scabicide	Several plants
Berberine	Bacillary dysentery	*Berberis vulgaris*, *Berberis asiatica*
Bergenin	Antitussive	*Ardisia japonica*
Betulinic acid	Anticancerous	*Betula alba*
Borneol	Antipyretic, analgesic, anti-inflammatory	Several plants
Bromelain	Anti-inflammatory, proteolytic	*Ananas comosus*
Caffeine	CNS stimulant	*Camellia sinensis*
Camphor	Rubefacient	*Cinnamomum camphora*
Camptothecin	Anticancerous	*Camptotheca acuminate*
Cephaeline	Bronchitis, emetic	*Cephaelis ipecacuanha*
Capsaicin	Analgesic in topical ointments, nasal sprays (Sinol-M) and dermal patches to relieve pain	*Capsicum* spp.
Beta-caryophyllene	Anti-inflammatory, antidepressant	*Syzygium aromaticum*, *Cannabis sativa*
(+)-Catechin	Haemostatic	*Potentilla fragarioides*
Charantin	Diabetes	*Momordica charantia*
Chymopapain	Proteolytic, mucolytic	*Carica papaya*
Cissampeline	Skeletal muscle relaxant	*Cissampelos pareira*
Cocaine	Local anaesthetic	*Erythroxylum coca*

(continued)

Table 8.1 (continued)

Active chemical	Potential use	Plant species
Codeine	Analgesic, antitussive	*Papaver somniferum*
Colchicine amide	Antitumour agent	*Colchicum autumnale*
Colchicine	Antitumour agent, anti-gout	*Colchicum autumnale, Gloriosa superba*
Convallatoxin	Cardiotonic	*Convallaria majalis*
Cuminaldehyde	Promising agent against alpha-synuclein aggregation (nerve stimulator)	*Cuminum cyminum*
Curcumin	Choleretic	*Curcuma longa*
Cynarin	Choleretic	*Cynara scolymus*
Danthron	Laxative	*Cassia species*
Demecolcine	Antitumour agent	*Colchicum autumnale*
Deserpidine	Antihypertensive, tranquilliser	*Rauvolfia canescens*
Deslanoside	Cardiotonic	*Digitalis lanata*
L-Dopa	Antiparkinsonism	*Mucuna sp*
Digitalin	Cardiotonic	*Digitalis purpurea*
Digitoxin	Cardiotonic	*Digitalis purpurea*
Digoxin	Cardiotonic	*Digitalis purpurea*
Diosgenin	Oestrogen replacement therapy, vaginal dryness, PMS (premenstrual syndrome), menstrual cramps, weak bones (osteoporosis)	*Dioscorea deltoidea, Trigonella foenum-graecum*
Dihydrocapsaicin	Analgesic in topical ointments, nasal sprays (Sinol-M) and dermal patches to relieve pain	*Capsicum* spp.
Emetine	Amoebicide, emetic	*Cephaelis ipecacuanha*
Ephedrine/pseudoephedrine	Sympathomimetic, antihistamine	*Ephedra sinica, E. gerardiana*
Etoposide	Antitumour agent	*Podophyllum peltatum*
Galantamine	Cholinesterase inhibitor	*Lycoris squamigera*
Gitalin	Cardiotonic	*Digitalis purpurea*
Glaucarubin	Amoebicide	*Simarouba glauca*
Glaucine	Antitussive	*Glaucium flavum*
Glaziovine	Antidepressant	*Ocotea glaziovii*
Glycyrrhizin	Sweetener, Addison's disease	*Glycyrrhiza glabra*
Gossypol	Male contraceptive	*Gossypium species*
Hemsleyadin	Bacillary dysentery	*Hemsleya amabilis*
Hesperidin	Capillary fragility	*Citrus species*
Hydrastine	Hemostatic, astringent	*Hydrastis canadensis*
Hyoscine	Cramps, spasms, renal or biliary colic	*Datura metel*
Hyoscyamine	Anticholinergic	*Atropa acuminate, Datura stramonium, Hyoscyamus niger*
Hydroxycitric acid	Antiobesity	*Garcinia cambogia*
Irinotecan	Anticancer, antitumour agent	*Camptotheca acuminata*
Kaempferol (3,5,7,4′-tetrahydroxyflavone)	Antiradical activity, anticancer	*Ageratum conyzoides*
Kainic acid	Ascaricide	*Digenea simplex*
Kawain	Tranquilliser	*Piper methysticum*

(continued)

Table 8.1 (continued)

Active chemical	Potential use	Plant species
Khellin	Bronchodilator	*Ammi visnaga*
Lanatosides A, B, C	Cardiotonic	*Digitalis lanata*
Lapachol	Anticancer, antitumour	*Tabebuia* sp.
Liquiritigenin, isoliquiritigenin	Antidiabetic	*Pterocarpus marsupium*
a-Lobeline	Smoking deterrent, respiratory stimulant	*Lobelia inflate*
Michellamine B	anti-AIDS	*Ancistrocladus korupensis* (found in Cameroon), *Ancistrocladus heyneanus*
Menthol	Rubefacient	*Mentha species*
Methyl salicylate	Rubefacient	*Gaultheria procumbens*
Monocrotaline	Antitumour agent (topical)	*Crotalaria sessiliflora*
Morphine	Analgesic	*Papaver somniferum*
Neoandrographolide	Dysentery	*Andrographis paniculata*
Nicotine	Insecticide	*Nicotiana tabacum*
Nordihydroguaiaretic acid	Antioxidant	*Larrea divaricata*
Noscapine	Antitussive	*Papaver somniferum*
Ouabain	Cardiotonic	*Strophanthus gratus*
Pachycarpine	Oxytocic	*Sophora pachycarpa*
Palmatine	Antipyretic, detoxicant	*Coptis japonica*
Papain	Proteolytic, mucolytic	*Carica papaya*
Papaverine	Smooth muscle relaxant	*Papaver somniferum*
Phyllodulcin	Sweetener	*Hydrangea macrophylla*
Physostigmine	Cholinesterase inhibitor	*Physostigma venenosum*
Picrotoxin	Analeptic	*Anamirta cocculus*
Pilocarpine	Parasympathomimetic	*Pilocarpus jaborandi*
Pinitol	Expectorant	*Several plants*
α-Pinene	Antibacterial	*Juniperus communis*
Podophyllotoxin	Antitumour anticancer agent	*Podophyllum peltatum, P. emodi*
Protoveratrines A, B	Antihypertensives	*Veratrum album*
Protodioscin	Diuretic, anabolic, aphrodisiac	*Tribulus terrestris*
Pseudoephedrine	Sympathomimetic	*Ephedra sinica*
Norpseudoephedrine	Sympathomimetic	*Ephedra sinica*
Psoralen	Vitiligo	*Psoralea corylifolia*
Quinidine	Antiarrhythmic	*Cinchona* spp.
Quinine	Antimalarial, antipyretic	*Cinchona* spp.
Quisqualic acid	Anthelmintic	*Quisqualis indica*
Rescinnamine	Antihypertensive, tranquilliser	*Rauvolfia serpentina*
Reserpine	Antihypertensive, tranquilliser, mental disorder, schizophrenia	*Rauvolfia serpentina*
Rhomitoxin	Antihypertensive, tranquilliser	*Rhododendron molle*
Rorifone	Antitussive	*Rorippa indica*
Rotenone	Pesticide, Insecticide	*Lonchocarpus nicou*
Rotundine	Analgesic, sedative, tranquilliser	*Stephania sinica*
Rutin	Capillary fragility	*Citrus species*
Salicin	Analgesic	*Salix alba*

(continued)

Table 8.1 (continued)

Active chemical	Potential use	Plant species
Sanguinarine	Dental plaque inhibitor	*Sanguinaria canadensis*
Santonin	Ascaricide	*Artemisia maritima*
Scillaren A	Cardiotonic	*Urginea maritima*
Scopolamine	Sedative	*Datura species*
Sennosides A, B	Laxative	*Cassia species*
Silymarin	Antihepatotoxic	*Silybum marianum*
Sparteine	Oxytocic	*Cytisus scoparius*
Stevioside	Sweetener	*Stevia rebaudiana*
Strychnine/brucine	CNS stimulant	*Strychnos nux-vomica*
Taxol	Antitumour agent	*Taxus wallichiana*
Teniposide	Antitumour agent	*Podophyllum peltatum*
Tetrahydrocannabinol (THC)	Antiemetic, decrease ocular tension	*Cannabis sativa*
Tetrahydropalmatine	Analgesic, sedative, tranquilliser	*Corydalis ambigua*
Tetrandrine	Antihypertensive	*Stephania tetrandra*
Theobromine	Diuretic, vasodilator	*Theobroma cacao*
Theophylline	Diuretic, bronchodilator	*Theobroma cacao and others*
Thymol	Antifungal (topical)	*Thymus vulgaris*
Tinosporic acid, cordifolioside	Immunomodulatory	*Tinospora cordifolia*
Topotecan	Antitumour, anticancer agent	*Camptotheca acuminata*
Trichosanthin	Abortifacient	*Trichosanthes kirilowii*
Trigonelline	Antidiabetic, lipid lowering	*Trigonella foenum-graecum*
Tubocurarine	Skeletal muscle relaxant	*Chondrodendron tomentosum*
Tylophorine	Bronchodilator	*Tylophora indica*
Valepotriates	Sedative, antisleep	*Valeriana jatamansi, V. officinalis*
Vasicine	Cerebral stimulant, increase brain metabolism, increase memory	*Vinca minor*
Vasicine	Bronchodilator	*Adhatoda vasica, A. beddomei*
Vincamine and vindoline	antiulcer property	*Catharanthus roseus*
Vinblastine	Antitumour, antileukemic agent	*Catharanthus roseus*
Vincristine	Antitumour, antileukemic agent	*Catharanthus roseus*
Withaferin A	Anti-inflammatory activity, antistress activity (adaptogenic activity)	*Withania somnifera*
Xanthotoxine	Vitiligo and psoriasis	*Ammi majus*
Xanthoxin	Phytohormone	*Heracleum candicans*
Yohimbine	Aphrodisiac	*Pausinystalia yohimbe*
Yuanhuacine	Abortifacient	*Daphne genkwa*
Yuanhuadine	Abortifacient	*Daphne genkwa*

the wellness of the world. Plants with wellness traits are so many which need detailed research on the pattern of *Hippophae rhamnoides* and *Morinda citrifolia*. The majority of the population in the world would rightly continue to depend on herbal remedies for primary health care as there is continuous improvement on the authenticity and availability of phytomedicines because of due emphasis on pharmacognosy. Phytomedicines would continue to serve the humanity to combat unknown diseases and disorders.

References

Ahmed I, Karukh A, Mohmad O (2006) Modern phytomedicine: turning medicinal plants into drugs. Wiley-VCH Verlog GmbH &Co. KGaA, Weinheim

Ahmedullah M, Nayar MP (1999) Red data book of Indian plants, vol 4 (Peninsular India). Botanical Survey of India, Calcutta

Bent S, Ko R (2004) Commonly used herbal medicines in the United States: a review. Am J Med 116:478–485

Chakravarty BK (1993) Herbal medicines. Safety and efficacy guidelines. Regul Aff J 4:699–701

Chatterjee A, Pakrashi SC (eds) (1995–1997) The treatise on Indian Medicinal Plants, Vol 1 to 5. Publications and Information Directorate, New Delhi

Dev S (2006) A selection of prime ayurvedic drugs-ancient- modern concordance. Anamaya Publishers, New Delhi

Dubey NK, Kumar R, Tripathi P (2004) Global promotion of herbal medicine: India's opportunity. Curr Med 86:37–41

Gogtay NJ, Bhatt HA, Dalvi SS, Kshirsagar NA (2002) The use and safety of non-allopathic Indian medicines. Drug Saf 25:1005–1019

Kamboj VP (2000) Herbal medicine. Curr Med 78:35–39

Khare CP (2012) Dictionary of synonyms: Indian medicinal plants. I K International Publishing House, New Delhi, 700p

Pal SK, Shukla Y (2003) Herbal medicine: current status and the future. Asian Pac J Cancer Prev 4:281–288

Seth SD, Sharma B (2004) Medicinal plants in India. Indian J Med Res 120:9–11

Singh B, Chaurasia OP (2000) Medicinal flora of Indian cold deserts. Acta Hortic 523(13):65–72

Taylor L (2000) Plant based drugs and medicines. http://www.rain-tree.com/plantdrugs.htm#.VY6lblIYHKR. Accessed 19 Aug 2014

Prospects for Development of Biomedicines from the Medicinal Plants of Northeastern India

9

Nongalleima Khumukcham, Dipak Biswas, Nameirakpam Surjit Singh, and Lokesh Deb

Abstract

Northeastern India has a great range of ecological habitats due to its edaphic, climatic and altitudinal variations, which made it the geographical "gateway" for much of India's flora and fauna. Therefore, the region, being one of the richest areas of India in terms of biological values, has been in the spotlight for its high biodiversity and traditional knowledge and has been a priority for leading conservation agencies of the world. Ethnomedicobotany is one of the tools that help to deal with the direct relationship of plants and man to prevent and cure ailments. The indigenous medicinal plants grown in Northeastern India are useful folk medicines for the people of this region. In the present chapter, 320 medicinal plants have been listed which includes their traditional use, local name, region and parts used. These cover important medicinal plants used for traditional healthcare practices in the region and are also components of the available commercial herbal products. The World Health Organization (WHO) Collaborating Centres for Traditional Medicine reported that a total of 122 compounds identified in their survey were derived from only 94 plant species and 80 % of those plants were used for the same or related ethnomedical purposes. This provides an array of information about the rich indigenous knowledge on traditional medicine and the medicinal potential of the varied plants used by the local healers. This would prove to be an important resource for discovery of many bioactive principles. A comprehensive approach using systems biology could be the suitable way

N. Khumukcham • D. Biswas
N.S. Singh • L. Deb (✉)
Pharmacology Laboratory, Institute of Bioresources and Sustainable Development, A National Institute of Department of Biotechnology, Government of India, Takyelpat Institutional Area, Imphal 795 001, Manipur, India
e-mail: lokeshdeb.ibsd@nic.in

© Springer India 2016

Vijay Veer, R. Gopalakrishnan (eds.), *Herbal Insecticides, Repellents and Biomedicines: Effectiveness and Commercialization*, DOI 10.1007/978-81-322-2704-5_9

to prove the efficacy and to obtain information that might lead to the understanding of the mode of action.

9.1 Introduction

Northeastern states of India, flanked in the north by the Himalayas and in the south by the Bay of Bengal, constitute a characteristic narrow passageway that connects the Indian subcontinent to East Asia and Southeast Asia (Chakraborty et al. 2012). The immense variety of the climatic, edaphic and altitudinal variations in Northeastern India have resulted in a great range of ecological habitats. Northeastern India characterizes the transition zone between the Indian, Indo-Malayan and Indo-Chinese biogeographic regions and a meeting place of the Himalayan Mountains and Peninsular India, and it was the part of the northward-migrating "Deccan Peninsula" (Bhutani 2008). Northeastern India is therefore the geographical "gateway" for much of India's flora and fauna, and as a consequence, the region is onc of the richest areas of India in biological values (Shankar 2006). The Northeastern region has been in the spotlight for its high biodiversity and traditional knowledge, and this region has been a priority for leading conservation agencies of the world. The region is affluent in medicinal plants and many other rare and endangered taxa. High endemism in higher plants, vertebrates and avian diversity in this region has qualified it to be a biodiversity "hotspot" (Chatterjee et al. 2006).

Arunachal Pradesh extends between latitude 26°30′N–29°30′N and longitude 91°30′E–97°30′E, with annual rain fall of 1500–3750 mm and temperature 0–31 °C. Assam extends between latitude 24°–28°N and longitude 90°–96°E with annual rain fall of 2000–8000 mm and temperature ranging from 5 to 32 °C (Chakraborty et al. 2012). Meghalaya is extended between latitude 25°1′–26°5′N and longitude 85°45′–92°52′E (Bhakta 1996), the altitude range from 500 to 2089 m, the temperatures ranges from 2 °C in winter to 21 °C in summer. Mizoram is extended between latitude 21°58′–24°35′N and longitude 92°15′–93°29′E. The total area of Mizoram is 21,087 km^2, the total forest area being 15,825 km^2. The altitude ranges from 500 to 2157 m. Temperature varies from 18 to 29 °C in summer and from 11 to 24 °C in winter (Sharma et al. 2001). Manipur is extended between latitude 23.83°N–25.68°N and longitude 93.03°E to 94.78°E with annual rain fall of 1250–2700 mm and temperature: 14.5 to 38 °C (Chakraborty et al. 2012). Nagaland (Chakraborty 1979), a state of North East India extended between latitude 25°69′–27°49′N and longitude 93°20′–95°15′E, has a total area of 16,579 km^2 with average rainfall of 250 cm.

Sikkim is a very small hilly state in the Eastern Himalayas with a total geographical area of 7096 sq. km. Sikkim, covering just 0.2 % of the geographical area of the country, harbours more 26 % flowering plants has tremendous biodiversity and has been identified as the hot spot in the Eastern Himalayas. Sikkim, situated in the eastern Himalayas patterned, is the 22 state of India occupies a total area of 7096 sq km and lies between 27°5′ and 28°10′N latitude and 88°4′ and 88°58′E longitude (Das et al. 2012). Tripura is a small hilly state of North-Eastern India, surrounded by Bangladesh on three sides with rich biodiversity hot spot with huge variety of flora and fauna. The total area of the state is 10,497,697 sq km and located in the bio-geographic zone of 9B-North East Hills between 220–56′ and 240–32′N latitude and between 900–09′ and 920–20′E longitudes (Das et al. 2014).

Northeastern (NE) states of India have different communities belonging to different tribes with their own unique culture and traditions. These tribes have rich indigenous traditional knowledge on herbal medicines. Sajem and Gosai (2006) reported that the Northeastern states of India harbour more than 130 major tribal communities of the total 427 tribal communities found in India (2001 census). In general, the tribes of Northeastern India have been categorized into two broad ethnic communities, Khasi and the Jaintia tribes of Meghalaya, who belong to the "Monkhemar" culture of Austric dialect, and the rest of the tribal groups are basically Mongoloid, who belong to the Tibeto-Burman subfamily of the Tibeto-Chinese group (Mukherjee 2005; Dutta and Dutta 2005; Ramakrishnan 1992). NE India, due to its tropical climatic condition, supports a diverse array of flora and fauna within its territory. Different rare flora and fauna are available in these regions of India, which are being used as biomedicine in traditional healthcare practices. These traditional healthcare practices arise due to necessity, observations and long practices, which pass orally as family secrets/traditions and from generations to generations. There are enormous numbers of medicinal plants being used by different ethnic tribes from these NE states for treatment and cure of mild to severe diseases. There are reports on a survey of medicinal plants used for treatment of aphthae, arthritis, boils and bruises, cancer, cough, cholera, dysentery/dysentery with bloody stool, diarrhoea, epilepsy, fever, heart disease, hypo-/hypertension, hysteria, pyorrhoea, snake bite and skin diseases; for gallbladder stone removal; to ease labour; to prevent miscarriage, kidney problems, nasal problems and ulcer; as abortifacient, antifertility and antihelminthic; etc. The biomedicine of a place is unique and endemic to the place. We have observed many documentation works from different states, which use different plants and animals separately as well as combinedly as a formulation to treat many ailments. In the Rig Veda there are documentations of about 99, the Yajur Veda 82 and the Atharva Veda 28 medicinal plants to cure different ailments (Bhattacharjya and Borah 2008). The *Sushruta Samhita* attributed to Sushruta in the sixth century BC describes 700 medicinal plants, 64 preparations from mineral sources and 57 preparations based on animal sources Dwived and Dwived 2007). Among the 120 active compounds currently isolated from the higher plants and widely used in modern medicine today, 80 % show a positive correlation between their modern therapeutic use and the traditional use of the plants from which they are derived (Fabricant and Farnsworth 2001). At least 7000 medical compounds in the modern pharmacopoeia are derived from plants (Interactive European Network for Industrial Crops and their Applications, 2000–2005).

9.2 The Medicinal Plants Used in Traditional Healthcare Practices in Northeastern India

A wide variety of plants have been used in Northeastern India for the treatment of mild to complex types of health problems or ailments since time immemorial. NE India, being a landlocked region, is underdeveloped than the rest of the country and thus in the medicine and healthcare system. The inhabitants of deeper nooks of the region, having dense forest cover, are surviving without medical doctor or modern medical facilities till date. Since they cannot afford the modern medical facilities, they are depending on plants as the alternative sources of medicine. Medicinal plants of the region are easily available and are not costly to procure as compared to the pharmaceuticals and modern synthetic drugs. Many pharmaceutical companies are using medicinal plants and their bioactive principles as components for manufacture of new drugs. The medicinal plants used in the NE states along with their scientific names, local names, parts of the plant being used for the treatment and kind of diseases treated are listed in Table 9.1, and the photographs of a few plants are provided in Fig. 9.1. The rich ethnic cultural diversity and biological diversity of the region can be a platform for advanced ethnopharmacological studies. The use of plants as medicine is accomplished due to the presence of secondary metabolites which are bioactive principles mediating their wide range of effects. These metabolites or bioactive compounds are working as effectively as conventional drugs. However, looking to the present scenario of the research, many herbal companies are exploiting the medicinal plants for new formulations, and pharmaceutical industries are targeting on those metabolites for the discovery of new drugs. Varied medicinal plants are grow-

Table 9.1 List of the medicinal plants used in traditional healthcare practices in Northeastern India

Sl. no.	Name of medicinal plants/ scientific name	Family	Local name	Parts used	Bioactivity	State	References
1.	*Andrographis paniculata*	Acanthaceae	Maha-tita, wild	Whole plant	Antimalarial activity	**Assam**	Namsaa et al. (2011)
2.	*Artemisia vulgaris* L.	Asteraceae	Chirota, cultivated	Stem and flowering tip	Antimalarial and repellent activity		
3.	*Azadirachta indica* (A. Juss.) L.	Meliaceae	Neem	Bark and leaf	Anti-plasmodial and repellent activity		
4.	*Alstonia scholaris* (L.)	Apocynaceae	Chotiana	Wild root and bark	Anti-plasmodial activity		
5.	*Adhatoda vasica* Nees.	Acanthaceae	Titabahak	Leaf and roots	Insecticidal activity		
6.	*Aegle marmelos* (L.) Correa ex. Roxb.	Rutaceae	Bel	Root and bark	Larvicidal activity	**Assam**	Namsaa et al. (2011)
7.	*Annona squamosa* L.	Annonaceae	Sita phol	Root and leaves	Larvicidal activity		
8.	*Aristolochia indica* L.	Aristolochiaceae	Iswarmool	Root	Larvicidal activity		
9.	*Caesalpinia volkensii* Harms.	Caesalpiniaceae	Swarnakanti	Leaf	Anti-plasmodial activity		
10.	*Coptis teeta* Wall.	Ranunculaceae	Mishmitita	Seeds, roots, rhizomes	Antimalarial activity		
11.	*Cissampelos pareira* L.	Menispermaceae	Goria lota	Root	Anti-plasmodial activity		
12.	*Cymbopogon citratus*	Gramineae	Gandhabiringa	Leaf	Larvicidal and repellent activity		
13.	*Clerodendrum infortunatum* (L.) Gaertn.	Verbenaceae	Dhopat-tita	Leaf	Antimalarial activity		

14.	*Gymnopetalum cochinchinensis* Kurz.	Cucurbitaceae	Kauri korola	Root	Larvicidal activity	**Assam**	Namsaa et al. (2011)
15.	*Lantana camara* L.	Verbenaceae	Goo-phool	Bark	Larvicidal and repellent activity		
16.	*Ocimum sanctum* L.	Lamiaceae	Tulasi	Leaf	Larvicidal and repellent activity		
17.	*Stephania hernandifolia* (Willd.) Walp.	Menispermaceae	Tubuki lota	Roots	Larvicidal activity		
18.	*Vitex peduncularis* Wall.	Verbenaceae	Osai, ahoi	Leaf, bark	Larvicidal activity		
19.	*Withania somnifera* (L.) Dunal	Solanaceae	Ashwagandha	Whole plant	Anti-plasmodial and antimalarial activity		
20.	*Zanthoxylum hamiltonianum* Wall.	Rutaceae	Tezmooi	Roots	Anti-plasmodial activity		
21.	*Abelmoschus esculentus* (L.)	Thelypteridaceae	Bilongly	Leaf	Anti-poison	**Arunachal Pradesh**	Tangjanga et al. (2011)
22.	*Acacia caesia* (L.) Willd.	Bromeliaceae	Hahpandong	Premature leaf	Cough		
23.	*Annona squamosa* L.	Oxalidaceae	Koapdoh	Fruit	Jaundice		
24.	*Ardisia pedunculosa* Wall.	Bignoniaceae	Kosy	Leaf and stem	Stomach ache and bee bite		
25.	*Averrhoa carambola* L.	Caricaceae	Kaadong	Fruit	Gastric, high pressure		
26.	*Azadirachta indica* Juss.	Meliaceae	Neem	Leaf	Antidiabetic efficacy		
27.	*Bauhinia purpurea* L.	Caesalpiniaceae	Sekang	Bark and seed	Anti-inflammatory activity of bioactive compounds: bauhinoxepin F and bauhinoxepin I		
28.	*Bidens pilosa* L. var. minor (Bl.) Scherff.	Asteraceae	Mutkein	Leaf	Antidiabetic activity and polyacetylenes		
29.	*Calamus tenuis* Roxb.	Rutaceae	Makri	Fruit	Loose motion		
30.	*Callicarpa arborea* Roxb.	Theaceae	khalap	Leaf	Wound washing and stomach ache		
31.	*Citrus aurantifolia* (Christm) Swingle	Caryophyllaceae	Pipi	Leaf	Boils		

(continued)

Table 9.1 (continued)

Sl. no.	Name of medicinal plants/ scientific name	Family	Local name	Parts used	Bioactivity	State	References
32.	*Clerodendrum colebrookianum* Walp.	Verbenaceae	Kuthaap (Manipuri), Ongiin (Arunachal Pradesh)	Leaf	Hypolipidemic effect in high-fat diet-fed rats		
33.	*Clerodendrum viscosum* Vent.	Verbenaceae	Dupatteeta	Leaf	Antimicrobial activity		
34.	*Coptis teeta* Wall.	Asteraceae	Wo pohing	Leaf	Blood coagulant		
35.	*Cudrania cochinchinensis* (Lour.) Kudo & Masam.	Malvaceae	Seloam	Leaf	Blood coagulant		
36.	*Curcuma caesia* Roxb.	Zingiberaceae	Humen	Rhizome	Anticancer, antioxidant		
37.	*Ficus hispida* L.f.	Moraceae	Mukongpong	Leaf	Hypoglycemic activity		
38.	*Ipomoea aquatica* Forsskal	Convolvulaceae	Humen	Leaf and young shoots	(1) Oral hypoglycaemic activity (2) Used against diabetic patient		
39.	*Lactuca gracilis* DC.	Crassulaceae	Rangkha pansi	Leaf	Bone fracture		
40.	*Millingtonia hortensis* L.f.	Mimosaceae	Reseh	Root	Tooth ache		
41.	*Momordica dioica* Roxb. ex. Willd.	Cucurbitaceae	Bhat karela	Fruit	Hypoglycemic and hypolipidemic activities of fruit pulp extracts on alloxan-induced diabetic rat		
42.	*Mikania scandens* (L.) Willd.	Asteraceae	Hipini	Leaf	Blood coagulant		
43.	*Moringa oleifera* (L.) Lam.	Moringaceae	Sajona	Pod	Hypoglycaemic activity in alloxan-diabetic rats		
44.	*Murraya koenigii* (L.) Spreng.	Rutaceae	Kari	Leaf	Hypoglycemic activity of fruit juice in diabetic mice model		
45.	*Oxalis griffithii* Edgew. et Hook.f.	Papaveraceae	Kanni	Capsule	Body ache		
46.	*Paederia scandens* (Lour.) Merr.	Acanthaceae	Kelong	Leaf	Stomach ache, scabies		

47.	*Panax pseudoginseng* Wall.	Araliaceae	Ginseng	Tuber	(1) Used for the treatment of diabetes (2) Trilinolein is a potent natural plant antioxidant		
48.	*Pongamia pinnata* (L.) Pierre	Papilionaceae	Yakopi	Seed	Antidiabetic activity of cycloart-23-ene-3, 25-diol in streptozotocin-nicotinamide-induced diabetic mice		
49.	*Saccharum spontaneum* L.	Poaceae	Thoamuh	Stem	Jaundice		
50.	*Scoparia dulcis* L.	Scrophulariaceae	Meeta boon	Leaf	Antidiabetic activity of scoparic acid D in streptozotocin-induced diabetic rats		
51.	*Solanum viarum* Dunal	Solanaceae	Soekhah	Fruit	Anti-germs		
52.	*Sonchus arvensis* L.	Asteraceae	Soeloe	Inflorescence	Tooth ache		
53.	*Spondias pinnata* (L.f.) Kurz.	Anacardiaceae	Pakang/amra	Fruit pulp	Hypoglycemic activity		
54.	*Stephania glandulifera* Miers.	Scrophulariaceae	Kobelang	Leaf	Jaundice and malaria		
55.	*Tamarindus indica* L.	Caesalpiniaceae	Teitli	Pod/seed	It reduces blood sugar level in streptozotocin-induced diabetic rats		
56.	*Terminalia arjuna* Beddome	Combretaceae	Arjun gash	Stem, bark	Antidiabetic effect in alloxan-induced diabetic rats		
57.	*Terminalia chebula* Retz.	Combretaceae	Hellica	Fruit	It reduces significantly the blood sugar level in alloxan-diabetic rats		
58.	*Tinospora cordifolia* (Willd.) Hook.f.	Menispermaceae	Kakyungha or sagunilota	Leaf	Antidiabetic activity in streptozotocin-induced diabetic rats		
59.	*Urena lobata* L.	Urticaceae	Oyik	Inner bark	Rheumatic arthritis		
60.	*Vitis barbata* Wall.	Vitaceae	Songramurich	Stem bark	Antidiabetic activity of *Vitis vinifera* in diabetic rats		

(continued)

Table 9.1 (continued)

Sl. no.	Name of medicinal plants/ scientific name	Family	Local name	Parts used	Bioactivity	State	References
61.	*Zanthoxylum acanthopodium* Wall. ex. Hook.f.	Poaceae	Sapa	Corn	Gall bladder problem		
62.	*Zanthoxylum nitidum* (Roxb.) DC.	Rutaceae	Ongear	Stem	Gastritis and diabetes		
63.	*Canarium strictum* Roxb.	Burseraceae	Gamsen	Bark	Skin diseases	**Meghalaya**	Doley et al. (2010)
64.	*Glochidion velutinum* Wight.	Phyllanthaceae	Byake	Fruits	Diarrhoea, digestion of food		
65.	*Zanthoxylum armatum* DC.	Rutaceae	Singlu	Fruits and bark	Cold, cough and relief from sore		
66.	*Stereospermum colais* DC.	Bignoniaceae	Damium	Bark	Jaundice, piles, relieve pain in sprain and stomach pain		
67.	*Stephania glandulifera* Miers.	Menispermaceae	Rabaka	Tuber and stem	Delivery, treatment of most of the stomach problems and quick fresh wound healing		
68.	*Schefflera venulosa* Harms.	Araliaceae	Paleh	Bark	Skin diseases		
69.	*Pothos scandens* L.	Araceae	Ridik	Whole plant	Bone fracture		
70.	*Portulaca oleracea* L.	Portulacaceae	Pashi-pali-echi	Whole plant	Digestion of food		
71.	*Pedilanthus tithymaloides* Poir	Euphorbiaceae	Kolokkali	Latex	Piles		
72.	*Michelia oblonga* Wall. ex. H.kf. & Thoms.	Magnoliaceae	Phulsopa	Fruits	Digestion and relief from the tonsillitis		
73.	*Meyna spinosa* L.	Rubiaceae	Kampur	Fruits and bark	Headache and hair washing		
74.	*Liparis nervosa* (Thun.) Lindl.	Orchidaceae	Honyum	Whole plant	Antimalarial drugs		
75.	*Gynocardia odorata* R. Br.	Flacourtiaceae	Tak	Fruit and bark	Diarrhoea, gastric, blood purification		

76.	*Musa paradisiaca* L.	Musaceae	Somomo	Whole plants	Treatment of severe diarrhoea or cholera	**Nagaland**	Jamir et al. (1999)
77.	*Mussaenda* spp.	Rubiaceae	Nokderang	Fresh root	Treatment of liver disorders and indigestion		
78.	*Paederia foetida* L.	Rubiaceae	Siizii	Root, leaves and shoot	Treatment of roundworm, abdominal colic and gastric ulcer		
79.	*Passiflora edulis* Sims.	Passifloraceae	Antsulashi	Fresh leaves and fruits	Acute dysentery and also in hypertension, constipation		
80.	*Piper betle* L.	Piperaceae	Patu	Fresh leaf	Acute abdominal spasm and abdominal colic		
81.	*Adhatoda vasica* Nees.	Acanthaceae	Sangtam tu	Leaves	Joint pain, lumber pain and sprains		
82.	*Aegle marmelos* Corr.	Rutaceae	Charamjang	Unripe dried fruit, leaves	Cough and mucous secretion		
83.	*Allium ascalonicum* L.	Liliaceae	Rupchi	Leaves	Wound of injured, antihelmintic		
84.	*Achyranthes aspera* L.	Amaranthaceae	Uthlengra	Whole plant	Contraceptive	**Tripura**	Das et al. (2014)
85.	*Adiantum philippense* L.	Adiantaceae	Khokochor	Fresh leaves	Abortifacient		
86.	*Allium cepa* L.	Alliaceae	Resunchao	Bulb	Contraceptive		
87.	*Allium sativum* L.	Alliaceae	Resun	Bulb	Contraceptive		
88.	*Aloe barbadensis* Mill.	Aloaceae	Ghretokumari, mussobbar	Whole plant	Contraceptive		
89.	*Ananas comosus* (L.) Merr.	Bromeliaceae	Birtung	Fresh leaves	Abortifacient	**Tripura**	Das et al. (2014)
90.	*Annona reticulata* L.	Annonaceae	Ata	Unripe fruit	Abortifacient		
91.	*Azadirachta indica* A. Juss.	Meliaceae	Neem	Stem bark, seed	Abortifacient		
92.	*Bombax ceiba* L.	Bombacaceae	Gochu	Fresh seeds	Abortifacient		
93.	*Butea monosperma* (Lam.) Taub.	Fabaceae	Polash	Dried seed	Abortifacient		
94.	*Caesalpinia pulcherrima* (L.) Sw.	Caesalpiniaceae	Radhachura	Seed	Abortifacient		

(continued)

Table 9.1 (continued)

Sl. no.	Name of medicinal plants/ scientific name	Family	Local name	Parts used	Bioactivity	State	References
95.	*Calotropis procera* (Aiton) W.T. Aiton	Asclepiadaceae	Akanda Angrapata	Root	Contraceptive	**Tripura**	Das et al. (2014)
96.	*Carica papaya* L.	Caricaceae	Hogeyegulo	Seed	Contraceptive		
97.	*Cassia alata* L.	Caesalpiniaceae	Thechou	Leaves, roots	Abortifacient Contraceptive		
98.	*Costus speciosus* Sm.	Costaceae	Khetoki	Seed	Abortifacient		
99.	*Cuscuta reflexa* Roxb.	Convolvulaceae	Chinailat	Whole plant	Contraceptive		
100.	*Cynodon dactylon* (L.) Pers.	Poaceae	Durpa	Whole plant	Abortifacient Contraceptive		
101.	*Datura metel* L.	Solanaceae	Dutra	Fresh root	Abortifacient		
102.	*Dioscorea bulbifera* L.	Dioscoreaceae	Khudupan	Whole plant	Contraceptive		
103.	*Drynaria quercifolia* (L.) J. Smith	Polypodiaceae	Banartola	Rhizome	Abortifacient	**Tripura**	Das et al. (2014)
104.	*Ferula assafoetida* L.	Apiaceae	Hing	Latex	Contraceptive		
105.	*Ficus religiosa* L.	Moraceae	Neao	Leaves	Contraceptive		
106.	*Gossypium herbaceum* L.	Malvaceae	Khol	Fresh root	Contraceptive		
107.	*Hibiscus rosa-sinensis* L.	Malvaceae	Nipui par, joba	Flower, stem, bark	Abortifacient		
108.	*Holarrhena antidysenterica* (L.) Wall.	Apocynaceae	Kuruchi	Seed	Abortifacient		
109.	*Lawsonia inermis* L.	Lythraceae	Mehendi	Leaves	Abortifacient		
110.	*Leucas aspera* Spreng.	Lamiaceae	Donkalas	Fresh stem and root	Abortifacient		
111.	*Lygodium flexuosum* (L.) Sw.	Schizaeaceae	Sakbangma, suilen	Leaves	Abortifacient Contraceptive		
112.	*Mimosa pudica* L.	Mimosaceae	Rajuriher	Whole plant	Abortifacient		
113.	*Momordica charantia* L.	Cucurbitaceae	Kangla	Seed	Contraceptive Abortifacient		
114.	*Moringa oleifera* Lam.	Moringaceae	Sajna	Root	Contraceptive		
115.	*Musa balbisiana* Colla	Musaceae	Mot munei	Seed	Contraceptive		

116.	*Oroxylum indicum* Vent.	Bignoniaceae	Bakilong	Bark	Contraceptive	**Tripura**	Das et al. (2014)
117.	*Phlogacanthus thyrsiflorus* (Roxb. ex. Hardw.) Mabb.	Acanthaceae	Ravanbasak	Leaves	Contraceptive		
118.	*Phyllanthus emblica* L.	Euphorbiaceae	Amlaki	Fruit	Contraceptive		
119.	*Piper betle* L.	Piperaceae	Khasiapatho, pangua	Stem	Abortifacient		
120.	*Plumbago indica* L.	Plumbaginaceae	Swetochita	Fresh root	Abortifacient		
121.	*Plumbago zeylanica* L.	Plumbaginaceae	Chichirimiri	Fresh root	Abortifacient		
122.	*Rauvolfia serpentina* Bail.	Apocynaceae	Chandoma	Root	Abortifacient	**Tripura**	Das et al. (2014)
123	*Ricinus communis* L.	Euphorbiaceae	Letao	Seed	Abortifacient		
124.	*Rubia cordifolia* L.	Rubiaceae	Manjistha	Seed	Abortifacient		
125.	*Solanum xanthocarpum* Schrad.	Solanaceae	Kantikari	Seed	Abortifacient		
126.	*Sapindus mukorossi* Gaertn.	Sapindaceae	Ritha	Seed	Contraceptive		
127.	*Saraca asoca* (Roxb.) W. J. De Wilde	Fabaceae	Asok	Seed	Abortifacient		
128.	*Stephania japonica* (Thunb.) Miers.	Menispermaceae	Samsota	Seed	Contraceptive	**Tripura**	Das et al. (2014)
129.	*Streblus asper* Lour.	Moraceae	Sheora	Leaves	Abortifacient		
130.	*Tacca laevis* Roxb.	Taccaceae	Tealkha	Fresh stem	Contraceptive		
131.	*Tamarindus indica* L.	Caesalpiniaceae	Tingtoi, thenthoi	Rhizome, fruit	Abortifacient		
132.	*Terminalia arjuna* (Roxb.) Wight. & Arn.	Combretaceae	Arjun	Fruit	Abortifacient		
133.	*Terminalia bellirica* Roxb.	Combretaceae	Boyra	Fruit	Contraceptive		
134.	*Terminalia chebula* Retz.	Combretaceae	Bakala	Fruit, bark	Contraceptive		
135.	*Thevetia peruviana* K. Schum.	Apocynaceae	Khumbarangcha	Seed	Abortifacient		
136.	*Vitex negundo* L.	Lamiaceae	Nisinda	Seed	Abortifacient		
137.	*Zingiber officinale* Roscoe	Zingiberaceae	Ada	Rhizome	Contraceptive		

(continued)

Table 9.1 (continued)

Sl. no.	Name of medicinal plants/ scientific name	Family	Local name	Parts used	Bioactivity	State	References
138.	*Allium sat um* L.	Liliaceae	Lasung	Rhizome	Chest pain, back pain and sore throat constipation and also hypertension	**Tripura**	Das et al. (2014)
139.	*Aloe vera* Tourn. ex. L.	Liliaceae	Alo naro	Leaves	Burns		
140.	*Amorphophallus campanulatus* Roxb.	Araceae	Shitsu nupang	Fresh stalks	Anthelmintic		
141.	*Azadirachta indica* A. Juss.	Meliaceae	Neem tu	Leaves	Vomiting, fever associated with headache and also in the treatment of hypertension		
142.	*Canna indica* L.	Cannaceae	Amjitera	Rhizomes	Gastric troubles	**Tripura**	Das et al. (2014)
143.	*Capsicum* spp.	Solanaceae	Chilli	Fruits	Antipruritic and counterirritant		
144.	*Carica papaya* L.	Caricaceae	Mamazu	Latex	Treatment of ringworm		
145.	*Caryota urens* L.	Palmae	Asang renra	Cotton-like tissue fibres	To stop bleeding and to join the cut edges		
146.	*Centella asiatica*	Apiaceae	Longshikok	Leaves	Treatment of diarrhoea		
147.	*Clerodendrum colebrookianum* Walp.	Verbenaceae	Umerem	Leaves	Treatment of hypertension, dizziness		

148.	*Costus speciosus*	Zingiberaceae	Moori	Fresh rhizome	Muscle cramp and muscle pain	**Tripura**	Das et al. (2014)
149.	*Cucumis sativus* L.	Cucurbitaceae	Zungi	Fruit	Relief from stomach discomfort due to flatulence and hyperacidity		
150.	*Dioscorea* spp.	Dioscoreaceae	Tsumgrem shii	Yam	Relief from back pain		
151.	*Emblica officinalis* Gaertn.	Euphorbiaceae	Lozu	Fruits	To relieve from bronchitis		
152.	*Eucalyptus globulus* Labill.	Myrtaceae	Malistu	Leaves and fruits	To remove dandruff and to treat scalp infection		
153.	*Ficus carica* L.	Moraceae	Mongozono	Ripe fruits	Treatment of intestinal ulcer		
154.	*Polyalthia longifolia*	Annonaceae	Mongmong	Leaves and fruits	Carminative and antihelmintic		
155.	*Polygonum hydropiper* L.	Polygonaceae	Jakremtsu	Whole plants	Treatment of fungal infections and itching skin	**Tripura**	Das et al. (2014)
156.	*Psidium guajava* L.	Myrtaceae	Motiram tu	Fresh leaves, bark	Treatment of severe diarrhoea		
157.	*Sapindus mukorossi* Gaertn.	Sapindaceae	Ering jang	Fruits	Antipruritic during foot and toe infection, as shampoo and as detergent		
158.	*Solanum nigrum* L.	Solanaceae	Likokji	Raw fruit, root	Relief from cough and treatment of asthma		
159.	*Spondias mangifera* Willd.	Anacardiaceae	Mezunglashi	Leaves	Treatment of foot and toe infection		
160.	*Stephania hernandifolia* (Willd.) Walp.	Menispermaceae	Takulaizu	Root	Treatment of diarrhoea, nausea, vomiting and abdominal colic		

(continued)

Table 9.1 (continued)

Sl. no.	Name of medicinal plants/ scientific name	Family	Local name	Parts used	Bioactivity	State	References
161.	*Terminalia chebula* Retz.	Combretaceae	Nangka jang	Fruits	Antispasmodic, antiemetic and also as cough suppressor treatment of constipation	**Tripura**	Das et al. (2014)
162.	*Urtica urens* L.	Urticaceae	Jaklemtsu	Leaves	Treatment of constipation and stomach disorder		
163.	*Verbena officinalis* L.	Verbenaceae	Shunutamtsu	Fresh herb	Treatment of high fever and malaria, used as bitter tonic and appetiser.		
164.	*Zingiber officinale* Rosc.	Zingiberaceae	Sung sung	Rhizome	Treatment of throat pain and cough, common cold, fever and as antidote in food poisoning		
165.	*Abrus precatorius* L.	Papilionaceae	Sentet	Root juice	Haemorrhoids and related problems	**Mizoram**	Sharma et al. (2001)
166.	*Acacia concinna* DC.	Mimosaceae	Khangthur	Leaf	Malaria		
167.	*Acacia pennata* L. Willd.	Mimosaceae	Khangsen	Leaf	Children in indigestion		
168.	*Achyranthes bidentata*	Amaranthaceae	Vangvat-tur	Leaves	Treatment of leech bites		
169.	*Adhatoda vasica* Nees.	Acanthaceae	Kawldawi	Leaves, root	Treatment of cough, bronchitis, asthma, fever and malaria		
170.	*Mikania micrantha* H.B.K.	Asteraceae	Lungthi	Whole plants	Treatment of various types of cancer and tuberculosis		
171.	*Aeschynanthus maculata* Lindl.	Gesneriaceae	Bawltehlantai	Flowers	Throat pain and tonsillitis		

172.	*Agave americana* L.	Agavaceae	Saidai	Leaves and roots	Scurvy	**Mizoram**	Sharma et al. (2001)
173.	*Ageratum conyzoides* L.	Asteraceae	Vaihlenhlo	Whole plants	Treatment of cholera		
174.	*Albizia chinensis(Osbeck) Merr.*	Mimosaceae	Vang	Bark	As lotion in scabies and other skin diseases		
175.	*Amorphophallus campanulatus* Roxb.	Araceae	Telhawng	Rhizome	To reduce fat in obesity. It is also used in heart diseases		
176.	*Asparagus racemosus* Willd.	Liliaceae	Arkebawk	Rhizome	To remove kidney stone, treatment of uterine disorders		
177.	*Bauhinia purpurea* L.	Caesalpiniaceae	Vaube	Bark	Diarrhoea		
178.	*Begonia inflata* Clarke	Begoniaceae	Sekhupthur	Whole plants	Treatment of haemorrhoids		
179.	*Benincasa hispida* (Thunb.) Cogn.	Cucurbitaceae	Maipawl	Fruit	Taken orally in diarrhoea, dysentery, cholera, food poisoning and diabetes. It is also given orally for epilepsy and other nervous disorders	**Mizoram**	Sharma et al. (2001)
180.	*Blumea balsamifera*	Asteraceae	Buarthau	Leaves	Asthma and cough. It is also given orally as a diuretic and externally during bath to reduce oedema		
181.	*Butea frondosa* Koen. ex. Roxb.	Papilionaceae	Fartuah or tuahpui	Bark	Typhoid fever and hypertension		
182.	*Callicarpa arborea* Roxb.	Verbenaceae	Hnahkiah	Bark	Applied on cuts and wounds as haemostatic		
183.	*Canarium resiniferum* Brace	Burseraceae	Berawthing	Fruit and bark	Abdominal colic		

(continued)

Table 9.1 (continued)

Sl. no.	Name of medicinal plants/ scientific name	Family	Local name	Parts used	Bioactivity	State	References
184.	*Capsicum frutescens* L.	Solanaceae	Anhling	Leaves, fruits and stem	To remove kidney stones	**Mizoram**	Sharma et al. (2001)
185.	*Carica papaya* L.	Caricaceae	Thingfanghma	Latex	Applied in tooth decay as an analgesic		
186.	*Chukrasia tabularis* A. Juss.	Meliaceae	Zawngtei	Fruit and bark	Hyperacidity, diarrhoea and fever		
187.	*Clerodendrum colebrookianum* Walp.	Verbenaceae	Phuihnum	Leaves	Hypertension and also in diabetes		
188.	*Clerodendrum infortunatum* L.	Verbenaceae	Phuihnam-chhia	Root	Antidandruff agent, scabies		
189.	*Colocasia esculenta* Linn. Schott.	Araceae	Dawl or bal	Juice of the stalk	On bee-sting, wounds and cuts		
190.	*Datura stramonium* L.	Solanaceae	Tawtawrawt-par	Leaf	Asthmatic problem and respiratory failure	**Mizoram**	Sharma et al. (2001)
191.	*Dendrocalamus hamiltonii* Nees.	Poaceae	Phulrua	Fruit	Treatment of hypotension		
192.	*Dichroa febrifuga* Lour.	Saxifragaceae	Khawsik-damdawi	Root	Treatment of fever, malaria and as emetic		
193.	*Dillenia indica* L.	Dilleniaceae	Kawrthingdeng	Leaves, bark and fruits	Treatment of cancer and diarrhoea		
194.	*Dioscorea alata* L.	Dioscoreaceae	Rambachim	Tuber	Haemorrhoids and syphilis		
195.	*Dysoxylum procerum* Hiern	Meliaceae	Thingthu-pui	Leaves	Dysentery		
196.	*Elaeagnus caudata* Schlecht ex. Momiyama	Elaeagnaceae	Sarzuk-pui	Root	Rheumatic pain		
197.	*Elaeagnus latifolia* L.	Elaeagnaceae	Sarzuk	Root	To remove retained placenta		
198.	*Ficus hispida* L.	Moraceae	Thei-pui	Root	Dysentery		

199.	*Ficus religiosa* L.	Moraceae	Hmawng or bung	Bark	Cough and various skin diseases	**Mizoram**	Sharma et al. (2001)
200.	*Gelsemium elegans* Gardner et. Champ.	Loganiaceae	Hanamtur	Root	Ringworm and in cases of tiger bites		
201.	*Gmelina arborea* L.	Verbenaceae	Thalmaung	Leaf juice	Coughs, gonorrhoea and ulcers		
202.	*Gynura conyza* Cass	Asteraceae	Buar-ze	Leaf juice	Wounds, ulcers, scabies, and as antidandruff agent		
203.	*Helicia excelsa* Bl.	Proteaceae	Sialhma	Leaves	Stomach pain		
204.	*Helicia robusta* Hk.f. et. T.	Proteaceae	Pasaltakaza	Rhizome	Uterine disorders, peptic ulcer and diabetes		
205.	*Hodgsonia heteroclita* Roxb.	Cucurbitaceae	Khaum	Seed	Uterine disorders		
206.	*Jatropha curcas* L.	Euphorbiaceae	Kang-damdawi	Stem	Treatment of burns	**Mizoram**	Sharma et al. (2001)
207.	*Kyllinga monocephala* Rottb.	Cyperaceae	Artelubawk	Root	An antidote for food poisoning, fever		
208.	*Lagerstroemia speciosa* L. Pers.	Lythraceae	Chawn-pui	Bark	Diarrhoea and dysentery		
209.	*Lantana camara* L.	Verbenaceae	Shilong flangsam	Whole plants	Antitetanus agent, rheumatism and malaria		
210.	*Melastoma malabathricum* L.	Melastomataceae	Builukhampa	Leaves	Treatment of diarrhoea and dysentery		
211.	*Mikania micrantha* H.B.K.	Asteraceae	Japan-hlo	Leaves	As antiseptic in cuts and wounds, treatment of diarrhoea, dysentery, fever and malaria		
212.	*Musa superba* Roxb.	Musaceae	Tumbu or changel	Whole plants	Treatment of chilling sensation, convulsion and cough		
213.	*Mussaenda macrophylla* Wall.	Rubiaceae	Vakep	Leaves	Coughs		

(continued)

Table 9.1 (continued)

Sl. no.	Name of medicinal plants/ scientific name	Family	Local name	Parts used	Bioactivity	State	References
214.	*Oryza sativa* L.	Poaceae	Buh or buh-pawl	Chopped rice straw	Removes kidney and gallbladder stones	**Mizoram**	Sharma et al. (2001)
215.	*Orthosiphon aristatus* Bl.	Lamiaceae	Zunthlum-kung	Leaf	Treatment of diabetes		
216.	*Osbeckia* spp.	Melastomataceae	Builukham	Root	Stomach troubles, pulmonary diseases and to remove kidney stones		
217.	*Oxyspora paniculata* DC.	Melastomataceae	Khampa	Root and rhizome	Heart diseases, kidney diseases and stomach troubles		
218.	*Paederia foetida* L.	Rubiaceae	Vawih-uih-hrui	Leaf juice	Toothache and gum ulcers		
219.	*Passiflora nepalensis* Wall.	Passifloraceae	Nau-awimu	Root	Fever and malaria	**Mizoram**	Sharma et al. (2001)
220.	*Passiflora* spp.	Passifloraceae	Sapthei	Fruits	Jaundice		
221.	*Pithecellobium angulatum* Benth.	Mimosaceae	Ardah-pu:	Leaves	Relief from toothache and gum boils		
222.	*Raphanus sativus* L.	Brassicaceae	Bul-uih	Root	Urinary complications and haemorrhoids		
223.	*Rhynchotechum ellipticum* A. DC.	Gesneraceae	Tiarrep	Leaves	Treatment of various types of cancer		
224.	*Sapindus mukorossi* Gaertn.	Sapindaceae	Hlingsi	Fruits	For gargle in cough and tonsillitis		
225.	*Saraca indica* L.	Caesalpiniaceae	Mualhawih	Bark	Uterine disorders and applied topically in scorpion sting		
226.	*Schima wallichii* Chois.	Ternstroemiacea	Khiang	Bark	Antiseptic		
227.	*Tagetes erecta* L.	Asteraceae	Derhken	Leafs	Eardrops in otorrhoea. It is also applied on boils and carbuncles		

228.	*Tamarindus indica* L.	Caesalpiniaceae	Tengtere or chimakelek	Fruits	Diarrhoea	**Mizoram**	Sharma et al. (2001)
229.	*Terminalia bellirica* Roxb.	Combretaceae	Thingvandawt	Flowers	Diarrhoea		
230.	*Uncaria laevigata* Wall.	Rubiaceae	Ralsamkuai	Roots	Tonsillitis		
231.	*Urena lobata* L.	Malvaceae	Sehnap	Leaves	In muscle pain and rheumatoid arthritis		
232.	*Viscum articulatum* Burm.	Loranthaceae	Lengpat	Whole plants	Fever, gout		
233.	*Vitex peduncularis* Wall.	Verbenaceae	Thingkhawi-hlu	leaf, root and bark	Treatment of malaria and blackwater fever		
234.	*Acorus calamus* Linn.	Araceae	Ok-hidak	Fresh rhizomes	Cough and Chest congestion	**Manipur**	Yumnam et al. (2012)
235.	*Adhatoda vasica* (L.) Nees.	Acanthaceae	Nongmangkha, Malabar nut	Fresh leaves, flowers	Treatment of cough and fever		
236.	*Ageratum conyzoides* L.	Compositae	Khongjai napi, goat weeds	Fresh whole parts of plant	Hair lotion		
237.	*Allium ascalonicum* Linn.	Amaryllidaceae	Meitei tilou macha, small onion	Leaves, bulb	Skin boil		
238.	*Allium hookeri* Thwaites	Amaryllidaceae	Maroi napakpi	Fresh whole parts of plant	High blood pressure, stomach ulcer		
239.	*Allium odorum* L.	Amaryllidaceae	Maroi nakoopi	Fresh whole parts of plant	Improving hair growth and help in reducing tension		

(continued)

Table 9.1 (continued)

Sl. no.	Name of medicinal plants/ scientific name	Family	Local name	Parts used	Bioactivity	State	References
240.	*Allium sativum* L.	Liliaceae	Chanam, garlic	Fresh leaves, bulb	Severe congestion of lungs due to cough	**Manipur**	Yumnam et al. (2012)
241.	*Alpinia allughas* (Retz.) Roscoe	Zingiberaceae	Poollei, shell ginger	Fresh bulb	Stomach pain		Yumnam et al. (2012)
242.	*Alpinia galanga* (Linn.) Willd.	Zingiberaceae	Kanghu, greater galangal	Fresh rhizomes	Ringworm and skin disease, to control high blood pressure		
243.	*Benincasa hispida* (Thunb.) Cogn.	Cucurbitaceae	Torbot, ash gourd	Fruits	Stomach ulcer, jaundice		
244.	*Blumea balsamifera* (L.) DC.	Asteraceae	Langthrei, elumea or Nagal camphor	Tender shoots, fresh leaves	Acidity problem, stone formation		
245.	*Chenopodium album* L.	Chenopodiaceae	Monsaobi lamb's quarters	Leaves, tender shoots	Liver enlargement problems		
246.	*Cyperus rotundus* L.	Cyperaceae	Sembang kaothoom, nutgrass	Fresh rhizomes	Fever, stomach disorder and bowel irritation		
247.	*Dactyloctenium aegyptium* (L.) Willd.	Cyperaceae	Poongphai, crowfoot	Whole parts of plants	Fever		
248.	*Datura stramonium* L.	Solanaceae	Sagoi daak, Jimson weed	Fresh or dried leaves	Severe asthma, chest pain		
249.	*Ficus hispida* Linn. f.	Urticaceae	Asi heibong, rough-leaved fig	Leaves, sticky latex	Skin in ringworm		
250.	*Fragaria indica* f.	Rosaceae	Heirongkaklab, Indian strawberry	Fresh whole parts of plant	Problem of urinary tract and stone	**Manipur**	Yumnam et al. (2012)
251.	*Hibiscus rosa-sinensis* L.	Malvaceae	Juba kusum, China rose	Sweat from the flowers	To cure soreness of the tongue and ulcer in the mouth		
252.	*Hibiscus sabdariffa* L.	Malvaceae	Shilosougri, red sorrel	Fresh leaves, dried fruits	Stone problem, gastric problem		
253.	*Leucas lavandulaefolia* Willd.	Labiatae	Mayang lemboom	Leaves, tender shoots	Relieve headache, sinusitis		
254.	*Melothria perpusilla* (Blume) Cogn.	Cucurbitaceae	Lamthabi	Whole parts of plant	Jaundice, kidney infection		
255.	*Meyna laxiflora* L.	Rubiaceae	Heibi, may-nuh	Fresh leaves, dried fruits	Blood purifier, boils and dysentery		

256.	*Nelumbo nucifera* Gaertn.	Nymphaeaceae	Thambal, east Indian lotus	Young leaves, dried leaves, seeds, roots	Diabetes, diarrhoea, cholera, liver, cardiac complaints, pile bleeding and leprosy	**Manipur**	Yumnam et al. (2012)
257.	*Nymphaea rubra* Roxb. ex. Andrews	Nymphaeaceae	Tharo angangba, red water lily	Rhizomes	To cure nose bleeding, piles, dysentery and as cardio tonic		
258.	*Polygonum posumba* Buch.	Polygonaceae	Fakpai, knotgrass	Fresh leaves	Gastric problems		
259.	*Sagittaria sagittifolia*	Alismataceae	Koukha, arrowhead, or Hawaii arrowhead	Leaves, tender shoots, rhizomes	Skin diseases and itch, to control high blood pressure		
260.	*Solanum nigrum* L.	Solanaceae	Leipung khangnga	Fruits	Fever, cough, mouth and tongue ulcer		
261.	*Trapa bispinosa* Roxb.	Trapaceae	Heikak, water chestnut	Fruits, tender shoots, leaves	Dysentery and diarrhoea, improving the blood circulation		
262.	*Zanthoxylum acanthopodium* DC.	Rutaceae	Mukthroobi	Leaves, shoots	Relief from cough and asthma, to cure tasteless and false smell		
263.	*Zingiber officinale* Roscoe	Zingiberaceae	Shing	Leaves, rhizomes	Used for cough and asthma, digestion		
264.	*Polygonum orientale* Linn.	Polygonaceae	Obuvii	Leaves	Diarrhoea and Dysentery	**Manipur**	Lokho (2012)
265.	*Melia azedarach* Linn.	Meliaceae	Siikhasii	Barks	Blood pressure, acidity and ringworm infection		
266.	*Chenopodium ambrosioides* Linn.	Chenopodiaceae	Nopuepro	Leaves	Headache, fever and blood pressure	**Manipur**	Lokho (2012)
267.	*Morus nigra* Linn.	Urticaceae	Khelosii	Leaves and roots	Jaundice		
268.	*Cannabis sativa* Linn.	Cannabaceae	Kanjapro	Leaves	Bone fracture, sprain and muscle pain		
269.	*Polytrichum juniperinum* Hewd.	Polytrichaceae	Shiipa	Whole plant	Cuts and nose bleeding		
270.	*Bidens pilosa* Linn.	Asteraceae	Shanghapiti-e	Leaves	Cold and fever, headache and blood pressure		
271.	*Centella asiatica* Linn.	Umbelliflorae	Korivii	Whole plant	Gastritis, ulcer and blood pressure		

(continued)

Table 9.1 (continued)

Sl. no.	Name of medicinal plants/ scientific name	Family	Local name	Parts used	Bioactivity	State	References
272.	*Erythrina variegata* Linn.	Papilionaceae	Letosii	Bark	Poison	**Manipur**	Lokho (2012)
273.	*Rhus semialata* Linn.	Anacardiaceae	Omoshii	Fruits	Dysentery and Diarrhoea		
274.	*Gynura bicolor* (Roxb. & Willd.) DC.	Asteraceae	Tabovii	Leaves and young stems	Ulcer, chronic acidity		
275.	*Clerodendrum colebrookianum* Walp.	Verbenaceae	Pijiivii	Leaves	Blood pressure and abdominal pain		
276.	*Houttuynia cordata* Thunb.	Saururaceae	Shakama	Whole plant	Stomach ache, gas formation and expulsion of worms		
277.	*Artemisia nilagirica* (Cl.) Camp.	Asteraceae	Shupriipriso	Leaves	Dandruff, cuts and wounds	**Manipur**	Lokho (2012)
278.	*Xanthosoma sagittifolium* (Linn.) Schott.	Araceae	Birovii	Stem	Bee-sting and insect bites		
279.	*Elsholtzia blanda* Bentham	Lamiaceae	Shiipriikholo	Leaves	Hypertension, headache and blistered lips		
280.	*Bombax ceiba* Linn.	Malvaceae	Pikriisii	Bark	Snake bites		
281.	*Brugmansia suaveolens* (Humb. & Bonpl. ex. Willd.) Bercht. & J. Presl.	Solanaceae	Mikrii taɔopro	Leaves and barks	Sprain, muscle pain and snake bites		
282.	*Oroxylum indicum* (Linn.) Benth. ex. Kurz.	Bignoniaceae	Kakidziihe	Bark and root	Cancer, diarrhoea, diabetes and hypertension		

283.	*Achyranthes aspera* Hook. F./Justicia adhatoda Linn.	Acanthaceae	Tohuopa	Leaves and roots	Malarial fever, abdominal pain, indigestion and urine disorder	**Manipur**	Lokho (2012)
284.	*Hibiscus sabdariffa* Linn.	Malvaceae	Okhrivii	Leaves and calyx	Tonic and skin allergy		
285.	*Physalis peruviana* Linn.	Solanaceae	Tsiibobopro	Leaves and fruit	Dysentery, diarrhoea, jaundice and tonic		
286.	*Solanum torvum* Sw.	Solanaceae	Modoro shiikhokha	Fruits	Blood pressure, headache and fever		
287.	*Momordica charantia* Linn.	Cucurbitaceae	Khenavii	Leaves	Fever, headache, blood pressure and cold		
288.	*Juglans regia* Linn.	Juglandaceae	Okhoshii	Bark	Tooth plague and bleeding gums		
289.	*Psidium guajava* (Linn.) Kuntze	Myrtaceae	Pungdol/pondal	Fruit and young tender leaves	Dysentery and diarrhoea	**Manipur**	Lokho (2012)
290.	*Emblica officinalis* Gaertn.	Euphorbiaceae	Heikru/shiihos hii	Fruit	Cold and cough		
291.	*Aconitum bisma* (Buch.-Ham.) Rapaics	Ranunculaceae	Bikhma	Tubers, roots	Tuber is used in food poisoning, asthma, cough and bronchitis	**Sikkim**	Panda and Misra (2010)
292.	*Aeschynanthus sikimensis* (C.B. Clarke) Stap	Gesneriaceae	Baklaypatay	Rhizome	Decoction of root is used in fever and throat pain		
293.	*Aesculus indica (Wall. ex Camb.) Hook. f.*	Sapindaceae	Pangra	Fruits	Seed oil used in rheumatism and mumps		
294.	*Aesandra butyracea* (Roxb.) Baehni	Sapotaceae	Chiuri	Fruits	Used in rheumatism		
295.	*Allium wallichii* var. Kunth	Alliaceae	Alliaceae	Leaves	Viral flue and used in high altitude sickness		

(continued)

Table 9.1 (continued)

Sl. no.	Name of medicinal plants/ scientific name	Family	Local name	Parts used	Bioactivity	State	References
296.	*Artemisia vulgaris* L.	Compositae	Titepati	Leaf decoction	Leaf decoction used on cuts and bruises to stop bleeding mostly in nose and measles and fever	**Sikkim**	Panda and Misra (2010)
297.	*Bergenia* ciliata (Royle)	Saxifragaceae	Pakhanbhed	Rhizomes, roots	Used in fever and applied to boils; rhizome in white discharge		
298.	*Betula utilis* D. Don	Betulaceae	Bhojpatra	Bark	Used to heal up wounds from bone fracture		
299.	*Bischofia javanica* Blume	Euphorbiaceae	Kainjal	Leaves, bark	Fruits are used in making wine; stem bark is used for irregular menstruation and pain		
300.	*Brugmansia suaveolens* (Humb. & Bonpl. ex. Willd.)	Solanaceae	Kolodhaturo	Leaves	Applied to cure swellings, sprain and rheumatism		
301.	*Buddleja asiatica* Lour.	Buddlejaceae	Bhimsen pate	Leaves, flowers, stem	Used for skin problems and as abortifacient		
302.	*Cordyceps sinensis* Berk.	Clavicipitaceae	Yarchagumba	Whole plant	Rejuvenates the liver and heart and retards aging processes in the immune system		
303.	*Daphne bholua* Ham. ex. D. Don	Thymelaeaceae	Kagatey	Bark and root	Bark decoction given to treat fever; root bark used for intestinal worms		

304.	*Dioscorea deltoidea* Wall. ex. Griseb.	Dioscoreaceae	Kurkurtarul	Bark and tuber	Tuber used in rheumatoid arthritis, asthma and fever	**Sikkim**	Panda and Misra (2010)
305.	*Eupatorium cannabinum* L.	Asteraceae	Banmara	Leaves/stem	Leaf and stem extract used on cuts and bruises to stop bleeding and infection		
306.	*Fraxinus floribunda* Wallich	Oleaceae	Lakuri	Bark	Bark boiled and applied for gout, sprain and used in fracture		
307.	*Heracleum wallichii* DC.	Apiaceae	Chimphing	Fruits and root	Fruits used orally during influenza, root as aphrodisiac		
308.	*Lindera neesiana* (Wall. ex. Nees.) Kurz	Lauraceae	Timbur	Bark and fruits	Flower used for excessive seminal discharge in dream; fruits are used to induce vomiting		
309.	*Marsdenia roylei* Wight.	Asclepiadaceae	Bahunilahara	Roots and leaves	Cooling effect in gonorrhoea		
310.	*Nardostachys jatamansi* DC.	Valerianaceae	Jatamansi	Root	Root used for hair loss, in epilepsy and hysteria		
311.	*Orchis latifolia* L.	Orchidaceae	Pachamala	Root tuber	Root tubers highly nutritious, used as aphrodisiac		
312.	*Oxalis corniculata* L.	Oxalidaceae	Chari amilo	Whole plant root	Leaf juice taken to cure dysentery and fever anaemia and tympanitis for appetite digestion		

(continued)

Table 9.1 (continued)

Sl. no.	Name of medicinal plants/ scientific name	Family	Local name	Parts used	Bioactivity	State	References
313.	*Panax pseudoginseng* Wall	Araliaceae	Mangan	Roots	Root taken to reduce fever, indigestion and vomiting and also used as tonic	**Sikkim**	Panda and Misra (2010), Maity et al. (2004)
314.	*Picrorhiza kurroa* Royle ex. Benth.	Scrophulariaceae	Kutki	Roots	Used as laxative, brain tonic, emetic, good in paralysis, jaundice		
315.	*Podophyllum Hexandrum* Royle	Berberidaceae	Papri	Whole plant	Cures septic wounds and diarrhoea		
316.	*Rubia manjith* Roxb. ex. Fleming	Rubiaceae	Manghito	Stem root	Cures jaundice, urinary tract infection, liver complaints and irregular menstruation, as general tonic, treatment of eye and ear diseases, blood purifier, treatment of joint pains, leucoderma and in skin diseases		
317.	*Saussurea gossypiphora* D. Don	Asteraceae	Kapisful	Plant/root	Plant paste used for cuts and bruises; root paste used to cure cough, asthma, fever and dysentery; inflorescence used for sexual dysfunction		
318.	*Stephania glabra* Lour.	Menispermaceae	–	Root bulb	Powder used in diabetes tuberculosis, asthma, fever	**Sikkim**	Panda and Misra (2010)
319.	*Swertia chirata* (Wall.) C.B. Clarke	Gentianaceae	Chirata	Plant	Plant juice used to cure malaria fever		
320.	*Taxus baccata* L.	Taxaceae	Dhengresalla	Leaf/bark	Leaf extracts used in breast and throat cancer		

Fig. 9.1 Some important medicinal plants. (**a**) *Adhatoda vasica*, (**b**) *Aloe barbadensis*, (**c**) *Clerodendrum colebrookianum*, (**d**) *Cuscuta reflexa*, (**e**) *Curcuma longa*, (**f**) *Prunus persica*, (**g**) *Cannabis sativa* (dried), (**h**) *Achyranthes aspera*, (**i**) *Zingiber officinale*, (**j**) *Phlogacanthus curviflorus*, (**k**) *Zingiber zerumbet*, (**l**) *Benincasa hispida*, (**m**) *Psidium guajava*, (**n**) *Datura metel*, (**o**) *Solanum nigrum*, (**p**) *Aegle marmelos*, (**q**) *Oroxylum indicum*, (**r**) *Artemisia nilagirica*, (**s**) *Nelumbo nucifera*, (**t**) *Phyllanthus emblica*

Fig. 9.1 (continued)

o

Solanum nigrum

p

Aegle marmelos

q

Oroxyllum indicum

r

Artemisia nilgarica

s

Nelumbo nucifera

t

Phyllanthus emblica

Fig. 9.1 (continued)

ing in NE India, and there are demands for those bioresources from across the world. Many pharmaceutical products are currently available in clinical use, which are having a long history of their uses as herbal remedies, including aspirin, digitalis, quinine and opium (Tony 1968). The World Health Organization (WHO) estimates that 80 % of the populations of some Asian and African countries are presently using herbal medicine for primary healthcare. Studies in the USA and Europe have shown that their use is less common in clinical settings, but has become increasingly more in recent years as scientific evidence about the effectiveness of herbal medicine has become more widely available. The annual global export value of pharmaceutical plants in 2011 accounted for over US$2.2 billion (Jennifer and Sherman 1998). The medicinal plants used in one state for treatment of a particular ailment are also found to be used by other different states for curing different diseases. The research status of the plants, scientific evaluation and traditional use reported in scientific communications are presented in Table 9.1.

Adhatoda vasica Nees. used as insecticidal in Assam were reported to be used as antidiabetic in Manipur and also for treatment of cough and fever (Table 9.1), while it was also reported for treatment of joint pain, lumber pain and sprains in Nagaland (Jamir et al. 1999). *Andrographis paniculata* used as antimalarial and *Azadirachta indica* (A. Juss.) L. used as anti-plasmodial and repellent activity in Assam were found to be used as antidiabetic in Manipur. *Aegle marmelos* is used for treatment of fever in Meghalaya, while it is used as larvicidal in Assam and for treatment of dysentery, cough and mucous secretion in Nagaland. In Mizoram, *Benincasa hispida* was taken orally for diarrhoea, dysentery, cholera, food poisoning and diabetes. It is also given orally for epilepsy and other nervous disorders where it was found to treat cough and fever in Meghalaya (Dolui et al. 2004). In Tripura *Carica papaya* and *Costus speciosus* were used as contraceptive and to treat muscle pain and cramp, respectively, while in Nagaland, they were applied topically to treat ringworm and as an abortifacient, respectively. *Zingiber officinale* was used for the treatment of throat pain and cough, common cold and fever and as antidote in food poisoning in Nagaland as well as in Manipur. *Paederia foetida* was used for toothache and gum ulcer in Mizoram, while it was used for colic and gastric ulcer in Nagaland. *Momordica charantia* was used for fever, headache, blood pressure and cold in Mizoram and Manipur, while it was used as contraception in Tripura.

Likewise, there are multiple uses of a plant in different states for treatment of different diseases; also, some plants were used for the same purpose in different states, indicating that the particular plant has indeed medicinal property commonly used as their ingredients of medicine in different cross-cultural tribes and regions of NE India.

9.3 Medicinal Plants of the Region as Components of the Available Commercial Herbal Products

As healthcare costs continue to escalate, the attraction for low-cost remedies has stimulated consumers to re-evaluate the potential of alternatives (Bouldin et al. 1999). Extracts and bioactive compounds of these medicinal plants are employed as composition in the manufacture of many herbal products by the leading companies. Ranging from small-scale to large-scale herbal companies, numerous medicinal plants and their extracts have been used as their components. The common plants used by traditional healers of Northeastern India which were found to be components of commercial products, along with their market price, are listed below (Table 9.2). The major bioactive principles are discussed below, which are in very much demand for either research or clinical uses or new drug discovery.

9.3.1 Catechin

The catechin of *Acacia caesia* belongs to a group of flavanols. (+)-Catechin inhibits intestinal tumour formation in mice (Weyant et al. 2001).

Table 9.2 The important medicinal plants used for traditional healthcare practices in Northeastern India, which are also components of the available commercial herbal products

Sl. No.	Plant name	Bioactive principle	Commercial product
1.	*Acacia caesia*	Catechin, catechol	Extract
2.	*Achyranthes aspera*	Triterpenoid saponins oleanolic acid	Tablets
3.	*Abelmoschus esculentus*	Isopentyl 2-methyl butanoate and heptanoic acid 2-methylbutyl ester	Powder
4.	*Alpinia galanga*	Myrcene, 1,8-cineole	Extract
5.	*Annona squamosa*	Atisine, oxophoebine, reticuline	Extract powder
6.	*Averrhoa carambola*	Oxalic acid	Starfruit powder
7.	*Azadirachta indica*	Azadirachtin	Natural azadirachtin, insecticide azadirachtin, neem seed extract
			Face wash, face scrub, face pack, body wash, skin cream
8.	*Bauhinia purpurea*	β-Sitosterol	Dietary supplements
9.	*Bombax ceiba*	Lupeol, saponins, gums and 4,5,7-trihydroxyflavone-3-O-β-D glucopyranosyl (1–4)-α-L-rhamnopyranoside	Acne/pimple cream
10.	*Capsicum frutescens*	Quercetin 3-O-α-L-rhamnopyranoside	Capsicum extract
11.	*Centella asiatica*	Brahmic and asiatic acid	Capsules
12.	*Citrus aurantifolia*	Limonin	Body oil, lip balm
13.	*Cudrania cochinchinensis*	6-p-hydroxybenzyl kaempferol (1) and 6-p-hydroxybenzyl quercetin	Capsules
14.	*Curcuma longa*	Curcumin	Curcumin
15.	*Cuscuta reflexa*	6,7-Dimethoxy-2H-1-benzopyran-2-one	Hair oil
16.	*Hibiscus sabdariffa*	Anthocyanins, protocatechuic acid (PCA)	Capsules
17.	*Houttuynia cordata*	Ethyl nonyl ketone	Herbal toothpaste, dietary supplement
18.	*Juglans regia*	Linoleic acid, oleic, linolenic and palmitic acids	Organic oil
19.	*Momordica charantia*	Charantagenins D	Herbal tea, capsules
20.	*Moringa oleifera*	Benzyl glucosinolate	Juice
21.	*Murraya koenigii*	β-caryophyllene, (E)-β-ocimene	Extract
22.	*Nelumbo nucifera*	Quercetin	Extract
23.	*Oroxylum indicum*	Naringin	Extract
24.	*Paederia scandens*	Iridoid glycoside	Extract
25.	*Panax pseudoginseng*	Ginsenosides	Capsules
26.	*Phyllanthus emblica*	Gallic acid	Fruit powder
27.	*Physalis peruviana*	Provitamin A, minerals, vitamin C, vitamin B complex and polysaccharides	Berries
28.	*Psidium guajava*	a-Selinene (23.7 %), p-caryophyllene (18.8 %) and 8-selinene	Toothpaste, lip balm
29.	*Rhus semialata*	6-Pentadecylsalicylic acid	Sunscreen
30.	*Raphanus sativus*	Rasatiol	Homoeopathic medicine
31.	*Scoparia dulcis*	Scoparic acid A–C, scopadulcic acid A and B	Dietary supplement

(continued)

Table 9.2 (continued)

Sl. No.	Plant name	Bioactive principle	Commercial product
32.	*Stephania glandulifera*	Alkaloids, coumarin	Pastille
33.	*Tamarindus indica*	2-Phenylacetaldehyde, 2-furfural and hexadecanoic acid	Skin cream, skin lotion
34.	*Terminalia arjuna*	(+)-Catechin, (+)-gallocatechin and (−)-epigallocatechin	Skin cream, cleanser
35.	*Terminalia chebula*	Tannins, anthraquinones, chebulinic acid, chebulagic acid	Fruit powder, tablets
36.	*Tinospora cordifolia*	Cordifolioside, tinocordifolin and tinocordifolioside	Capsules, powder
37.	*Zingiber officinale*	Gingerol	Capsules
38.	*Zingiber zerumbet*	Zerumbone	Ayurvedic soap

Catechol was first isolated in 1839 by H. Reinsch by distilling it from the solid tannic preparation catechin, which is the residuum of catechu, the boiled or concentrated juice of *Mimosa catechu* (*Acacia catechu* L.f.). Upon heating the catechin above its decomposition point, a substance Reinsch first named "pyrocatechol" was distilled and condensed as a white solid ("pyro" referring to heat) (n.wikipedia.org/wiki/Catechol). Myrcene is a monoterpene having analgesic property. It has anti-inflammatory properties through prostaglandin E2 (Lorenzetti et al. 1991).

9.3.2 1,8-Cineole

It is also known as eucalyptol. In 1870, F.S. Cloez identified and ascribed the name *eucalyptol* to the dominant portion of *Eucalyptus globulus* oil (Boland et al. 1991) Eucalyptol is used as an insecticide and insect repellent (Klocke et al. 1987; Sfara et al. 2009).

9.3.3 Atisine

Atisine is the most abundant alkaloid found in the roots of *Annona squamosa* and is used as a pesticide. Oxophoebine showed selective toxicity against DNA repair and recombination-deficient mutants of the yeast *Saccharomyces cerevisiae*. Reticuline found in *Annona squamosa* is one of the alkaloids found in opium, and experiments in rodents suggest it possesses potent central nervous system-depressing effects (Morais et al. 1998).

9.3.4 Rotundifolone

Rotundifolone, of *Ardisia pedunculosa*, also known as lippione, is a monoterpene known to have antinociceptive activity (Sousa et al. 2007).

9.3.5 Oxalic Acid

Oxalic acid is a dicarboxylic acid and is found in *Averrhoa carambola.* It is mainly used in dyeing industries. Oxalic acids are not carcinogenic but, if inhaled, are toxic to health. In humans, ingested oxalic acid has an oral LD_{Lo} (lowest published lethal dose) of 600 mg/kg (Oxalic Acid Material Safety Data Sheet 2014). It has been reported that the lethal oral dose is 15–30 g (IDLHs 2015).

9.3.6 Azadirachtin

Azadirachtin is a limonoid group present in neem, which is commonly used as an insecticide and a pesticide. Many herbal companies use neem and its extracts in the production of many herbal products.

9.3.7 β-Sitosterol

β-Sitosterol of *Bauhinia Purpurea* is one of several phytosterols (plant sterols) with chemical structures similar to that of cholesterol. β-Sitosterol supplements or enriched foods should be avoided during pregnancy and breastfeeding, since not enough is known about its effects on unborn and newborn children (http://en.wikipedia.org/wiki/Beta-Sitosterol). High levels of β-sitosterol concentrations in blood have been correlated with increased severity of heart disease in men having previously suffered from heart attacks (Assmann et al. 2006).

9.3.8 Calliterpenone

Calliterpenone is a natural plant growth promoter. Calliterpenone (CA), a stereo-isomer of abbeokutone, in comparison to gibberellic acid (GA3) attributes on growth of, trichomes, essential oil biosynthesis and expression of some oil biosynthetic pathway genes (Bose et al. 2013).

9.3.9 Torvpregnanosides

Torvpregnanosides isolated from aerial parts of *Solanum torvum* have anti-neutrophilic inflammatory steroidal glycosides from *Solanum torvum.*

9.3.10 Zerumbone

Zerumbone, a sesquiterpene, is a major compound of *Zingiber zerumbet.* Zerumbone markedly suppresses free radical generation, pro-inflammatory protein production and cancer cell proliferation accompanied by apoptosis (Murakami et al. 2002).

9.3.11 α-Cyperone

α-Cyperone, isolated from the rhizomes of *Cyperus rotundus*, inhibits LPS-induced COX-2 expression and PGE2 production through the negative regulation of NFκB signalling in RAW 264.7 cells (Jung et al. 2013).

9.3.12 Charantagenins D

Charantagenins D from the fruit of *Momordica charantia* L. with an -OMe substituent group in the side chain exhibited significant cytotoxic activities against cancer cells (Wang et al. 2012).

9.3.13 Naringin

Naringin isolated from *Oroxylum indicum* is a flavanone glycoside. Naringin is an inhibitor of vascular endothelial growth factor (VEGF) release, which causes angiogenesis (Schindler and Mentlein 2006).

9.3.14 Rasatiol

Rasatiol isolated from *Raphanus sativus* accelerated fibroblast growth in a dose-dependent manner and increased the production of type 1 collagen, fibronectin and elastin. Phosphorylation of p42/44 extracellular signal-regulated kinase, p38 mitogen-activated protein kinase and Akt was remarkably increased by rasatiol, indicating that enhanced ECM production is linked to the activation of intracellular signalling cascades (Roh et al. 2013).

9.3.15 Ginsenoside

Ginsenoside is the major compound of *Panax pseudoginseng.* It is also known as panaxosides, and it belongs to a class of steroid glycosides and triterpene saponins, found exclusively in the plant genus *Panax* (ginseng). In one study on breast cancer and different ginsenosides, it was found that ginsenoside-Rc was capable of inhibiting the growth of these cancer cells. This suggests that there is a possibility that ginsenoside-Rc may have effects that prevent or limit the development of breast cancer (Murphy 2000).

According to PubMed, there are 48 scientific reports on the zerumbone of *Zingiber zerumbet*, 27 results on 1,8-cineole of *Alpinia galanga*, and one result on *Annona squamosa* and its compounds oxophoebine and reticuline. *Ardisia pedunculosa* has no results with its rotundifolone and (R)-limonene in PubMed. However, there are 755 research documents on the Google database for this plant and its activity. Altogether two documents on the research of oxalic acid of *Averrhoa carambola* and 48 documents on the research of azadirachtin of *Azadirachta indica* Juss. are available in the Google database.

9.3.16 Okanin

Okanin is one of the most abundant chalcone [1,3-diaryl-2-propen-1-one] compounds found in the genus *Bidens* (Asteraceae) that has been used as various folk medications in Korea and China for treating inflammation, malaria, hypertension, diabetes, peptic ulcer, snake bite and small pox (Yuan et al. 2008; Li et al. 2005; Hoffmann and Holzl 1988). Across the world researchers are working on the okanin for exploring its medicinal uses for treatment as antidiabetic, analgesic and anti-inflammatory (Chien et al. 2009; Fotso et al. 2014), and so far four scientific literatures are available in the public domain.

9.3.17 Ethyl Caffeate

Ethyl caffeate is an ester of a hydroxycinnamic acid, a naturally occurring organic compound. Ethyl caffeate suppresses NF-kappa B activation and its downstream inflammatory mediators, iNOS, COX-2 and PGE2, in vitro or in mouse skin (Chiang et al. 2005). It is administered intraperitoneally in rats previously which is able to prevent the dimethylnitrosamine-induced loss in body and liver weight, as well as to reduce the degree of liver injury. It can be considered as a promising natural compound for future application in chronic liver diseases (Boselli et al. 2009), and so far one scientific literature was available in the public domain.

9.3.18 Quercetin

Quercetin is a plant pigment (flavonoid). It is found in many plants and foods, such as red wine, onions, green tea, apples, berries and *Ginkgo biloba*. People use quercetin as a medicine for treating conditions of the heart and blood vessels including "hardening of the arteries" (atherosclerosis), high cholesterol, heart disease and circulation problems. It is also used for diabetes, cataracts, hay fever, peptic ulcer, schizophrenia, inflammations, asthma, gout, viral infections and chronic fatigue syndrome (CFS), in preventing cancer and for treating chronic infections of the prostate. Quercetin is also used to increase endurance and improve athletic performance (Fabjan et al. 2003; Mammela et al. 2000; Konishi et al. 1996). The bioactive constituent quercitrin present in *Capsicum* has shown an antioxidant activity (Materska and Perucka 2005), and so far two scientific literatures were available in the public domain.

9.3.19 Brahmic and Asiatic Acid

The brahmic and asiatic acid of *Centella asiatica* (L.) is a pentameric triterpenoid. *C. asiatica*, commonly known as "gotu kola", "asiatic pennywort", "Indian pennywort", "Indian water navelwort", "wild violet" and "tiger herb" in English, is a tropical plant, which is cultivated due to its medical importance in some countries including Turkey, and it has a long history of utilization in Ayurvedic and Chinese traditional medicines since centuries ago (Meulenbeld and Wujastyk 2001). Across the world researchers are working on the brahmic and asiatic acids for exploring their medicinal uses as antioxidant, anti-inflammatory and protection against glutamate- or beta-amyloid-induced neurotoxicity (Krishnamurthy et al. 2009), and so far 68 scientific literatures were available in the public domain.

9.3.20 Limonin

Limonin is a limonoid and a bitter, white, crystalline substance found in citrus and other plants. It is also known as limonoate D-ring lactone and limonoic acid di-delta-lactone. Ongoing research programmes are examining the effects of limonin in human diseases. Citrus seed extracts reportedly have antiviral properties, inhibiting replication of retroviruses like HIV-1 and HTLV-I (Balestrieri et al. 2011). Neuroprotective effects of limonin have also been described (Yoon et al. 2010). Limonin reduces proliferation of colon cancer cells (Chidambara et al. 2011) and has been tested as an anti-obesity agent in mice (Ono et al. 2011), and so far one scientific literature was available in the public domain.

9.3.21 Saponins

Saponins are a class of chemical compounds found in particular abundance in various plant species. Colebroside A (1), a new diglucoside of fatty acid ester of glycerin, was isolated from the aerial parts of *Clerodendrum colebrookianum* Walp., along with nine known compounds (2–10). Their structures were elucidated by spectroscopic and chemical methods. Compounds 2, 3, 4, 5, 7, 8, 9 and 10 have been obtained from this plant for the first time (Yang et al. 2000), and so far one scientific literature was available in the public domain.

9.3.22 Viscosene

Across the world researchers are working on the viscosene for exploring its medicinal uses for treatment of antimicrobial and cytotoxic activities (Amin et al. 2012) and anti-cervical cancer (Sun et al. 2013), and so far six scientific literatures were available in the public domain.

9.3.23 Berberine and Coptine

Berberine and coptine are compounds of *Coptis teeta* known as Yunnan goldthread (Chinese), its rhizome is used as an antimicrobial and anti-inflammatory. Five scientific literatures were available in the public domain.

9.3.24 Kaempferol

Kaempferol is a natural flavonol, a type of flavonoid, that has been isolated from tea, broccoli, *Delphinium*, witch hazel, grapefruit, cabbage, kale, beans, endive, leek, tomato, strawberries, grapes, Brussels sprouts, apples and other plant sources. Kaempferol is a yellow crystalline solid with a melting point of 276–278 °C. It is slightly soluble in water but soluble in hot ethanol and diethyl ether (Mustafa et al. 2010). It is found in fruits, vegetables, leaves and grains. It can be used as an ingredient in supplements, beverages or foods (*WordNet.*) Across the world researchers are working on the viscosene and quercetin for exploring their medicinal uses for treatment of antioxidant activity (Zhou et al. 2014) and anti-inflammatory activity. Five scientific literatures were available in the public domain.

9.3.25 Curcumin

Curcumin is a diarylheptanoid. It is the principal curcuminoid of the popular South Asian spice turmeric, which is a member of the ginger family (*Zingiberaceae*). Turmeric's other two curcuminoids are desmethoxycurcumin and bis-desmethoxycurcumin. The curcuminoids are natural phenols that are responsible for the yellow colour of turmeric. The enol form is more energetically stable in the solid phase and in solution. Curcumin can be used for boron quantification in the curcumin method. It reacts with boric acid to form a red-colour compound, rosocyanine (Dorland 2011). Across the world researchers are working on curcumin for exploring its medicinal uses for treatment of osteoarthritis (Appelboom et al. 2014) and anti-inflammatory cytokine gene promoter activity (McCann et al. 2014), melanoma (Jiang et al. 2015) and breast cancer

(Thulasiraman et al. 2014). 1131 scientific literatures were available in the public domain.

9.3.26 6,7-Dimethoxy-2H-1-benzopyran-2-one

6,7-Dimethoxy-2H-1-benzopyran-2-one from *Cuscuta reflexa* showed strong inhibitory activity against alpha-glucosidase. One scientific literature was available in the public domain. *Cuscuta reflexa* is a genus of about 100–170 species of yellow, orange or red (rarely green) parasitic plants. Formerly treated as the only genus in the family Cuscutaceae, it now is accepted as belonging in the morning glory family, Convolvulaceae, on the basis of the work of the Angiosperm Phylogeny Group (Stefanovic and Olmstead 2014).

9.3.27 Dilenetin and Betulinic Acid

Many researchers are exploring the medicinal uses of dilenetin and betulinic acid for the treatment of antidiabetic activity (Kumar et al. 2013) and antileukemic activity (Mallick et al. 2010), melanoma (Jiang et al. 2015) and breast cancer (Thulasiraman et al. 2014). 1131 scientific literatures were available in the public domain. *Dillenia indica* is a species of *Dillenia* native native to southeastern Asia, from India, Bangladesh andSri Lanka east to southwestern China (Yunnan) and Vietnam, and south through Thailand to Malaysia and Indonesia. (Germplasm Resources Information Network 2015).

9.3.28 Hortensin

Hortensin, a major compound of *Millingtonia hortensis*, is a medicinal plant widely used in many Asian countries. An aqueous crude extract of this plant has been shown the apoptosis induction on RKO colon cancer cells (Tansuwanwong et al. 2008; Chulasiri M et al. 2006). Across the world researchers are exploring hortensin and its medicinal uses for the treatment of cancer and, so far four scientific literatures were available in the public domain.

9.3.29 Benzyl Glucosinolate: A Compound of Moringa (*Moringa oleifera* Lam.)

Benzyl glucosinolate, a compound of moringa (*Moringa oleifera* Lam.), is an edible plant used as both a food and medicine throughout the tropics. A moringa concentrate (MC), made by extracting fresh leaves with water, utilizes naturally occurring myrosinase to convert four moringa glucosinolates into moringa isothiocyanates (Waterman et al. 2014). Across the world researchers are working on the benzyl glucosinolate for exploring its medicinal uses for treatment of inflammations and bacterial and microbial infections, and so far seven scientific literatures were available in the public domain.

9.3.30 β-Caryophyllene and (E)-β-Ocimene

Murraya koenigii (L.) Spreng., "curry leaf tree", is a popular spice and condiment of India. Sixty-one compounds were identified, of which eleven were present in all the specimens analysed. The two major volatile metabolites were identified as beta-caryophyllene (16.6–26.6 %) and alpha-humulene (15.2–26.7 %) along with nine minor compounds identified as beta-elemene (0.3–1.3 %), aromadendrene (0.5–1.5 %), beta-selinene (3.8–6.5 %), spathulenol (0.6–2.7 %), caryophyllene oxide (0.7–3.6 %), viridiflorol (1.5–5.5 %), 2-naphthalenemethanol (0.7–4.8 %), trivertal (0.1–1.0 %) and juniper camphor (2.6–8.3 %). The results suggest that beta-caryophyllene and alpha-humulene could be used as chemotaxonomical markers for Malaysian *M. koenigii*; hence, these specimens could be of the same stock and different from the ones in India, Thailand and China. Beta-caryophyllene and alpha-humulene could be used as chemotaxonomical markers for Malaysian *M. koenigii*; hence, these specimens could be of the same stock and different from the ones in India, Thailand and China (Nagappan et al. 2012).

β-Caryophyllene and (E)-β-ocimene are being explored for their medicinal uses for treatment of malaria and bacterial infections and used as chemotaxonomical markers, and so far four scientific literatures were available in the public domain.

9.3.31 Quercetin

The DPPH scavenging effect, the inhibition of human low-density lipoprotein oxidation and antioxidative contents were employed for the activity-guided purification to identify the antioxidant components of lotus leaves (leaves of *Nelumbo nucifera* Gaertn the result indicate that the antioxidant capacity of lotus leaves is partially relevant to its flavonoids. (Lin et al. 2009). (+)-1(R)-Coclaurine (1) and (−)-1(S)-norcoclaurine, together with quercetin 3-O-beta-D-glucuronide (4), were isolated from the leaves of *Nelumbo nucifera* (Nymphaeaceae) and identified as anti-HIV principles (Kashiwada et al. 2005).

Across the world researchers are working on the quercetin for exploring its medicinal uses for treatment of HIV and inflammatory diseases and as secretagogue, and so far 24 scientific literatures were available in the public domain.

9.3.32 Iridoid Glycoside

Iridoid glycoside is a compound of *Paederia scandens* which is used to treat aches, jaundice, dysentery and dyspepsia as a folk medicine in the southern region of China, Vietnam, India and Japan. Phytochemical studies revealed the presence of iridoids, flavonoids, volatile oil and other metabolites in these species, which possess versatile bioactivities like antinociceptive, anti-inflammatory, antidiarrhoeal, antitussive and antitumor activities (Wang et al. 2014).

Across the world researchers are working on iridoid glycoside for exploring its medicinal uses for treatment of analgesic and inflammatory, diarrhoeal, antinociceptive activity, and so far seven scientific literatures were available in the public domain.

9.3.33 P-Caryophyllene (18.8 %) and 8-Selinene

Psidium guajava has been known for its anti-inflammatory, antimicrobial, antioxidant, antidiarrhoeal, and antimutagenic properties (Yang et al. 2007), and 348 scientific literatures were available in the public domain.

9.3.34 6-Pentadecylsalicylic Acid

Bioassay-directed fractionation of the n-hexane extract of the stem of *Rhus semialata* Murr. var. roxburghii DC. (Anacardiaceae) has led to the isolation of 6-pentadecylsalicylic acid. It showed antithrombin activities. It also prolonged the clotting time in a dose-dependent manner in the clotting assay of thrombin-fibrinogen interaction (Kuo et al. 1991).

9.3.35 Scoparic Acid A–C and Scopadulcic Acid A and B

A β-glucuronidase-inhibitory diterpene called scoparic acid A (SA) was isolated from *Scoparia dulcis* L. together with scoparic acid B, scoparic acid C and the aphidicolin-like tetracyclic diterpenes scopadulcic acid A (SDA) and scopadulcic acid B (SDB). Scopadulcic acid B (SDB) and scopadulciol (SDC) were found to show multi-faceted pharmacological effects such as inhibitory effects on gastric acid excretion, bone resorption, replication of herpes simplex virus type 1 (HSV-1), etc. In addition, SDC was suggested to be applicable to cancer gene therapy using ganciclovir or acyclovir and the HSV-1 thymidine kinase gene called the suicide gene (Hayashi et al. 2011; Rial et al. 2002).

9.3.36 Tinocordifolin and Tinocordifolioside

Cordifolioside, tinocordifolin and tinocordifolioside are being studied for their medicinal uses for treatment of Parkinsonism and as radioprotective and cardioprotective, antispasmodic, anti-inflammatory, antiarthritic, antiallergic and antidiabetic uses, and so far 12 scientific literatures were available in the public domain

9.3.37 (+)-Catechin, (+)-Gallocatechin and (−)-Epigallocatechin

High-performance liquid chromatography and liquid chromatography-mass spectrometry analysis of *Terminalia arjuna* confirmed that it contains flavan-3-ols such as (+)-catechin, (+)-gallocatechin and (−)-epigallocatechin. Phenolic acids such as gallic acid and ellagic acid and its derivatives were also found in *T. arjuna* extract. Ellagic acid derivatives were isolated and their spectral studies indicated that isolated compounds were 3-O-methyl-ellagic acid 4-O-β-D-xylopyranoside, ellagic acid and 3-O-methyl ellagic acid 3-O-rhamnoside. Hydrolysis and thiolysis studies of high-molecular-weight polyphenols indicated that they are proanthocyanidins. Given these results, it may be possible to attribute the heart-health effects of *T arjuna* to these polyphenols which may be responsible for the endothelial benefit functions like tea (Saha et al. 2012; Singh and Chauhan 2014).

9.3.38 Gingerol

Gingerol is a major pungent compound of *Zingiber officinale*. It is reported to be an antioxidant and anticancer compound and is commonly used in curing many ailments in different places.

9.4 Conclusion

States in NE India have many ethnic groups with rich cultures and traditional knowledge. There is need of extensive survey and documentation of the health care practices of these tribes. Currently, there are many reports on scientific experimentations justifying the aspects attributing to the medicinal properties of plants used by them. These medicinal plants may have single use or multiple uses. Overexploitation is becoming a major ground for the extinction of the medicinal plants. Plants, which were not accessible or documented earlier for their therapeutic potential, can be a new source of research. Consequently, a sustainable utilization as well as conservation of the significant medicinal plants of NE India must be obligatory responsibility of each state and the people in it; else they would be lost before proper documentation and utilization in of ethnomedicine. Understanding the scenario of the uses and importance in the traditional healthcare system, we strongly draw the attention for future research on the prospect of development of biomedicines from the medicinal plants of Northeast India.

References

Amin MR, Mondol R, Habib MR et al (2012) Antimicrobial and cytotoxic activity of three bitter plants enhydrafluctuans, andrographispeniculataand-clerodendrumviscosum. Adv Pharm Bull 2:207–211

Appelboom T, Maes N, Albert A (2014) Anew curcuma extract (flexofytol) in osteoarthritis: results from a belgian real-life experience. Open Rheumatol J 8:77–81

Assmann G, Cullen P, Erbey J et al (2006) Plasma sitosterol elevations are associated with an increased incidence of coronary events in men: results of a nested case–control analysis of the Prospective Cardiovascular Münster (PROCAM) study. Nutr Metab Cardiovasc Dis 1:13–21

Balestrieri E, Pizzimenti F, Ferlazzo A et al (2011) Antiviral activity of seed extract from *Citrus bergamia* towards human retroviruses. Bioorg Med Chem 19(6):2084–2089

Bhakta GP (1996) Outline of geography of Meghalaya. Raj Publishing Concern, India. http://nopr.niscair.res.in/bitsream/123456789/9370/1/IJTK%203(4)%20358-364.pdf

Bhutani KK (2008) Herbal wealth of North-East India database and appraisal. National Instituteof Pharmaceutical Science and Research, Punjab

Boland DJ, Brophy JJ, House PN (1991) Eucalyptus leaf oils: use, chemistry, distillation and marketing. Inkata Press, Melbourne, p 6. ISBN 0-909605-69-6

Borah PC, Bhattacharjya DK (2008) Medicinal weeds of crop fields and role of women in rural health and hygiene in Nalbari district, Assam. IJTK 3:501–504

Bose SK, Yadav RK, Mishra S (2013) Effect of gibberellic acid and calliterpenone on plant growth attributes, trichomes, essential oil biosynthesis and pathway gene expression in differential manner in *Mentha arvensis* L. Plant Physiol Biochem 66:150–158
Boselli E, Bendia E, Di Lecce G et al (2009) Ethyl caffeate from Verdicchio wine: chromatographic purification and in vivo evaluation of its antifibrotic activity. J Sep Sci 21:3585–3590
Bouldin AS, Smith MC, Garner DD et al (1999) Pharmacy and herbal medicine in the US. Social Sci Med 49:279–289
Chakraborty N (1979) Nagaland -Nagamese-te Nagaland State LagaKotha. Assam Sahitya Sabha, Jorhat, pp 1–6
Chakraborty R, De B, Sen S et al (2012) North-East India an ethnic storehouse of unexplored medicinal plants. J Nat Prod Plant Res 1:143–152
Chatterjee S, Dutta P, Ghosh D et al (2006) Biodiversity significance of North East India. Forest Conservation Programme, New Delhi
Chiang Y-M, Lo C-P, Chen Y-P et al (2005) Ethyl caffeate suppresses NF-κB activation and its downstream inflammatory mediators, iNOS, COX-2, and PGE2 in vitroor in mouse skin. Br J Pharmacol 3:352–363
Chidambara M, Jayaprakasha K, Kumar GK et al (2011) Citrus limonin and its glucoside inhibit colon adenocarcinoma cell proliferation through apoptosis. J Agric Food Chem 6:2314–2323
Chien SC, Young PH, Hsu YJ, Chen CH, Tien YJ, Shiu SY, Li TH, Yang CW, Marimuthu P, Tsai LFL, Yang WC (2009) Anti-diabetic properties of three common Bidens pilosa variants in Taiwan. Phytochemistry 70:1246–1254
Chulasiri M, Bunyapraphatsara N, Moongkarndi P (2006) Mutagenicity and antimutagenicity of hispidulin and hortensin, the flavonoids from Millingtonia hortensis L. Environ Mol Mutagen 20(4):307–312
Das T, Mishra SB, Saha D, Agarwal S (2012) Ethnobotanical survey of medicinal plants used by ethnic and rural people in Eastern Sikkim Himalayan region. AJBAS 4(1):16–20
Das B, Talukdar AD, Choudhury MD (2014) A few traditional medicinal plants used as antifertility agents by ethnic people of Tripura, India. Int J Pharm Pharm Sci 6(3):47–53
Documentation for Immediately Dangerous to Life or Health Concentrations (IDLHs), Oxalic acid (2015) The National Institute for Occupational Safety and Health (NIOSH) http://www.cdc.gov/niosh/idlh/144627.html. Accessed 26 Mar 2015
Doley B, Gajurel PR, Rethy P et al (2010) Lesser known ethno medicinal plants used by the Nyshi community of Papumpare district, Arunachal Pradesh. IJB 1:34–36
Dolui AK, Sharma HK, Marein TB, Lalhriatpuii TC (2004) Folk herbal remedies of Meghalaya. IJTK 4:358–364
Dorland (2011) Dorland's illustrated medical dictionary. Elsevier, Philadelphia
Dutta BK, Dutta PK (2005) Potential of ethnobotanical studies in North East India: an overview. Indian J Tradit Knowl 4:7–14
Dwivedi G, Dwivedi S (2007) History of Medicine: Sushruta – the Clinician – Teacher par Excellence. Indian J Chest Allied Sci 49:243–244, National Informatics Centre
Fabricant DS, Farnsworth NR (2001) The value of plants used in traditional medicine for drug discovery. Environ Health Perspect 1:69–75. doi:10.1289/ehp.01109s169
Fabjan N, Rode J, Kosir IJ, Wang Z, Zhang Z, Kreft I (2003) Tartary Buckwheat (Fagopyrum tataricum Gaertn.) as a Source of Dietary Rutin and Quercitrin. J Agric Food Chem 51(22):6452–6455
Fotso AF, Longo F, Djomeni PD et al (2014) Analgesic and antiinflammatory activities of the ethyl acetate fraction of Bidenspilosa (Asteraceae). Inflammopharmacology 2:105–114
Germplasm Resources Information Network: Dillenia indica (2015) Wikipedia, http://en.wikipedia.org/wiki/Germplasm_Resources_Information_Network. Accessed 25 Mar 2015
Hayashi T et al (2011) Investigation on traditional medicines of Guarany Indio and studies on diterpenes from Scopariadulcis. YakugakuZasshi 131(9):1259–1269
Hoffmann B, Holzl J (1988) New chalcones from *Bidens pilosa*. Planta Med 54:52–54
Huang CF, Chen YW, Yang CY (2011) Extract of lotus leaf (Nelumbonucifera) and its active constituent catechin with insulin secretagogueactivity. J Agric Food Chem 4:1087–1094
Jamir TT, Sharma HK, Dolui AK (1999) Folklore medicinal plants of Nagaland, India. Fitoterapia 70:395–401
Jennifer B, Sherman PW (1998) Antimicrobial functions of spices: why some like it hot. Q Rev Biol 73(1):3–49. doi:10.1086/420058
Jiang AJ, Jiang G, Li LT, Zheng JN (2015) Curcumin induces apoptosis through mitochondrial pathway and caspases activation in human melanoma cells. Mol Biol Rep 1:267–275
Jung SH, Kim SJ, Jun BG (2013) α-Cyperone, isolated from the rhizomes of Cyperusrotundus, inhibits LPS-induced COX-2 expression and PGE2 production through the negative regulation of NFκB signalling in RAW 264.7 cells. J Ethnopharmacol 147:208–214
Kashiwada Y, Aoshima A, Ikeshiro Y et al (2005) Anti-HIV benzylisoquinoline alkaloids and flavonoids from the leaves of Nelumbonucifera, and structure-activity correlations with related alkaloids. Bioorg Med Chem 2:443–448
Kaur M, Singh A, Kumar B (2014) Neuroprotective effect of Tinosporacordifolia ethanol extract on 6-hydroxy dopamine induced Parkinsonism. Indian J Pharmacol 2:176–180
Klocke JA, Darlington MV, Balandrin MF (1987) 8-Cineole (eucalyptol), a mosquito feeding and ovipositional repellent from volatile oil of Hemizoniafitchii (Asteraceae). J Chem Ecol 12:2131
Konishi T, Nishio T, Kiyosawa S et al (1996) The constituents of Taxilluskaempferi and the host, Pinusthunbergii. Catechins and flavones of Taxilluskaempferi. Yakugaku Zasshi 116(2):148–157

Krishnamurthy RG, Senut MC, Zemke D (2009) Asiatic acid, a pentacyclic triterpene from Centellaasiatica, is neuroprotective in a mouse model of focal cerebral ischemia. J Neurosci Res 11:2541–2550

Kumar S, Kumar V, Prakash O (2013) Enzymes inhibition and antidiabetic effect of isolated constituents from Dilleniaindica. Biomed Res Int 2013:382063. doi:10.1155/2013/382063

Kuo SC, Teng CM, Lee LG et al (1991) 6-Pentadecylsalicylic acid: an antithrombin component isolated from the stem of Rhussemialata var. roxburghii. Planta Med 3:247–249

Li S, Kuang HX, Okada Y, Okuyama T (2005) New flavanone and chalcone glucosides from *Bidens bipinnata* Linn. J Asian Nat Prod Res 7:67–70

Lin HY, Kuo YH, Lin YL, Chiang W (2009) Antioxidative effect and active components from leaves of Lotus (Nelumbo nucifera) J Agric Food Chem 12; 57(15):6623-9

Lokho A (2012) The folk medicinal plants of the Mao Naga in Manipur, North East India. IJSRP 2(6):1–8

Lorenzetti BB, Souza GR, Sarti SL et al (1991) Myrcene mimics the peripheral analgesic activity of lemongrass tea. J Epharmacol 1:43–48

Maity D, Pradhan N, Chauhan AS (2004) Folk uses of some medicinal plants from North Sikkim. Indian J Tradit Knowl 3(1):66–71

Mallick S, Kumar D, Vedasiromoni JR, Pal BC (2010) Anti-leukemic activity of *Dillenia indica* L. fruit extract and quantification of betulinic acid by HPLC. Phytomedicine 17(6):431–435

Mammela P, Savolainenb H, Lindroosa L, Kangasd J, Vartiainen T (2000) J Chromatogr A 891(1):75–83

Materska M, Perucka I (2005) Antioxidant activity of the main phenolic compounds isolated from hot pepper fruit (Capsicum annuum L). J Agric Food Chem 53(5):1750–1756

McCann MJ, Johnston S, Reilly K, Men X, Burgess EJ, Perry NB, Roy NC (2014) The effect of turmeric (Curcuma longa) extract on the functionality of the solute carrier protein 22 A4 (SLC22A4) and interleukin-10 (IL-10) variants associated with inflammatory bowel disease. Nutrients 6(10):4178–4190

Meulenbeld GJ, Wujastyk D (2001) Studies on Indian medical history. Motilal Banarsidas, New Delhi

Morais D, Liana CL, José M et al (1998) Central depressant effects of reticuline extracted from Ocoteaduckei in rats and mice. J Ethnopharmacol 1:57–61

Mukherjee TK (2005) Editorial. Indian J Tradi Knowl 4:3–4

Murakami A, Takahashi D, Kinoshita T, Koshimizu K, Kim HW, Yoshihiro A, Nakamura Y, Jiwajinda S, Terao J, Ohigashi H (2002) Zerumbone, a Southeast Asian ginger sesquiterpene, markedly suppresses free radical generation, proinflammatory protein production, and cancer cell proliferation accompanied by apoptosis: the α, β-unsaturated carbonyl group is a prerequisite. Carcinogenesis 23(5):795–802

Murphy L (2000) American ginseng in the prevention and treatment of human breast cancer. Southern Illinois University Carbondale, Carbondale

Mustafa RA, Abdul Hamid A, Mohamed S, Bakar FA (2010) Total phenolic compounds, flavonoids, and radical scavenging activity of 21 selected tropical plants. J Food Sci 75(1):C28–C35

Nagappan T, Ramasamy P, Vairappan CS (2012) Chemotaxonomical markers in essential oil of Murraya koenigii. Nat Prod Commun 7(10):1375–1378

Ono E, Inoue J, Hashidume T, Shimizu M, Sato R (2011) Anti-obesity and anti-hyperglycemic effects of the dietary citrus limonoid nomilin in mice fed a high-fat diet. Biochem Biophys Res Commun 410(3):677–681

Oxalic Acid Material Safety Data Sheet (2014) Radiant Indus Chem. Accessed 20 May 2014.

Panda AK, Misra S (2010) Health traditions of Sikkim Himalaya. J Ayurv Integ Med 3:183–189

Ramakrishnan PS (1992) Tropical forests, exploitation, conservation and management. Impact Sci Soc 42:149–162

Roh SS, Park SB, Park SM, Choi BW, Lee MH, Hwang YL, Kim CH, Jeong HA, Kim CD, Lee JH (2013) Novel compound rasatiol isolated from Raphanus sativus has a potential to enhance extracellular matrix synthesis in dermal fibroblasts. Ann Dermatol 25(3):315–320

Saha A, Pawar VM, Jayaraman S (2012) Characterisation of polyphenols in Terminalia arjuna bark extract. Indian J Pharm Sci 74(4):339–347

Sajem AL, Gosai K (2006) Traditional use of medicinal plants by the Jaintia tribes in North Cachar Hills district of Assam, northeast India. J Ethnobiol Ethnomed 2:33

Schindler R, Mentlein R (2006) Flavonoids and vitamin E reduce the release of the angiogenic peptide vascular endothelial growth factor from human tumor cells. J Nut 136(6):1477–1482. PMID 16702307

Sfara V, Zerba EN, Alzogaray RA (2009) Fumigant insecticidal activity and repellent effect of five essential oils and seven monoterpenes on first-instar nymphs of Rhodnius prolixus. J Med Entomol 46(3):511–515

Sharma HK, Chhangte L, Ashoke DK (2001) Traditional medicinal plants in Mizoram, India. Fitoterapia 72:146–161

Singh PP, Chauhan SM (2014) Activity-guided isolation of antioxidants from the leaves of Terminalia arjuna. Nat Prod Res 28(10):760–763

Sousa DP, Junior EV, Oliveira FS et al (2007) Antinociceptive activity of structural analogues of rotundifolone: structure-activity relationship. Z Naturforsch C 62(1–2):39–42

Stefanovic S, Olmstead RG (2014) Testing the phylogenetic position of a parasitic plant (*Cuscuta,* Convolvulaceae, Asteridae): Bayesian inference and the parametric bootstrap on data drawn from three genomes. Syst Biol 53(3):384–399

Sun C, Nirmalananda S, Jenkins CE, Debnath S, Balambika R, Fata JE, Raja KS (2013) First ayurvedic approach towards green drugs: anti cervical cancer-cell properties of Clerodendrum viscosumroot extract. Anticancer Agents Med Chem 13(10):1469–1476

Tangjanga S, Namsab ND, Arana C, Litin A (2011) An ethnobotanical survey of medicinal plants in the Eastern Himalayan zone of Arunachal Pradesh. India J Ethnopharmacolo 134:18–25

Tansuwanwong S, Yamamoto H, Imai K, Vinitketkumnuen U (2008) Antiproliferation and apoptosis on RKO colon cancer by Millingtonia hortensis. Plant Foods Hum Nutr (Formerly Qualitas Plantarum) 64(1):11–17

Thulasiraman P, McAndrews DJ, Mohiudddin IQ (2014) Curcumin restores sensitivity to retinoic acid in triple negative breast cancer cells. BMC Cancer 27;14:724

Tony S (ed) (1968) Plants in the development of modern medicine. Harvard University Press, Cambridge. ISBN 0-674-67330-1. http://www.traffic.org/medicinal-plants/

Wang X, Sun W, Cao J, Qu H, Bi X, Zhao Y (2012) Structures of new triterpenoids and cytotoxicity activities of the isolated major compounds from the fruit of Momordica charantia L. J Agric Food Chem 60(15):3927–3933

Wang L, Jiang Y, Han T, Zheng C, Qin L (2014) A phytochemical, pharmacological and clinical profile of Paederia foetida and P. scandens. Nat Prod Commun 9(6):879–886

Waterman C, Cheng DM, Rojas-Silva P, Poulev A, Dreifus J, Lila MA, Raskin I (2014) Stable, water extractable isothiocyanates from *Moringa oleifera* leaves attenuate inflammation *in vitro*. Phytochemistry 103:114–122

Weyant MJ, Carothers AM, Dannenberg AJ, Bertagnolli MM (2001) (+)-Catechin inhibits intestinal tumor formation and suppresses focal adhesion kinase activation in the Min/+ Mouse. Cancer Res 61:118

Yang H, Jiang B, Hou AJ, Lin ZW, Sun HD (2000) Colebroside A, a new diglucoside of fatty acid ester of glycerin from Clerodendrum colebrookianum. J Asian Nat Prod Res 2(3):177–185

Yang XL, Hsieh KL, Liu JK (2007) Guajadial: an unusual meroterpenoid from guava leaves Psidium guajava. Org Lett 9(24):5135–5138

Yoon JS, Yang H, Kim SH, Sung SH, Kim YC (2010) Limonoids from Dictamnus dasycarpus protect against glutamate-induced toxicity in primary cultured rat cortical cells. J Mol Neurosci 42(1):9–16

Yuan LP, Chen FH, Ling L (2008) Protective effects of total flavonoids of *Bidens bipinnata* L. against carbon tetrachloride-induced liver fibrosis in rats. J Pharm Pharmacol 60:1393–1402

YumnamRS DCO, Abujam SK, Chetia D (2012) Study on the ethnomedicinal system of Manipur. Int J Pharm Biol Arch 3(3):587–591

Zhou Q, Chen L, Yin HJ, Li B, Tian Y, Chen HW, Xiao YH, He XH, Zeng YL, Dong JX (2014) Two new flavonols from Cudrania cochinchinensis. J Asian Nat Prod Res 16(10):976–981

10 Traditional Anti-poison Plants Used by the Monpa Tribe of Arunachal Pradesh

Jambey Tsering, Hui Tag, Baikuntha Jyoti Gogoi, and Vijay Veer

Abstract

Arunachal Pradesh is a treasure house of biodiversity as well as traditional knowledge. The state harbours over 800 medicinal plants. Rising demands of significant medicinal plants such as *Taxus baccata*, *Paris polyphylla*, *Swertia chirayita*, *Neopicrorhiza scrophulariiflora*, etc. for developing pharmaceutical drugs have drastically reduced the wild population of these species. Some important medicinal plants of the state, their bioactive efficacy and conservation status are mentioned in this paper. This paper also describes the traditional use of 13 ethnomedicinal plants used by the *Monpa* tribe of Arunachal Pradesh as antidote against food poisoning, snake bite, scorpion bite and insect bites. Some of the potent antidote plants claimed by the *Monpa* tribe are *Aconitum heterophyllum*, *Asparagus racemosus*, *Ligularia amplexicaulis*, *Rhododendron hodgsonii*, *Swertia hookeri* and *Verbascum thapsus*. Out of the 13 traditional antidote plants, six are new to science as traditional medicine and are yet to be pharmaceutically validated for their bioactivity. Given the strong traditional cultural use of the plants, it is expected that biochemical and pharmacological studies would reveal the bioactive compounds, which could be further developed as holistic evidence-based drugs to cure some complicated ailments/diseases.

10.1 Introduction

Arunachal Pradesh is a treasure house of medicinal plants because of its climatic and ecological diversities and favourable microclimatic condition (Mani 1974; Rodgers and Panwar 1988; Singh et al. 2007). The entire territory forms a complex hill system with varying elevations ranging from 50 m to about 7000 m, traversed

J. Tsering • H. Tag (✉)
Department of Botany, Rajiv Gandhi University, Rono Hills, Doimukh 791112, Arunachal Pradesh, India
e-mail: huitag2008rgu@gmail.com

B.J. Gogoi • Vijay Veer
Defence Research Laboratory (Defence Research and Development Organisation), Tezpur 784 001, Assam, India

© Springer India 2016

Vijay Veer, R. Gopalakrishnan (eds.), *Herbal Insecticides, Repellents and Biomedicines: Effectiveness and Commercialization*, DOI 10.1007/978-81-322-2704-5_10

throughout by a number of rivers and rivulets. Forest of different climatic types from tropical, subtropical, pine, temperate subalpine to alpine is prevalent in the state. The state is known as the hub of medicinal plants and has been identified as one of the biodiversity hotspot areas in the world (Myers et al. 2000). The state is inhabited by 26 major tribes and 110 sub tribes, which have their own traditional ways of using the forest resources for sustaining livelihood (Tag and Das 2004). The tribes inhabiting the state recognise over 800 species of plants as having medicinal properties (Gajural et al. 2003). These plants are found in different vegetation and altitude gradient.

The *Monpa* tribe is one of the largest tribes among the 26 major tribal communities found in the state. *Monpas* inhabit two districts of the state (Tawang and West Kameng) with a total area of 9507 km^2 (Census 2011). Anthropological reports have confirmed the fact that the *Monpas* undoubtedly are belonging to mongoloid racial stock and they are of Tibeto-Burman linguistic groups (Bradley 1997). *Monpas* are thought to have migrated from Tibet and Bhutan over past many centuries. They share close similarities in the utilisation of biological resources as well as cultural and religious similarities with both Bhutan and Tibet. Due to the lack of traditional Tibetan medicine system (*Sowa-rigpa*) in the area, *Monpas* have developed reliable traditional botanical-based medicinal knowledge over the centuries while living in close association with nature. Their medicinal and cultural knowledge have been proven effective for mitigation of various types of ailments or diseases that often plagued the community in the earlier centuries (Tag et al. 2014). The use of antidote plants by different tribes of India has been reported in recent decades, which is evident from the works of Sikdar and Dutta (2008), Kumar and Singh (2014), Gnanavel and Franklin (2014) and Soman (2014). However, no literature evidence which focused on antidote plants used by tribal peoples of Arunachal Pradesh was found. Efforts have been made in the present study to highlight the folklore claims of antidote plants used by the *Monpa* tribe administered through traditional *Bonpo* (priest) and *Menpa* (medicine man) in the treatment of snake bites, scorpion bites, food poisoning and contamination with poisonous plants.

10.2 Medicinal Plant Wealth of Arunachal Pradesh

Arunachal Pradesh with a total forest cover of 81.22 % is enriched by over 800 species of medicinal plants, 650 species of orchids, 85 species of terrestrial mammals and 500 species of birds (Tag et al. 2005). Such a rich biodiversity in the state coupled with rich cultural diversity have provided an initial advantage to its inhabitants for observing and scrutinising the rich flora and fauna for developing their own traditional knowledge in curing various ailments. Medicinal plants continue to provide health security to millions of rural people all over the world. According to WHO's estimates, over 80 % of the people in the developing countries and 65 % of India's population depend upon traditional medicines for sustenance and healthcare needs (Farnsworth and Soejarto 1991). India is the world's second largest exporter of medicinal plants after China. The vast majority of medicinal plants are harvested from the natural forests. A bulk of it is traded, generating a turnover of US$2.5 billion annually.

The demand for medicinal plants is ever increasing as people are more and more fascinated towards herbals. The demand for herbal crude drugs from pharmaceutical companies is another major factor, leading to extinction of significant medicinal plants from its wild habitat. The rising demand and the destructive harvesting practices are not only threatening the survival of many species, but also the livelihood of the people who depend on the products. An estimated 316 species in India are under the threat of extinction. Since 2008, UNDP has been supporting the state medicinal plant boards in the three biodiversity-rich states (Arunachal Pradesh, Uttarakhand and Chhattisgarh) to devise and implement strategies that promote sustainable use and conservation of medicinal plants. One such strategy that has worked wonders for the communities is the setting up of medicinal plant conservation areas (MPCAs). MPCAs are natural forest areas established and managed by the state forest departments in collaboration with local communities to conserve threatened medicinal plants. So far, 21 MPCAs have been set up, conserving 32 globally significant medicinal plants and numerous other plants (Tag, 2012). Below are enumerated 20 GSMPs found in the state (Table 10.1).

Table 10.1 Some globally significant medicinal plants of Arunachal Pradesh and their bioactivity

Sl. No.	Plants	Family	Part used	Bioactivity/bio-constituents	References
1.	*Aconitum heterophyllum* Wall. ex Royle	Ranunculaceae	Rt	Dihydroatisine, heterophyllinine, lycoctonine[1], antibacterial, anti-inflammatory[2], antioxidant[3]	1: Nisar et al. 2009; 2: Verma et al. 2010; 3: Prasad, et al. 2012
2.	*Bergenia ciliata* (Haw.) Sternb.	Saxifragaceae	Rt	Bergenin, catechin, gallic acid[4], antibacterial[5], anti-inflammatory[6], cytoprotective[7]	4: Dhalwal et al. 2008; 5: Sinha et al. 2001a; 6: Sinha et al. 2001b;
3.	*Cinnamomum tamala* (Buch.-Ham.) T. Nees & Eberm.	Lauraceae	Lf	Myrcene, α-pinene, germacrene A[8], antihyperlipidemic, antidiabetic, antibacterial, antioxidant, antimicrobial, anti-inflammatory, antidiarrhoeal[9]	8: Dhulasavant et al. 2010; 9: Mir et al. 2004
4.	*Coptis teeta* Wall.	Ranunculaceae	Rt	Berberine, coptine[10]	10: Latif et al. 2010
5.	*Embelia ribes* Burm. f.	Primulaceae	Fr	Embelin, anticonvulsant[11]; antioxidant, neuroprotective[12]	11: Mahendran et al. 2011; 12: Ansari et al. 2008
6.	*Fritillaria cirrhosa* D. Don	Liliaceae	Rt	Imperialine[13], anti-inflammatory, antitussive[14]	13: Chan et al. 2000; 14: Wang et al. 2011
7.	*Gentiana kurroo* Royle	Gentianaceae	Wh	Lupeol, ursolic acid[15], anti-inflammatory, immunomodulatory[16], antibacterial, antioxidant[17], analgesic[18]	15: Maurya et al. 2012; 16: Latif et al. 2006; 17: Baba and Malik 2014; 18: Wani et al. 2011;
8.	*Illicium griffithii* Hook. f. & Thomson	Schisandraceae	Br	α-Pinene, limonene, linalool[19], antibacterial, antifungal[20]	19: Dung et al. 1995; 20: Vijayakumar et al. 2012
9.	*Mahonia napaulensis* DC.	Berberidaceae	Fr	Antifungal[21]	21: Bajpai and Vankar 2007
10.	*Nardostachys jatamansi* (D. Don) DC.	Caprifoliaceae	Rt	Jatamansin[22], jatamansinol, angelicin, visnadin[23], norseychelanone, α- and β-patchoulenes[24], antioxidant[25], hepatoprotective, neurotoxic[26]	22. Shanbhag et al. 1964; 23: Shanbhag et al. 1965; 24: Rucker et al. 1976; 25: Sharma and Singh 2012; 26: Rao et al. 2005
11.	*Neopicrorhiza scrophulariiflora* (Pennell) D.Y. Hong	Plantaginaceae	Rt	Iridoid[27], picroside I, scroneoside A[28], antioxidant, antineoplastic[29], immunomodulatory, anti-inflammatory[30]	27: Li et al. 1998; 28: Wang et al. 2013; 29: Rajkumar et al. 2011; 30: Smit et al 2000
12.	*Oroxylum indicum* (L.) Kurz	Bignoniaceae	Sd, Br	Baicalein, chrysin, oroxylin, anticancer[31], anti-inflammatory, antimicrobial[32], antioxidant[33]	31: Chen et al. 2003; 32: Ali et al. 1998; 33: Yan et al. 2011
13.	*Paris polyphylla* Sm.	Melanthiaceae	Rt	Diosgenin, polyphyllin[34]; anticancer[35]	34: Singh et al 1980; 35:Yan et al. 2009

(continued)

Table 10.1 (continued)

Sl. No.	Plants	Family	Part used	Bioactivity/bio-constituents	References
14.	*Piper pedicellatum* C. DC.	Piperaceae	Fr	Pedicellamide, antifungal, antioxidant[36]	36: Tamuly et al. 2013
15.	*Rheum nobile* Hook. f. & Thomson	Polygonaceae	Rt	Quercetin[40]	40: Iwashina et al. 2004
16.	*Rhododendron anthopogon* D. Don	Ericaceae	Lf, Fl	Epicatechin, quercitrin, syringic acid[41], α-pinene, β-pinene, limonene, inflammatory, antimicrobial, antiproliferative[42]	41: Sharma et al. 2010; 42: Innocenti et al. 2010
17.	*Sinopodophyllum hexandrum* (Royle) T.S. Ying	Berberidaceae	Wh	Picropodophyllone, isopicropodophyllone, dehydropodophyllotoxin, antifungal[37], radioprotective, antioxidant[38], cytotoxic, insecticides[39]	37: Rahman et al. 1995; 38: Arora et al. 2005; 39: Miyazawa et al. 1999
18.	*Swertia chirayita* (Roxburgh) H. Karst.	Gentianaceae	Wh	Swerchirin, swertianin, anticarcinogenic, antidiabetic, antioxidant, hepatoprotective[43]	43: Joshi and Dhawan 2005
19.	*Taxus baccata* L.	Taxaceae	Lf	Taxanes, taxicin-I, taxicin-II[44], taxol, anticancer[45], antibacterial[46]	44: Appendino et al. 1992; 45: Malik et al. 2011; 46: Reddy et al. 2001
20.	*Valeriana jatamansi* Jones	Caprifoliaceae	Rt	Bakkenolide[47], iridoid[48], neuroprotective[47, 48], antioxidant, apoptosis[49]	47: Xu et al. 2011a; 48: Xu et al. 2011b; 49: Bhatt et al. 2012

Rt root; *Fr* fruit; *Lf* leaf; *Fl* flower; *Br* bark; *Sd* seed; *Wh* whole plant

10.3 Antidote Plants Used by *Monpa* Tribe and Their Bioactivity

The *Monpa* tribe uses a good number of medicinal plants to treat a variety of diseases and ailments using pre-Buddhist local healing methods called *Bon*. The people residing in the remote and far-flung areas, particularly, the *Brokpas* (Yak herder), are the main ethomedicinal knowledge holders. *Brokpas* have their own *Bonpo* or *Menpa* who treats the ailing patients of the villages. *Bonpo* (priest with supernatural attributes) treats the patient with herbs by chanting of mantra, addressing the *Phu* (the mountain gods, the supreme soul) for early recovery of the patients suffering from ailments due to food poisoning, snakebite, inflammation, contagious diseases and insect bites. In the absence of modern medical facilities in remote mountain ecosystem, plant-based herbal medication is the first choice for treatment of diseases/ailments among the *Monpa* living in Himalaya and sub-Himalayan region since many centuries.

During the study conducted between 2010 and 2014 in the *Monpa*-inhabited regions of Arunachal Pradesh, 13 plants were reported to be used as antidote against snake bite, scorpion bites, body toxication, etc. Altogether, about 220 informants were interviewed during the study. Information regarding the habits and habitats of antidote plants, parts used, time of harvesting, dose preparation, mode of treatment and vernacular name were collected directly from the traditional healers (*Bonpo* and *Menpa*), village heads, elderly people, cattle herder and hunters who occasionally act as *Menpa* (herbalists). The plant were identified and authenticated by Botanical Survey of India, Arunachal Regional Centre, Itanagar. Voucher specimens of each species were collected and deposited at Plant Systematic and Pharmacognosy Research Laboratory,

Department of Botany, Rajiv Gandhi University, Doimukh, Arunachal Pradesh. Enumerated below are the 13 plants used as antidote agents by the *Monpa* tribe of Arunachal Pradesh. Photographs of six selected best known antidotes plants used by the tribe are presented in Fig. 10.1.

[**Coll. no.**, Collection number; **local nm.**, local name]

10.3.1 *Aconitum heterophyllum* Wall. ex Royle [Ranunculaceae]

Coll. no. and local nm JT/HT/149/2012; Gonga-karpu.

Habit and habitat Herbs; mountain slopes between 3500 and 3600 m.

Plant characters Erect, up to 1.5 m tall with taproots. Leaves

Fig. 10.1 (**a**) *Aconitum heterophyllum* Wall. ex Royle; (**b**) *Asparagus racemosus* Willd.; (**c**) *Ligularia amplexicaulis* DC.; (**d**) *Rhododendron hodgsonii* Hook. f.; (**e**) *Swertia hookeri* C.B. Clarke; (**f**) *Verbascum thapsus* L

palmately divided, dentate, acute. Inflorescence corymbose raceme; pedicel 7–10 cm with 2 bracteoles. Flowers bisexual, zygomorphic. Sepals 5, petaloid, greenish-yellow. Petals 5, creamy colour. Stamens numerous, ca. 5–8 mm; anthers ellipsoid-globose. Carpels 2; style long, ca. 3–5 mm.

Traditional uses Water decoction or a small part of dried tuber is given to the patient to treat aconite contamination, snake and scorpion bite. The root infusion is given during stomachache, high fever and hereditary diseases and also used for any unknown disease.

The plant is known as 'Atees' (Sanskrit) and considered as an effective Ayurvedic medicine. It is naturally found in the Himalayan region at an altitude between 2000 and 4000 m and cultivated throughout the world. The tuber is reported as anti-inflammatory (Verma et al. 2010), antioxidant (Prasad et al. 2012), antibacterial (Ahmad et al. 2008), antifungal, cytotoxic (Anwar et al. 2003), antiviral (Pandey et al. 2004) and immune-stimulant properties (Atal et al.1986). The major bioactive constituents isolated so far are dihydroatisine, heterophyllinine and lycoctonine (Nisar et al. 2009).

10.3.2 *Anaphalis busua* (Buch.-Ham.) DC. [Asteraceae]

Coll. no. and local nm JT/HT/446/2013; Tayong.

Habit and habitat Herbs; grasslands and roadsides between 2100 and 2800 m.

Plant characters Perennial, up to 1 m high; stems erect, woolly pubescent. Leaves linear-lanceolate to oblanceolate, 1-nerved. Heads in terminal corymbose clustered; flowers yellowish-green.

Traditional uses Low concentrated infusion of shoots is given during food poisoning. Also used to treat body inflammation, blood discharge and contaminated diseases. The leaves are used for external blood clotting.

The plant is known as 'tall pearly everlasting'. The plant is reported to be grown from Bhutan, south China, north India and Nepal. In many countries, the plant is considered as invasive weeds. Phytochemical and pharmacological activities are yet to be analysed.

10.3.3 *Anaphalis triplinervis* (Sims) Sims ex C.B.Clarke [Asteraceae]

Coll. no. and local nm JT/HT/285/2012; Dhakoi.

Habit and habitat Herbs; grassy slopes (2000–2300 m).

Plant characters Perennial, up to 50 cm high; stem woody. Leaves broadly elliptic to oblong, woolly hair abaxially, 3 nerved. Heads solitary or few in terminal corymb; flowers white.

Traditional uses	Infusion of shoots is given during food poisoning, inflammation, blood discharge and contaminated diseases.

Common name of *A. triplinervis* is 'triple-nerved pearly everlasting'. It grows wild throughout the Himalayan region up to 2300 m altitude. No phytochemical and pharmacological analyses have been carried out.

10.3.4 *Arisaema utile* Hook.f. ex Schott [Araceae]

Coll. no. and local nm	JT/HT/406/2013; Dawa.
Habit and habitat	Herbs; under *Abies* forests (3500–3600 m).
Plant characters	Dioecious; tuber depressed, globose, 3–10 cm in diam., with small tubercles. Leaf solitary; petiole green with dark purple spots, 20–40 cm, leaf blade 3-foliolate; leaflets sessile, leaflet ovate-rhombic, 7–15×6–11 cm, cuneate, acuminate; spathe tube purplish brown cylindric, 4–5×1.5–2 cm, spadix unisexual.
Mode of uses	Water decoction of dried fruits is consumed to remove all types of poison contamination in body. Tuber is used in reproductive and bone-related diseases. Fruit is also believed to improve blood circulation.

This very intriguing plant has a flower that is shaped like an extended hood of a cobra and hence called as 'Sikkim cobra lily'. The plant grows well in Bhutan, south China, north India, Myanmar, Nepal and Pakistan at an altitude between 2800 and 3100 m. Very limited studies have been carried out on its pharmacological activities. Antimicrobial, cytotoxic and antioxidant activities were confirmed by Mubashir and Shah (2012) and antiproliferative activity by Dhuna (2010).

10.3.5 *Asparagus racemosus* Willd. [Asparagaceae]

Coll. no. and local nm	JT/HT/418/2013; Nge, Ngalangma.
Habit and habitat	Climbers; in forests (1400–1500 m).
Plant characters	Hermaphroditic; stems branched. Cladodes in fascicles of 3–6, linear; leaf spur spinescent; spine straight or subrecurved, 1.5–2 cm on main stems, 5–10 mm on branches. Inflorescences developing after cladodes, axillary, many-flowered raceme 1–4 cm. Stamens equal, ca. 0.7 mm; anthers yellow, minute.
Mode of uses	Raw/infusion of roots is taken during food poisoning. Root is also used in reproductive and body's water storage diseases.

A. racemosus is commonly found throughout the Himalayan region, Malaysia, Myanmar, Nepal, Pakistan, and in some parts of Africa and Australia. The plant is commonly known as 'shatavari' which means curer of a hundred diseases. Due to its multiple uses, the demand for *A. racemosus* is constantly on the rise. Due to destructive harvesting, combined with habitat destruction, and deforestation, the plant is now considered endangered in its natural habitat. The tuberous root is experimentally proven as antioxidant, anticancer, anti-inflammatory, antiageing (Hayes 2008a, b) and antidepressant (Singh et al. 2009). Two novel compounds were isolated from the tuber extract: shatavarins and sarsasapogenin (Sidiq et al. 2011).

10.3.6 *Berberis angulosa* Wall. ex Hook.f. & Thomson [Berberidaceae]

Coll. no. and local nm	JT/HT/196/2012; Lae-kanchan.
Habit and habitat	Shrubs; scrublands and forests (3500–3800 m).
Plant characters	Up to 2 m tall. Spines 3-fid, 7–12 mm, slender. Leaves obovate, both surfaces glabrous, inconspicuously veined, cuneate, entire, acute. Flowers solitary. Sepals in 2 whorls. Petals obovate, ca. 6×4.8 mm. Berry shiny, red, subglobose, 10–12×9–12 mm.
Mode of uses	The decoction of flowers is consumed during food poisoning. Sometimes fruits are also used. The flowers are eaten raw during diarrhoea and are applied on eye during eye infection.

B. angulosa is reported from some parts of China, Northeast India and Nepal. It grows well between 3500 and 4500 m asl. Berberine, which has antibacterial and antitumour effects, is universally present in rhizomes of all *Berberis* species. However, no specific study on B. angulosa has been carried out.

10.3.7 *Drynaria propinqua* (Wallich ex Mett.) Bedd. [Polypodiaceae]

Coll. no. and local nm	JT/HT/068/2011; So.
Habit and habitat	Herbs; epiphytic on rocks and trees (1800–2800 m).
Plant characters	Rhizomes long, terete, 1–2 cm in diam.; scales appressed, peltate, 3–6×1–1.5 mm; fronds dimorphic, glabrous; orbicular or ovate, 10–20×7–18 cm, pinnatifid up to 3–13, irregularly dentate, acute. Sori in 1 regular row between costa and margin.
Mode of uses	The extract of rhizomes is used as antidote against toxicities resulting from meat consumption. Decoction of rhizomes is consumed to reduce fever arising from poisoning. It is also taken orally to treat cold. The rhizomes along with other plants are used to treat kidney-related problems.

D. propinqua is an epiphytic herb native to Nepal, Bhutan, China, India and Southeast Asia. The plant is commonly used in Chinese system of medicine; however, no clinical tests have yet reported. Two studies have identified 4 chemical compounds: propinqualin, 4-O-β-D-glucopyranosyl caffeic acid, β-sitosterol-3-O-β--D-glucopyranoside (Liu et al. 1992) and (-)-epiafzelechin 3-O-β-D-allopyranoside (Liu et al. 1994).

10.3.8 *Ligularia amplexicaulis* DC. [Asteraceae]

Coll. no. and local nm	JT/HT/236/2012; Rihu.
Habit and habitat	Herbs; grasslands (2900–3100 m).
Plant characters	A perennial herb, up to 1 m high; rhizome short, stout. Leaves green, large, 20–30 × 15–20 cm, obovate, rounded, dentate. Inflorescence terminal, 10 × 15 cm in diam. Flowers shaggy, yellow, 5–6 long-slender yellow rays.
Mode of uses	Infusion of root is used as antidote and to treat jaundice and liver- and lung-related diseases. Leaves and tender shoots are applied on open wound for healing.

L. amplexicaulis is commonly called as 'Leopard plant' which grows in the Himalayas on moist rocky slopes, generally besides streams, at elevations ranging between 2700 and 4800 m. Till now, no scientific research has been reported.

10.3.9 *Rhododendron hodgsonii* Hook. f. [Ericaceae]

Coll. no. and local nm	JT/HT/483/2014; Laah.
Habit and habitat	Shrubs; *Abies* forests, thickets (3600–3900 m).
Plant characters	Up to 3–7 m tall. Petiole terete, 40–50 mm. Leaves leathery, oblong-elliptic to oblanceolate, 16–30 × 5–12 cm, cuneate to rounded. Inflorescence 15–20 flowered. Corolla tubular-campanulate, magenta to purple, 3.5–4 cm; stamens 15–18, 2.5–3.5 cm, filaments glabrous; style 2–2.5 cm, glabrous. Capsule curved, 30–40 × ca. 7 mm.
Mode of uses	The tender leaves are eaten raw, which helps immediate vomiting during poison ingestion.

R. hodgsonii is known as 'Hodgson's Rhododendron' and is found between 3500 and 4000 m above sea level in Bhutan, China, North India and Nepal. No phytochemicals have been isolated so far.

10.3.10 *Solanum viarum* Dunal [Solanaceae]

Coll. no. and local nm	JT/HT/274/2012; Tag-tsa.
Habit and habitat	Herbs; in open forests, wastelands, grasslands (1600–2200 m).
Plant characters	Up to 1 m high; much branched, prickly. Leaves ovate, shallowly lobed, petiole stout 3–7 cm. Flowers in axillary cymes, ca. 2 cm across, white

	calyx, prickly. Berries globose, covered with green lines, pale yellow when ripe, 2–3 cm in diam.
Mode of uses	The root infusion is used as antidote during scorpion bite. Leaves are used as diuretic. Fruits are given as contraceptive. Fruits heated over fire are used as analgesic by crushing between the infected teeth.

S. viarum known as 'the tropical soda apple' is a perennial shrub native to Asia, Africa and South America. It is considered as an invasive species in the lower eastern coastal states of the United States and recently on the mid-north coast of Australia. No pharmaceutical research has been reported yet on this plant.

10.3.11 *Swertia hookeri* C.B. Clarke [Gentianaceae]

Coll. no. and local nm	JT/HT/346/2012; Gudue-serp.
Habit and habitat	Herbs; alpine and subalpine meadows (4000–4300 m).
Plant characters	Prostrate, up to 1 m high; stems terete, hollow, pinkish. Radical leaves spatulate-elliptic; cauline leaves elliptic, 5-nerved. Flowers in panicles, dull yellow with violet tip on outer side. Capsule oblong. Seeds discoid.
Mode of uses	The roots of plant having 8 internodes are collected and decoction is made to treat body poisoning.

S. hookeri is native to Bhutan, China, India and Nepal. A quercetin and some xanthone compounds were identified to be present: 3-*O*-stearyl-3′,4′,5,7-tetra-*O*-methylquercetin, 1-*O*-stearyl-3,5-dimethoxyxanthone, 1-hydroxy-3,5-dimethoxyxanthone, tetraoxygenated xanthones, 8-*O*-β-D-glucosyl-1,3-dihydroxy-5-methoxyxanthones and 8-*O*-β-D-glucosyl-1,5-dihydroxy-3-methoxyxanthone (Ghosal et al. 1980).

10.3.12 *Triplostegia glandulifera* Wall. ex DC. [Caprifoliaceae]

Coll. no. and local nm	JT/HT/424/2013; Gonga-karpu.
Habit and habitat	Herbs; forests, grassy slopes (2000–2300 m).
Plant characters	Perennial, erect, 15–40 cm tall; taproots nearly fusiform, 20–40 × 4–10 mm, slightly fleshy; stems angular, puberulent. Leaves opposite, obovate-lanceolate, 2–6 × 1.5–4 cm, pinnatifid, serrate. Inflorescences paniculiform; pedicels ca. 1 mm. Corolla white or rose, funnelform, 3–5 mm. Anthers white. Style slightly longer than stamens.
Mode of uses	Tuber is used as a substitute of *A. heterophyllum*. A small piece of dry tuber is eaten to stop blood discharge in dysentery.

T. glandulifera is a lesser known medicinal plant used in the Tibetan system of medicine. The plant is well distributed between 1800 and 4000 m in Himalayan region (Garhwal to Bhutan to Arunachal); south, west and central China; Burma; Taiwan; Celebes; and New Guinea. The plant is known to have anti-stimulant and hypoglycaemic activities (Liu et at. 2008a, b).

10.3.13 *Verbascum thapsus* L. [Scrophulariaceae]

Coll. no. and local nm	JT/HT/370/2012; Jyugpa-serji.
Habit and habitat	Herbs; grassy slopes, roadsides (2200–2900 m).
Plant characters	Erect, up to 1.5 m tall, densely stellate-tomentose. Basal leaves in rosette, oblanceolate, crenate, petiolate, densely tomentose; stem leaves sessile. Flowers in dense, tomentose, spicate racemes, yellow. Capsule globose
Mode of uses	The fruit extract is used during excessive antidote medication. The fruits are also used to treat liver enlargement and blood discharge from reproductive organs.

V. thapsus (great mullein) is a species native to Europe, northern Africa and Asia and introduced in the Americas and Australia. It is widely used throughout the world for herbal remedies, with well-established emollient and astringent properties. Mullein remedies are especially recommended for coughs and related problems but also used in topical applications against a variety of skin problems. Roots possess strong antiviral property (McCutcheon et al. 1995). Many investigations on the whole plant of V. thapsus could lead to isolation of verbascoside, vatic acid, α-spinasterol, siakogenins (Mehrotra et al. 1989), iridoidglycosides (Seifert et al. 1985), sterones, sesquiterpene acid (Khuroo et al. 1988) and phenylethanoid and lignan glycosides (Warashina et al. 1991).

10.4 Commercial Prospects of Medicinal Plants Used by *Monpa* Tribe

One of the most significant antidote plants found in the study area was *A. heterophyllum*, locally known as *Gonga-karpu*. The tuber of *A. heterophyllum* is highly valued in the *Monpa* culture and used in the treatment of a variety of ailments like stomachache, high fever, hereditary diseases, antidote, etc. The dried root powder is also mixed with other crude medicines to increase the efficiency of crude drugs. Due to its high demand in the local area as well as from outside pharmaceutical company agents, the wild population of *A. heterophyllum* is dwindling fast; however, the plant is now successfully cultivated in nurseries in some villages. Another rare species recorded was *Asparagus racemosus*, but it is also successfully cultivated.

Out of the 13 traditional antidote plants, six species have no record of any pharmacological studies. These include *Anaphalis busua*, *A. triplinervis*, *Berberis angulosa*, *Ligularia amplexicaulis*, *Rhododendron hodgsonii* and *Solanum viarum*. However, *Aconitum* and *Taxus* are still widely hunted as medicinal raw materials by the pharmaceutical agents through unorganised market channel. The commercial prospects of the 13 species could boost up the level of sustainable income source for the rural community if proper cultivation and harvesting protocols are developed for each species through agro-horticultural research trial. Prioritisation and robust policy approach is needed at all levels to preserve and promote the traditionally used significant medicinal plants the sub-Himalayan region.

10.5 Traditional Healing Practices of *Monpa* Tribe

Antidote plants are still widely used by the residents of Tawang and West Kameng districts and are the first choice of medication for the *Brokpa* community and other interior *Monpa* villagers residing in remote and far-flung areas. The *Monpa* tribe is very rich in ethnomedicinal knowledge. They used a good number of plants to treat various types of diseases and ailments. Among the *Monpas*, common ailments, which they treat, include body toxicity, respiratory problem, digestive system malfunction, jaundice and reproductive-related ailments. The major causes of toxicities include aconite contamination, food poisoning, snake bite and scorpion bite. As per the information of *Bon* and *Menpa*, food poisoning may be either intentional or unintentional. Intentional food poisoning is generally treated by mantras through *Bonpo*, but in some villages, plant-based drug (decoction) is also used which is generally administered through herbalist (*Menpa*).

The tribal communities of Arunachal Pradesh use a variety of ethnomedicine forms, but are gradually being replaced by modern medicine. The present study revealed that antidote plants are still widely used by the traditional healers of the *Monpa* tribe. The plant medicines are administered to the patient in two alternative processes. (1) Treatment of poisoned patient means person accidentally or intentionally poisoned through consumption of herbal food, caused by insects or scorpion bites. (2) Another form of treatment is through *Menpa* (medicine men or simple herbalists) who have deep knowledge about the properties and function of such anti-poison plants found in their mountain ecosystem. It is concluded that although *Monpa* embraced Buddhism way back, in the ethnomedication process, they still heavily rely on age-old indigenous medication methods ingrained in their traditional faith and belief system called *Bon*, and the people who are experts in such local spiritual healing paths of pre- and post-Buddhism *Monpa* of Arunachal Pradesh are called *Bonpo*. Despite rapid advances of western medicine in some localities, the holistic *Bon* legacy of indigenous medication system treasured through pre-Buddhism and post-Buddhism healing experts called *Bonpo* of the *Monpa* continues to play a central role in treatment of patients residing in interior and far-flung localities of the *Monpas* who have placed themselves in high altitude and sub-Himalayan mountain region of Arunachal Pradesh.

10.6 Conclusion

On the basis of strong cultural claim attached with efficacy over certain medicinal plants used by the *Monpas* in their local *Bon* therapeutic practices, these anti-poison plants could be investigated through reverse pharmacological approach. Such investigation means investigation at biochemical and pharmacological level on those potential anti poisonous plants. Literature revealed that almost all of the 13 species have been partly reported to have been investigated phytochemically, which contains certain antioxidant and wound healing and biological activities which validate the traditional claim of the *Bonpo*. Our investigation has further confirmed the fact that socioreligious beliefs attached with medicinal plants used by the local communities have some scientific basis, which in fact offers a clue for scientific investigation of the biological materials. Such ethnobotanical studies can easily lead the researchers to successful discovery of noble compounds from the plants, which could prove to be the panacea for human ailments in the future.

Acknowledgement The first and second authors express their thankfulness to Defence Research Laboratory, Tezpur (Assam), for financial support through extramural project, and Centre with Potential for Excellence in Biodiversity (CPEB-II), Rajiv Gandhi University, Rono Hills, Doimukh, Arunachal Pradesh, for their financial and logistic support funded by University Grant Commission (UGC), New Delhi. The third and fourth authors are thankful to the Ministry of Defence under DRDO, New Delhi, for financial support to carry out this project work.

References

Ahmad M, Ahmad W, Ahmad M, Zeeshan M, Obaidullah, Shaheen F (2008) Norditerpenoid alkaloids from the roots of *Aconitum heterophyllum* wall with antibacterial activity. J Enzyme Inhib Med Chem 23(6):1018–1022

Ali RM, Houghton PJ, Raman A, Hoult JRS (1998) Antimicrobial and anti-inflammatory activities of extracts and constituents of *Oroxylum indicum* (L.) Vent. Phytomedicine 5(5):375–381. doi:10.1016/S0944-7113(98)80020-2

Ansari MN, Bhandari U, Islam F, Tripathi CD (2008) Evaluation of antioxidant and neuroprotective effect of ethanolic extract of *Embelia ribes* burm in focal cerebral ischemia/reperfusion-induced oxidative stress in rats. Fundam Clin Pharmacol 22(3):305–314. doi:10.1111/j.1472-8206.2008.00580.x

Anwar S, Ahmad B, Sultan M, Gul W (2003) Biological and pharmacological properties of Aconitum chasmanthum. J Biol Sci 3:989–993

Appendino G, Gariboldi P, Pisetta A, Bombardelli E, Gabetta B (1992) Taxanes from *Taxus baccata*. Phytochemistry 31(12):4253–4257. doi:10.1016/0031-9422(92)80454-M

Arora R, Chawla R, Puri SC et al (2005) Radioprotective and antioxidant properties of low-altitude *Podophyllum hexandrum*. J Environ Pathol Toxicol Oncol 24(4):299–314. doi:10.1615/JEnvironPatholToxicolOncol.v24.i4.70

Atal CK, Sharma ML, Koul A (1986) Immunomodulating agents of plant origin-I: preliminary screening. J Ethnopharmacol 18:133–141

Baba SA, Malik SA (2014) Evaluation of antioxidant and antibacterial activity of methanolic extracts of *Gentiana kurroo* Royle. Saudi J Biol Sci. doi:10.1016/j.sjbs.2014.06.004

Bajpai D, Vankar PS (2007) Antifungal textile dyeing with *Mahonia napaulensis* D.C. Leaves extract based on its antifungal activity. Fiber Polym 8(5):487–494

Bhatt ID, Dauthal P, Rawat S, Gaira KS, Jugran A, Rawal RS, Dhar U (2012) Characterization of essential oil composition, phenolic content, and antioxidant properties in wild and planted individuals of *Valeriana jatamansi* Jones. Sci Hortic-Amsterdam 136(1):61–68. doi:10.1016/j.scienta.2011.12.032

Bradley D (1997) Tibeto-Burman languages and classification. In: Bradley D (ed) Papers in Southeast Asian Linguistics no. 14: Tibeto-Burman languages of the Himalayas. Pacific Linguistics, Canberra, pp 1–72

Census (2011) District statistical report. Government of Arunachal Pradesh, Tawang District

Chan S, Li S, Lin G, Li P (2000) Pharmacokinetic study and determination of imperialine, the major bioactive component in antitussive *Fritillaria cirrhosa*, in rat by high-performance liquid chromatography coupled with evaporative light-scattering detector. Anal Biochem 285(1):172–175. doi:10.1006/abio.2000.4756

Chen LJ, Games DE, Jones J (2003) Isolation and identification of four flavonoid constituents from the seeds of *Oroxylum indicum* by high-speed counter-current chromatography. J Chromatogr A 998(1):95–105. doi:10.1016/S0021-9673(02)01954-4

Dhalwal K, Shinde VM, Biradar YS, Mahadik KR (2008) Simultaneous quantification of bergenin, catechin, and gallic acid from *Bergenia ciliata* and *Bergenia ligulata* by using thin-layer chromatography. J Food Compos Anal 21(6):496–500. doi:10.1016/j.jfca.2008.02.008

Dhulasavant V, Shinde S, Pawar M, Naikwade NS (2010) Antihyperlipidemic activity of *Cinnamomum tamala* Nees. on high cholesterol diet induced hyperlipidemia. Int J PharmTech Res 2(4):2517–2521

Dhuna V (2010) P27 Purification and characterization of an N-acetyllactosamine specific lectin from tubers of *Arisaema utile* having anti-proliferative effect on human cancer cell lines. Eur J Cancer Suppl 8(2):19

Dung XN, Chinh ND, Leclercq PA (1995) Volatile constituents of the fruit oil of *Illicium griffithii* Hook. f. et Thoms. from Vietnam. J Essent Oil Res 7(4):451–452. doi:10.1080/10412905.1995.9698561

Farnsworth NR, Soejarto DD (1991) Global importance of medicinal plants. In: Akerele O, Heywood V, Synge H (eds) Conservation of medicinal plants. Cambridge University Press, Cambridge

Gajural PR, Rethy P, Singh B (2003) Preliminary studies on wild edible plants resources of Dihang Debang biosphere reserve. In: Abstract & souvenir of 13th annual conference of IAAT and international symposium on plant taxonomy: advances & relevance, TM Bhagalpur University, Bihar, 14–15 Nov 2003, p 79

Ghosal S, Biswas K, Jaiswal DK (1980) Xanthone and flavonol constituents of *Swertia hookeri*. Phytochemistry 19(1):123–126. doi:10.1016/0031-9422(80)85027-8

Gnanavel R, Franklin CJ (2014) Medicinal plant based antidote against snake bite by Irula tribes of Tamil Nadu, India. World J Pharm Sci 2(9):1029–1033

Hayes PY, Jahidin AH, Lehmann R et al (2008a) Steroidal saponins from the roots of *Asparagus racemosus*. Phytochemistry 69:796–804

Hayes PY, Jahidin AH, Lehmann R, Penman K, Kitching W, Voss JJD (2008b) Steroidal saponins from the roots of *Asparagus racemosus*. Phytochemistry 69(3):796–804. doi:10.1016/j.phytochem.2007.09.001

Innocenti G, Acqua SD, Scialino G, Banfi E, Sosa S, Gurung K, Barbera M, Carrara M (2010) Chemical composition and biological properties of *Rhododendron anthopogon* essential oil. Molecules 15(4):2326–2338. doi:10.3390/molecules15042326

Iwashina T, Omori Y, Kitajima J, Akiyama S, Suzuki T, Ohba H (2004) Flavonoids in translucent bracts of the Himalayan *Rheum nobile* (Polygonaceae) as ultraviolet shields. J Plant Res 117(2):101–107

Joshi P, Dhawan V (2005) *Swertia chirayita* – an overview. Curr Sci India 89(4):635–640

Khuroo MA, Qureshi MA, Razdan TK, Nichols P (1988) Sterones, iridoids and a sesquiterpene from *Verbascum thapsus*. Phytochemistry 27(11):3541–3544. doi:10.1016/0031-9422(88)80764-7

Kumar V, Singh PK (2014) Ethnomedicinal plants used as antidote for snake- bite and scorpion-sting in Bundelkhand (U.P.), India. IOSR-JESTFT 8(1):52–55

Latif A, Khan TF, Afaq SH (2006) Anti-inflammatory activity of flower tops of *Gentiana kurroo* Royle extract. Pharmacology 3:575–580

Latif A, Razique A, Asadullah RRS, Zuberi RH (2010) Phytochemical and physico-chemical study of *Coptis teeta* Wall.: an effective drug of choice in ocular ailments. Eur J Integ Med 1(1):22–23. doi:10.1016/j.eujim.2008.08.128

Li JX, Li P, Tezuka Y, Namba T, Kadota S (1998) Three phenylethanoid glycosides and an iridoid glycoside from *Picrorhiza scrophulariiflora*. Phytochemistry 48(3):537–542. doi:10.1016/S0031-9422(98)00030-2

Liu SQ, Xiao ZY, Feng R (1992) Chemical constituents of *Drynaria propinqua* (Wall) J. Sm. Zhongguo Zhong Yao Za Zhi 17:737–739

Liu S, Xiao Z, Feng R (1994) A flavanol glycoside from *Drynaria propinqua*. Phytochemistry 35(6):1595–1596. doi:10.1016/S0031-9422(00)86903-4

Liu X, Guo M, Li L et al (2008a) Study on the hypoglycemic effect of Triplostegia glandulifera wall. Mod Med Health 3:6–37, 2008–09, Tsinghua Tongfang Knowledge Network Technology Co., Ltd. (Beijing)

Liu X, Guo M, Xu J et al (2008b) Study on the anti-stimulation action of Triplostegia glandulifera wall. Mod Med Health 3:6–37, 2008–09, Tsinghua Tongfang Knowledge Network Technology Co., Ltd. (Beijing)

Mahendran S, Thippeswamy BS, Veerapur VP, Badami S (2011) Anticonvulsant activity of embelin isolated from *Embelia ribes*. Phytomedicine 18(2–3):186–188. doi:10.1016/j.phymed.2010.04.002

Malik S, Cusido RM, Mirjalili MH, Moyano E, Palazon J, Bonfill M (2011) Production of the anticancer drug taxol in *Taxus baccata* suspension cultures: a review. Process Biochem 46(1):23–34

Mani MS (1974) Ecology and biogeography in India. Dr. Junk b. v. Publishers, Hague

Maurya A, Khan F, Bawankule DU, Yadav DK, Srivastava SK (2012) QSAR, docking and in vivo studies for immunomodulatory activity of isolated triterpenoids from *Eucalyptus tereticornis* and *Gentiana kurroo*. Eur J Pharm Sci 47(1):152–161. doi:10.1016/j.ejps.2012.05.009

McCutcheon AR, Roberts TE, Gibbons E, Ellis SM, Babiuk LA, Hancock REW, Towers GHN (1995) Antiviral screening of British Columbian medicinal plants. J Ethnopharmacol 49(2):101–110. doi:10.1016/0378-8741(95)90037-3

Mehrotra R, Ahmed B, Vishwakarma RA, Thakur RS (1989) Verbacoside: a new luteolin glycoside from *Verbascum thapsus*. J Nat Prod 52(3):640–643. doi:10.1021/np50063a028

Mir SR, Ali M, Kapoor R (2004) Chemical composition of essential oil of *Cinnamomum tamala* Nees et Eberm. leaves. Flavour Frag J 19(2):112–114. doi:10.1002/ffj.1236

Miyazawa M, Fukuyama M, Yoshio K, Kato T, Ishikawa Y (1999) Biologically active Components against Drosophila melanogaster from *Podophyllum hexandrum*. J Agric Food Chem 47(12):5108–5110. doi:10.1021/jf9903509

Mubashir S, Shah WA (2012) Antimicrobial, cytotoxic and antioxidant activities of *Arisaema utile*. J Pharm Res 5(3):1368

Myers N, Mittermeier RA, Mittermeier CG, da Fonseca GAB, Kent J (2000) Biodiversity hotspots for conservation priorities. Nature 403:853–858

Nisar M, Obaidullah AM, Wadood N, Lodhi MA, Shaheen F, Choudhary MI (2009) New diterpenoid alkaloids from *Aconitum heterophyllum* Wall: selective butyrylcholinestrase inhibitors. J Enzyme Inhib Med Chem 24(1):47–51

Pandey H, Nandi SK, Kumar A, Palni UT, Chandra B, Palni L (2004) In-vitro propagation of Aconitum balfourii Stapf: an important aconite of the Himalayan alpines. J Hortic Sci Biotech 79:34–41

Prasad SK, Kumar R, Patel DK, Sahu AN, Hemalatha S (2012) Physicochemical standardization and evaluation of in-vitro antioxidant activity of *Aconitum heterophyllum* Wall. Asian Pac J Trop Biomed 2(2):S526–S531. doi:10.1016/S2221-1691(12)60266-4

Rahman A, Ashrafa M, Choudhary MI, Rehman H, Kazmi MH (1995) Antifungal aryltetralin lignans from leaves of *Podophyllum hexandrum*. Phytochemistry 40(2):427–431. doi:10.1016/0031-9422(95)00195-D

Rajkumar V, Guha G, Kumar RA (2011) Antioxidant and anti-neoplastic activities of *Picrorhiza kurroa* extracts. Food Chem Toxicol 49(2):363–369. doi:10.1016/j.fct.2010.11.009

Rao VS, Rao A, Karanth KS (2005) Anticonvulsant and neurotoxicity profile of *Nardostachys jatamansi* in rats. J Ethnopharmacol 102(3):351–356. doi:10.1016/j.jep.2005.06.031

Reddy PS, Jamil K, Madhusudhan P, Anjani G, Das B (2001) Antibacterial activity of isolates from *Piper longum* and *Taxus baccata*. Pharm Biol 39(3):236–238

Rodgers WA, Panwar HS (1988) Planning wildlife protected area network in India, vol I & II. A report prepared for the Department of Environment, Forests & Wildlife, Government of India, Wildlife Institute of India, Dehradun

Rucker G, Tautges J, Maheswari ML, Saxena DB (1976) Norseychelanone, α- and β-patchoulenes and patchouli alcohol from *Nardostachys jatamansi*. Phytochemistry 15(1):224. doi:10.1016/S0031-9422(00)89094-9

Seifert K, Schmid J, Lien NT, Johne S (1985) Iridoide aus *Verbascum arten*. Planta Med 51:409–411

Shanbhag SN, Mesta CK, Maheshwari ML, Paknikar SK, Bhattacharyya SC (1964) Terpenoids-LII: jatamansin, a new terpenic coumarin from *Nardostachys jatamansi*. Tetrahedron 20(11):2605–2615. doi:10.1016/S0040-4020(01)90839-7

Shanbhag SN, Mesta CK, Maheshwari ML, Bhattacharyya SC (1965) Terpenoids-LXXV: constituents of *Nardostachys jatamansi* and synthesis of (±) dihydrosamidin and visnadin from jatamansin. Tetrahedron 21(12):3591–3597. doi:10.1016/S0040-4020(01)96975-3

Sharma SK, Singh AP (2012) In-vitro antioxidant and free radical scavenging activity of *Nardostachys jatamansi* DC. J Acupunct Merid Stud 5(3):112–118. doi:10.1016/j.jams.2012.03.002

Sharma N, Sharma UK, Gupta AP, Sinha AP (2010) Simultaneous determination of epicatechin, syringic acid, quercetin- 3-O-galactoside and quercitrin in the leaves of *Rhododendron* species by using a validated HPTLC method. J Food Compos Anal 23(3):214–219. doi:10.1016/j.jfca.2009.11.003

Sidiq T, Khajuria A, Suden P, Singh S, Satti NK, Suri KA, Srinivas VK, Krishna E, Johri RK (2011) A novel sarsasapogenin glycoside from *Asparagus racemosus* elicits protective immune responses against HBsAg. Immunol Lett 135(1-2):129–135. doi:10.1016/j.imlet.2010.10.013

Sikdar M, Datta U (2008) Traditional phytotherapy of the Nath people of Assam. Ethnomedicine 2(1):39–45

Singh SB, Thakur RS, Schulten HR (1980) Spirostanol saponins from *Paris polyphylla*, structures of polyphyllin C, D, E and F. Phytochemistry 21(12):2925–2929

Singh RK, Singh A, Sureja AK (2007) Traditional foods of Monpa tribe of West Kameng, Arunachal Pradesh. Indian J Tradit Knowl 6(1):25–36

Singh GK, Garabadu D, Muruganandam AV, Joshi VK, Krishnamurthy S (2009) Antidepressant activity of *Asparagus racemosus* in rodent models. Pharmacol Biochem Behav 91(3):283–290. doi:10.1016/j.pbb.2008.07.010

Sinha S, Murugesan T, Maiti K, Gayen JR, Pal B, Pal M, Saha BP (2001a) Antibacterial activity of *Bergenia ciliata* rhizome. Fitoterapia 72(5):550–552. doi:10.1016/S0367-326X(00)00322-1

Sinha S, Murugesan T, Maiti K, Gayen JR, Pal M, Saha BP (2001b) Evaluation of anti-inflammatory potential of *Bergenia ciliata* Sternb. rhizome extract in rats. R Pharm Soc Great Brit 53(2):193–196. doi:10.1211/0022357011775398

Smit HF, Kroes BH, Berg AJJ, Wal D, Worm E, Beukelman CJ, Dijk H, Labadie RP (2000) Immunomodulatory and anti-inflammatory activity of *Picrorhiza scrophulariiflora*. J Ethnopharmacol 73(1-2):101–109. doi:10.1016/S0378-8741(00)00268-3

Soman G (2014) Some medicinal plants of Panhala taluka used as antidotes. Int J Life Sci 2(3):276–278

Tag H, Das AK (2004) Ethnobotanical notes on the Hill Miri tribe of Arunachal Pradesh, India. Indian J Tradit Knowl 3(1):80–85

Tag H, Das AK, Kalita P (2005) Plants used by the Hill Miri tribe of Arunachal Pradesh in ethnofisheries. Indian J Tradit Knowl 4(1):57–64

Tag H (2012) Mainstreaming conservation and sustainable use of medicinal plant diversity in three Indian states: botanical and ecological survey of GSMP & Associated flora in Lumla-MPCA (Tawang FD). Final technical report, UNDP Cell, APSMPB, SFRI, Itanagar

Tag H, Tsering J, Hui PK, Gogoi BJ, Veer V (2014) Nutritional potential and traditional uses of high altitude wild edible plants in Eastern Himalayas, India. WASET 8(2):395–400

Tamuly C, Dutta PP, Bordoloi M, Bora J (2013) Antifungal and antioxidant pyrrole derivative from *Piper pedicellatum*. Nat Prod Commun 8(10):1451–1454

Verma S, Ojha S, Raish M (2010) Anti-inflammatory activity of *Aconitum heterophyllum* on cotton pellet-induced granuloma in rats. J Med Plant Res 4(15):1566–1569. doi:10.5897/JMPR09.502

Vijayakumar A, Duraipandiyan V, Jeyaraj B, Agastian P, Raj MK, Ignacimuthu S (2012) Phytochemical analysis and in vitro antimicrobial activity of *Illicium griffithii* Hook. f. & Thoms extracts. Asian Pac J Trop Dis 2(3):190–199. doi:10.1016/S2222-1808(12)60045-0

Wang D, Zhu J, Wang S, Wang X, Ou Y, Wei D, Li X (2011) Antitussive, expectorant and anti-inflammatory alkaloids from bulbus *Fritillariae cirrhosae*. Fitoterapia 82(8):1290–1294. doi:10.1016/j.fitote.2011.09.006

Wang H, Zhao W, Choomuenwai V, Andrews KT, Quinn RJ, Feng Y (2013) Chemical investigation of an antimalarial Chinese medicinal herb *Picrorhiza scrophulariiflora*. Bioorg Med Chem Lett 23(21):5915–5918. doi:10.1016/j.bmcl.2013.08.077

Wani BA, Ramamoorthy D, Ganai BA (2011) Preliminary phytochemical screening and evaluation of analgesic activity of methanolic extract of roots of *Gentiana kurroo* Royle in experimental animal models. Int J Pharm Pharm Sci 3(4):164–166

Warashina T, Miyase T, Ueno A (1991) Iridoid glycosides from *Verbascum thapsus* L. Chem Pharm Bull 12:3261–3264

Xu J, Yang B, Guo Y, Jin D, Guo P, Liu C, Hou W, Zhang T, Gui L, Sun Z (2011a) Neuroprotective bakkenolides from the roots of *Valeriana jatamansi*. Fitoterapia 82(6):849–853. doi:10.1016/j.fitote.2011.04.012

Xu J, Zhao P, Guo Y, Xie C, Jin D, Ma Y, Hou W, Zhang T (2011b) Iridoids from the roots of *Valeriana jatamansi* and their neuroprotective effects. Fitoterapia 82(7):1133–1136. doi:10.1016/j.fitote.2011.07.013

Yan LL, Zhang YJ, Gao WY, Man SL, Wang Y (2009) In-vitro and in vivo anticancer activity of steroid saponins of *Paris polyphylla* var. yunnanensis. Exp Oncol 31(1):27–32

Yan R, Cao Y, Chen C, Dai H, Yu S, Wei J, Li H, Yang B (2011) Antioxidant flavonoids from the seed of *Oroxylum indicum*. Fitoterapia 82(6):841–848. doi:10.1016/j.fitote.2011.04.006

Molecular Farming for Production of Biopharmaceuticals and Edible Vaccines in Plants

11

Ajitabh Bora, Hemanta Kumar Gogoi, and Vijay Veer

Abstract

Production of recombinant proteins such as edible vaccines, recombinant subunit vaccines, antibodies and other medical proteins in plants is referred to as molecular farming. Plants, as expression platform, have several advantages over mammalian or microbial systems such as low cost of production, fast scalability, absence of pathogenic microbes and their capability to synthesise complex proteins. This involves identification and isolation of the gene of a pathogen encoding antigenic protein and preparation of a suitable construct followed by its introduction to a suitable plant host system either through *Agrobacterium*-mediated or direct gene transfer technique for expression of the protein of interest. The plant parts of the transgenic plant containing the antigen are fed raw or the protein is extracted and administered. The cultivation of these plants needs to be done in a greenhouse or under in vitro condition in order to negate environmental and biosafety issues. Biofarming offers great opportunity to the pharmaceutical industry for production and supply of medicines at an affordable cost, particularly to the developing countries.

11.1 Introduction

Heterologous gene expression technology has received great attention since the recent past for production of edible vaccines and biopharmaceuticals. Several in vitro and in vivo systems have been developed for expression of foreign proteins, which is likely to increase in the near future. This has led to the development of the concept of molecular farming which refers to the production of pharmaceutically important and commercially valuable proteins in plants (Franken et al. 1997). Its aim is to provide a cheap and safer ways for large-scale production of recombinant proteins. This is a new area of science which combines biotechnology and plants for production of valuable products of biomedical importance.

Humans have been using plants as medicines since the early days of civilisation. The Egyptians

A. Bora (✉) • H.K. Gogoi • Vijay Veer
Defence Research Laboratory (Defence Research and Development Organisation), Tezpur 784 001, Assam, India
e-mail: ajitabhbora@rediffmail.com

© Springer India 2016
Vijay Veer, R. Gopalakrishnan (eds.), *Herbal Insecticides, Repellents and Biomedicines: Effectiveness and Commercialization*, DOI 10.1007/978-81-322-2704-5_11

listed around 700 medicinal plants during 1600 B.C. In India, Ayurvedic system of medicine is prevalent since time immemorial, which are based on *Charaka Samhita*, an early textbook on Ayurveda (Indian traditional medicine), which is one of the two authoritative texts, the other being *Sushruta Samhita* in the field of medicine and human health. Plants are the most abundant and cheap source of proteins on Earth. Plants have provided several beneficial molecules to mankind since centuries, but only in the past few decades they have been employed for heterologous gene expression. Plants offer an efficient and economic platform for large-scale production of recombinant proteins to meet the global demand with minimum risk of contamination of human pathogenic microbes. Human serum albumin was the first recombinant plant-derived pharmaceutical protein (PDP) produced in 1990 in transgenic tobacco and potato plants (Sijmons et al. 1990). Since then several human and other important proteins have been produced in a variety of crops such as vaccines, enzymes, growth factors, etc. (Goldstein and Thomas 2004).

Proteins are being used as vaccines, drugs and diagnostic reagents, which have led to an upsurge for production of recombinant proteins on a commercial scale. Proteins find several applications in the field of medicine, research and industry, but the conventional method for extraction of proteins from their native sources is costly and difficult, for example, proteins such as single-chain fragments (scFvs) have no known natural sources. Moreover, proteins from their natural sources may pose risk to human health, for example, several people contracted diseases after the use of contaminated blood products. Moreover, other systems for production of proteins through the use of mammalian cell culture and microbial systems have demerits in terms of cost of production, safety and scalability (Schwartz 2001; Chu and Robinson 2001; Houdebaine 2000).

Research findings indicate that molecular farming in plants has several advantages in comparison to other traditional systems; due to which, the use of plants for mass production of recombinant proteins is becoming popular worldwide (Fischer and Emans 2000; Gidding 2001). Production of recombinant proteins and other secondary metabolites in plants by means of cultivation and extraction of the protein of interest is referred to as molecular farming (Wilde et al. 2002). This technology, first demonstrated in the 1980s, is based on the genetic transformation capability of plants (Bevan et al. 1983). The human growth hormone is the first recombinant protein produced in transgenic plants in the year 1986, while the first recombinant antibody was produced in 1989 through expression in the F1 plant resulting from crossing of two transgenic parents (Hiatt et al. 1989; Barta et al. 1986). However, the production of recombinant proteins for commercial purpose started in 1997 with the expression of *avidin* (an egg protein) in transgenic maize (Hood et al. 1997). Thus, plants are potential candidate for use as bio-factories for mass production of different types of recombinant proteins which is due to their capability for performing post-translational changes leading to proper folding of the proteins as well as maintaining their structural and functional integrity.

Molecular farming in plants started since the successful transformation of the first higher plants (Fraley et al. 1983). Every protein has the potential for use as a product, for example, the marker gene *uids*, which was used to develop transformation systems in plants, is presently a product (Jefferson et al. 1987; Kusnadi et al. 1998; Witcher et al. 1998). The concept of using plants for production of human proteins was initially seen with scepticism. There was no much support for the concept of using plants for the production of therapeutic proteins before publication of the production of the first recombinant antibodies in tobacco in 1989. Thereafter, it was realised that transgenic plants can be an excellent platform for producing a variety of pharmaceutical proteins (Schillberg et al. 2003). Plants as the system for producing protein products have several advantages, such as it has low cost of production in comparison to other conventional systems; the method for cultivation of these plants already exists; they are free from human pathogenic microbes, which nullifies the chance of contamination of the protein product and produces proteins with proper folding; and the proteins thus produced have more stability (Horn et al. 2004).

11.2 Production of Edible Vaccines

Conventional vaccines are preparations that normally contain killed or attenuated pathogenic microbes or an antigen from the pathogen. They incite immune response but do not cause any disease. Successful vaccination programmes worldwide have eliminated several dreadful diseases from this Earth with very fewer instances of reoccurrence. Thus, oral immunisation is an economic and effective way of combating diseases on a wider scale.

Vaccines which are administered orally in the form of food are called edible vaccines. These are produced by expressing an orally active antigenic component of the target pathogen in a plant system. They are thereafter fed to human beings or animals in order to immunise them against the pathogen (Singh 2009). Such type of vaccines is safe for consumption as they contain only the antigen of a pathogen but do not have genes that could lead to the formation of the whole pathogen which will lead to disease condition (Marsa 1994). Edible vaccines are now being developed for several major diseases of both human beings and animals such as cholera, measles, hepatitis B and C and foot and mouth disease (Gidding et al. 2000). The plant system has the capacity for expression of more than one transgene, thereby making it possible for delivering multiple antigens for repeated inoculations (Conrad and Fiedler 1994).

Plant-based edible vaccines, unlike other oral vaccines, which get degraded in the stomach (because of gastric enzymes and low pH) and gut before inducing immune responses, are resistant to intestinal degradation due to the presence of the cell wall (Daniell et al. 2001; Webster et al. 2002) which makes it a preferred system for delivering vaccines.

In 1990 a patent was filed for expression of a surface protein from *Streptococcus* in tobacco (0.02 % of the total leaf protein) which was the first known record of edible vaccine (Mason and Arntzen 1995). The idea of developing edible vaccine became popular with the successful expression of hepatitis B surface antigen in tobacco plants (Mason et al. 1992). This invention has given much impetus and excitement to the biotechnologies which have led to several attempts for development of edible vaccines.

The need of the hour is to develop novel vaccines, particularly the edible ones, which are cheap and easy in administration and storage without refrigeration which are very essential for the success of any oral immunisation programme particularly in the developing countries. With the development in the field of molecular biology, newer techniques were employed for production of subunit vaccines (Washam 1997). Multicomponent vaccines are also being produced that provide immunisation against several diseases.

11.3 Advantages of Edible Vaccines

Edible vaccines have several advantages over the conventional vaccines which are listed below (Shah et al. 2011):

1. Edible vaccines offer an effective means for immunisation since they do not require adjuvants for enhancement of the immune response.
2. They can elicit mucosal immunity unlike the conventional vaccines.
3. They are economical in terms of cost of production, storage and delivery.
4. They find easy acceptance as these are not administered through injection like the conventional vaccines.
5. Such type of vaccines may be a source of novel and multicomponent vaccines which will provide immunisation against several diseases.
6. They are known to be safe for administration since there is no chance of proteins reforming to the pathogenic microbes against which immunisation is sought as these do not contain attenuated pathogens.

11.4 Limitations of Edible Vaccines

Edible vaccines, though easy to access, pose few challenges which are discussed below (Shah et al. 2011):

1. There is a chance for development of immunotolerance to the vaccine protein.
2. Due to the lack of standard methods for quantification of the plant material to be consumed, dosage differs with plant type, generation, age of the individual, ripeness and protein content. Low doses shall generate less antibodies, while higher doses will lead to immunotolerance.
3. Vaccine stability varies with the type of plant.
4. Those edible vaccines, which need to be cooked such as potatoes, may lead to denaturation of the protein.
5. Fresh fruits and vegetables containing a vaccine antigen are different and need to be stored properly in order to prevent microbial spoilage during storage.
6. There is a need for distinguishing a vaccine fruit from a normal one in order to avoid misadministration.
7. The functioning of these vaccines may be affected due to the differences in the glycosylation pattern of plants and humans.
8. They cannot be used as a regular component of a human or animal diet to avoid development of tolerance towards the antigen.

11.5 Mode of Action of Edible Vaccines

Edible vaccines activate both induced and mucosal immunity in the administered organism, which shall impart immunisation against pathogens entering through the mucosa, such as those causing tuberculosis, pneumonia, diarrhoea, STDs, etc. (Lal et al. 2007). The mucosal immune system provides first-degree protection to an organism against invasion of potential pathogens and is an important region for vaccination. The edible vaccines acts near the Peyer's patch (PP) of the intestine, which contains IgA-producing plasma cells having potentiality for populating mucosal tissue and acts as immune effector sites. The edible vaccine generally breaks down and releases antigens at the PP (Rudzik et al. 1975), where the M cells take up the released antigens and hand over to the B cells with the help of APC (antigen-presenting cells). The B cells then get differentiated into plasma cells and secrete IgA antibodies (Pant and Sanjana 2014), thereby inducing immunity.

11.6 Development of an Edible Vaccine

For development of an edible vaccine, the gene of a pathogen, which encodes for an orally active antigenic protein, is isolated following standard molecular biology techniques, and a suitable construct is prepared for expression in the plant system. The gene construct is introduced into the selected host plant either by means of *Agrobacterium*-mediated or direct gene transfer technique for integration with the host genome for the expression of the antigenic proteins (Fig. 11.1). The plant parts (fruits/leaves, tubers) of the transgenic plant which contain the antigenic protein are provided in the raw form to the humans or animals for consumption to elicit immunisation in the target organism (Singh 2009). The selection of the plant species for transformation and production of the desired antigen will depend on the target organism. For immunisation of human beings, fruits and vegetables which are eaten raw such as banana, tomato, etc., are suitable for use as edible vaccines. In the case of animals, feeds such as alfalfa and other fodder crops are suitable as the delivery system.

11.7 Plant-Based Recombinant Subunit Vaccines

Subunit vaccines are those vaccines which introduce an antigen into a human or animal system in order to induce immunity without introducing the pathogen, whole or otherwise. One way of

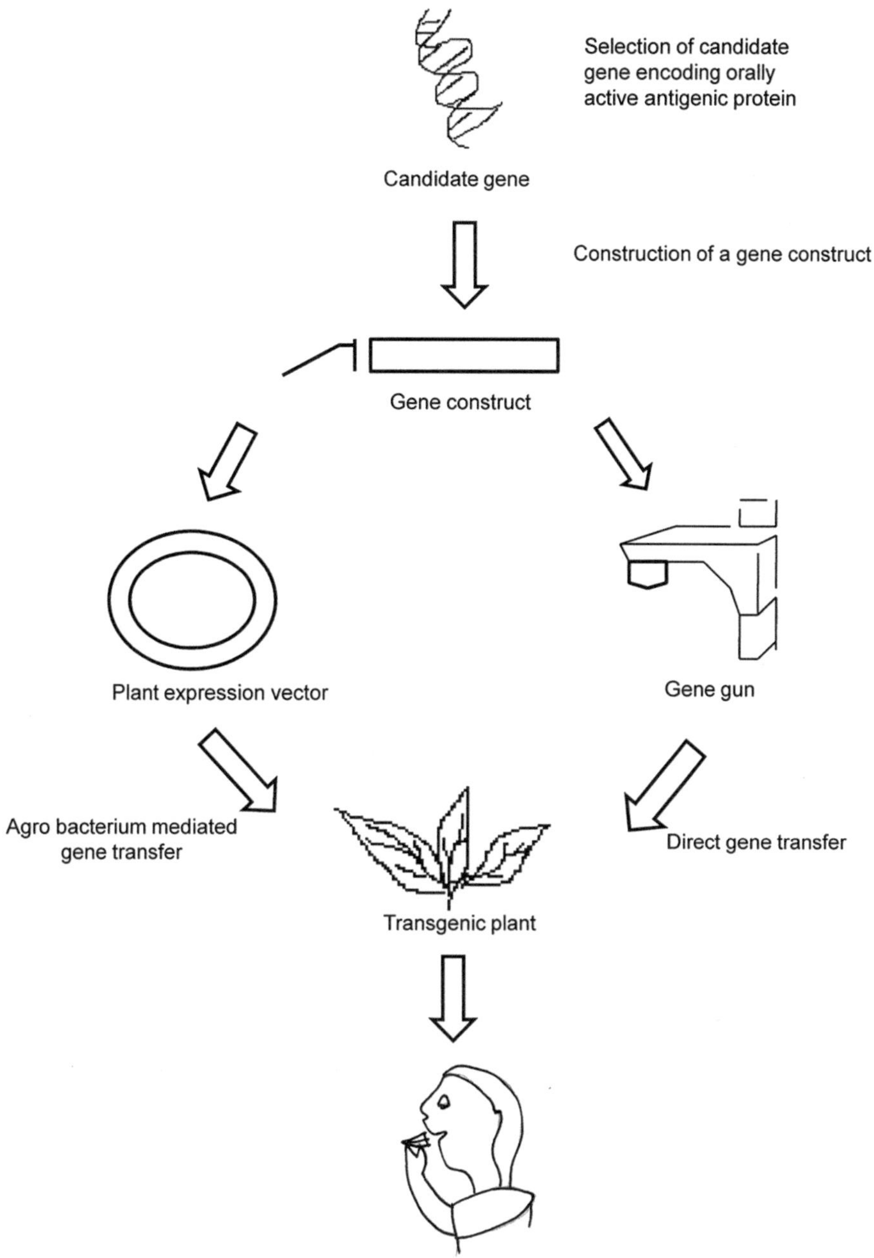

Fig. 11.1 Steps involved in the production of an edible vaccine

producing such type of vaccine is by extracting a specific protein from the pathogen and injecting it into the target immune system. This has a major disadvantage in that the isolated protein gets denatured and then will combine with antibodies other than the targeted ones. The other method is the identification and isolation of the gene coding the antigen from the pathogen and introducing it into a non-pathogenic host such as a virus, yeasts or a plant which serve as recombinant subunit vaccines.

The application of recombinant DNA technology has made subunit vaccine preparation efficient and safe. They can be provided in the form of a pure recombinant protein or as proteins delivered through live non-pathogenic vectors or as nucleic acid molecules which encode for the antigen (Singh 2009). There are several merits for using recombinant subunit vaccines such as pathogen-free production and purification procedure and high yields of a desired antigenic product.

Plant system can be a suitable platform for expression of recombinant subunit vaccines which can be extracted and purified for use as vaccines. The production cost of vaccines in plants is expected to be lower than other microbial systems. There are two approaches for production of recombinant subunit vaccines in plants, viz., (a) integration of the transgene, containing the antigenic gene, into plant genome and (b) expression as a coat protein fusion of a plant virus (Singh 2009).

11.8 Applications of Edible Vaccines

11.8.1 Malaria

Malaria, caused by a parasitic protozoan belonging to the genus *Plasmodium*, is a mosquito-borne disease infecting humans and animals. It is widespread in tropical and subtropical zones of the world. According to the WHO malaria fact sheet, March 2014, there were 207 million cases of malaria in 2012 which caused casualties of around 473,000–789,000 people, which were mostly children in Africa. This disease is mostly linked to poverty which has a profound negative impact on the economy of a nation. Malaria control measures such as spraying of insecticides did not have much impact on prevention of multiplication of the vector and resistance development in malarial parasites. Presently three antigens, viz., merozoite surface protein (MSP) 4, (MSP) 5 from *Plasmodium falciparum* and MSP 4/5 from *P. yoelii*, are being utilised for the development of a plant-based vaccine against malaria. When these recombinant proteins were co-administered with CTB as mucosal adjuvant to mice for oral immunisation, an effective antibody response against blood stage parasite was induced (Wang et al. 2004). These test antigenic proteins, produced in *E. coli*, could cause immunisation only when high dose of the antigens was administered which raises apprehension for its efficacy in inducing a significant immune response in humans. The expression levels of these antigens in plants are low for which more amount of the edible vaccine plant material needs to be taken in order to have immunity against malaria (Mishra et al. 2008). Thus, the transgenic technology has to be improved to enhance antigenic expression in plants for use as edible vaccines.

11.8.2 Measles

Measles, caused by paramyxoviruses, infects the respiratory system, skin and immune system. Global human death due to measles is estimated to be around 8 lakhs annually. Currently available vaccines against measles are live attenuated having certain limitations. With the purpose of development of an edible vaccine against measles, the MV-H antigen was expressed in tobacco plants. When these plants were fed to mice, the concentration of the antibodies attained five times the concentration considered protective for human beings. Besides, IgA antibodies were present in the faecal matter of the test mice (Huang et al. 2001). Transgenic tomato (Marquet-Blouin et al. 2003) and potato (Webster et al. 2002) have also been used for delivering viral antigens for the development of edible measles viruses.

11.8.3 Hepatitis B

Hepatitis B, caused by the hepatitis B virus (HBV), is a disease of the liver which infects humans world over, and vaccination is the only known means of preventing this disease. With an attempt to develop an edible vaccine using potato, hepatitis A surface antigen (HBsAg) subtype *ayw* was cloned into CaMv (cauliflower mosaic virus) plasmid which was used for transformation of the potato plant. The transgenic plants were shown to produce HBsAg, with higher antigen expression in the roots than in the leaves, due to which transgenic potatoes were not sufficient to induce the desired level of immunity (Domansky 1995). Further work is in progress to enhance HBsAg expression level by employing promoters, for example, patatin promoter and various other elements regulating transcription.

Transgenic tobacco plants, expressing HBsAg, when fed to animals as parenteral vaccine exhibited primary response similar to those achieved in case of traditional vaccine. However, on oral feeding of the tobacco plant, better primary response is observed over administration of conventional vaccines (Daniell et al. 2002).

11.8.4 Diabetes Mellitus Type 1

Diabetes mellitus type 1 (also called type 1 diabetes) is a form of diabetes mellitus which results from the autoimmune destruction of the beta cells that produce insulin in the pancreas. Nonobese diabetic mice developed insulin-dependent diabetes when fed with transgenic potato and tomato plants containing GAD67 gene-encoding diabetes-related proteins. Of the test population, 20 % of the prediabetic while 70 % of the control mice developed diabetics. Besides, a high level of Ig1 (cytokine-associated antibody that suppresses harmful immunity) was found in the treated mice. The antigen generated in the transgenic plants seems to maintain immunogenicity and thereby preventing diabetes in animals (Blanas et al. 1996; Ma et al. 1997).

11.8.5 Cholera

Cholera, caused by *Vibrio cholerae*, is an infection of the small intestine. It affects people world over and causes deaths. Transgenic potatoes, expressing B subunit of the *E. coli* heat-liable enterotoxin (LT-B), when fed to mice induced both induced and secretory antibodies. On boiling these potatoes for 5 minutes, nearly half of the antigenic proteins were found to be in its native state which shows that edible vaccines do not always undergo inactivation on cooking, thereby providing evidence that selection of plants for development of edible vaccines may go beyond raw foods (Leben et al. 1993; Richter et al. 1996; Mason et al. 1998a).

11.8.6 Bovine Pneumonia Pasteurellosis

Bovine pneumonic pasteurellosis is an economically important disease of ruminants with a wide prevalence throughout the continents. A transgenic white clover plant expressing *Mannheimia haemolytica* A1 leukotoxin 50 fusion protein was fed to rabbits for oral immunisation which gave encouraging results (Lee et al. 2001).

11.9 Clinical Trials

Several plant-based vaccines have been evaluated in various animal models for studying their immunogenic response. Clinical trials of few of the potential vaccine candidates are in progress. Tacket et al. (1998) for the first time conducted a clinical trial in volunteers who were fed with raw transgenic potatoes containing 3.7–15.7 μg/g of LT-B antigen. Ninety-one percent of the volunteers developed IgG anti-LT serum, half of which responded after the first dose.

The National Institute of Allergy and Infectious Diseases (NIAID) conducted clinical trials of transgenic potato, producing part of the *E. coli* toxin causing diarrhoea, in volunteers and found that it could trigger immune response in the volunteers without any serious side effects (Ball et al. 1999). Tacket et al. (1998) fed volunteers with potatoes expressing NVCP to study the immune response, of which 95 % developed immunity with few cases of modest response.

11.10 Production of Biopharmaceuticals

11.10.1 Selection of Host Plant System

Plants are gaining importance as potential system for production of recombinant proteins. Several recombinant proteins derived from plants are undergoing clinical trials and are in their final stage of development, and several others are in the pipeline. Majority of these are produced in transgenic plants, developed through a genetic transformation technique followed by extraction of the protein of interest.

Tobacco is one of the most commonly used plant systems for production of recombinant proteins. The popularity of tobacco as a well-established expression host is because of the availability of established transformation system and regulatory elements for the expression of the transgene (Fischer and Emans 2000). Tobacco, owing to its high biomass yield and ease of scalability, has become a popular system for molecular farming on a commercial scale. Tobacco, being a nonedible crop, there is low probability of food chain contamination by the transgenic materials (Stoger et al. 2000). Due to the presence of nicotine in tobacco which needs to be completely removed in the process of isolation and purification, plants such as lettuce and alfalfa are being used for molecular farming of recombinant proteins. Lettuce has been used for the production of hepatitis B virus subunit vaccine (Kapusta et al. 1999), and alfalfa was being used for protein expression by a Canadian biotech company Medicago Inc.

Although leafy crops are good expression systems due to their high biomass yield, the expressed proteins used to be unstable resulting in short shelf life which should be subjected to early processing after harvest. In contrast, seeds expressing proteins could be stored up to 3 years under refrigeration as these proteins do not undergo degradation (Larrick and Thomas 2001). Several cereal crops such as wheat, rice, maize, and barley have been explored for their use as recombinant protein expression systems (Stoger et al. 2000; Hood 2002).

Oleaginous crops such as oilseed rape or safflower are also being used for protein production, e.g. SemBioSys Genetics Inc. developed oleosin-fusion platform which involves production of the target protein as a fusion with oleosin. UniCrop, a Finnish biotech company, has also developed an oilseed-based platform, but they used it for isolation of the target protein from fast-growing sprouts in bioreactors (Fischer et al. 2004). Besides, few other plants have also been utilised for molecular farming with an aim to develop a better platform than the existing ones. Microalgae have also emerged as a potential system for molecular farming due to its rapid growth rate. A protein expression system has been developed in the microalgae *Chlamydomonas reinhardtii* for expression of an antibody against glycoprotein D of herpes simplex virus (Mayfield et al. 2003). Plants such as *Lemna* (duckweed) have also been used as expression platforms by Biolex Inc. and the moss *Physcomitrella patens* by Greenovation Inc., Freiburg, Germany (Fischer et al. 2004).

11.10.2 Transformation Methods

Transgenic plants, producing recombinant proteins, are being developed through two methods, viz., *Agrobacterium*-mediated transformation and particle bombardment. With both the techniques having its both merits and demerits, the choice depends upon the plant species selected for the transformation, available skill and other issues related to IP. Techniques such as electroporation, protoplast transformation or Whisker transformation have not been yet used for development of transgenic plants (Ma et al. 2003).

The soil pathogen *Agrobacterium tumefaciens* has been widely used as a vector for transferring the gene of interest to the host plant system. This is mostly used for molecular farming in dicotyledonous species such as tomato, tobacco, pea, potato and alfalfa (Gelvin 2003; Veluthambi et al. 2003). Particle bombardment is advantageous due to its less dependency on the genotype and could be a desired transformation technique for crops such as rice, wheat, maize and soya bean (Christou 1996). This is also the choice method for transformation of plastids unlike the *Agrobacterium*-mediated transfer which targets the nuclease (Gelvin 2003).

11.10.3 Antibody Production in Plants

An antibody is a protein, which is generated by the plasma cells and is used by the immune system against any foreign bodies such as pathogenic microbes. It is also known as an immunoglobulin (Ig). The foreign target bodies have a unique part called antigen which is recognised by the antibody. This specific binding affinity of the antibodies towards an antigen has led to its application in the diagnosis, prevention and treatment of diseases. Several recombinant antibodies having therapeutic properties have been developed or in the pipeline and a few are under trials.

Antibody production in plant system poses a unique challenge, as folding and assembling of the protein molecules should be proper in order that they can correctly detect corresponding antigens (Ma et al. 1995). Tobacco has been used as a platform for antibody production in earlier studies. Presently, cereals are being used for the production of antibodies as they have the advantage of long-term storage of the synthesised protein without significant degradation at normal temperature. A few other plant-based expression platforms, such as cell suspension cultures, agroinfiltrate leaves and virus-infected plants, are also being used for the synthesis of antibodies. These systems have a few advantages such as short time for its development and quick onset of protein synthesis and are unlike the transgenic or transplastomic plants (Fischer and Emans 2000).

Six plant-derived antibodies, viz. , *Avicidin*, *CaroRx*, *T84.66*, *AntiHSV* and *anti-RSV*, *38C13* and *PIPP*, have been developed for use as human therapeutics, which are under different phases of clinical trials. These products will be a great boon to the pharmaceutical industry once they successfully undergo clinical trials. A technology for mass production of vaccine against the dreaded Ebola virus in transgenic tobacco plants has been developed by a company Icon Genetics in Halle, Germany (Dohutia 2014).

11.10.4 Medical Proteins

Plants are being used for production of proteins having medical importance such as milk protein lysozyme and β-casein essential for improving health of children (Chong et al. 1997; Chong and Langridge 2000) and protein polymers having applications in surgery and tissue replacement. Human collagen protein has been expressed in transgenic tobacco plants (Ruggiero et al. 2000), and synthetic spider silk protein was produced in potato and tobacco (Scheller et al. 2001).

11.11 Regulatory and Safety Issues

Molecular farming involves production of genetically modified plants for production of biopharmaceuticals for which it falls under the purview of environmental regulatory and safety laws. Plant-based vaccines, particularly edible vaccines, have to face several ecological safety issues which need to be addressed to get environmental clearance for its successful commercialisation. A few risks related to the molecular farming products include spread of transgene through seed dispersal, horizontal gene transfer and transfer of pollens, besides their effects on pollinators, herbivorous animals and rhizospheric microbes and possible risks of contamination of food chain by plant materials containing recombinant proteins (Ma et al. 2003). Some concerns with the mass production of the transgenic plants containing the transgene are (i) gene transfer to other plant species, (ii) their possibility of becoming weeds and (iii) contamination of the plant vaccine itself in plant debris which might pollute drinking water. These issues may be dealt with through controlled cultivation of the transgenic plants in greenhouses or plant tissue culture facilities in order to prevent environmental release of the pharmaceuticals.

The use of innocuous plant-based markers and removal of selectable marker genes are a prerequisite to prevent incorporation of superfluous gene sequences. The probability of gene transfer through dispersal of the pollens could be prevented by adopting certain measures like cultivating the transgenic plants inside a greenhouse and by selecting those plants as an expression platform which do not outcross with its wild relatives.

The percentage risk of horizontal gene transfer has been low. The possible adverse implication of the products of molecular farming on the organisms, other than the targeted ones, may be averted by the application of regulated promoters which restricts the expression of the transgene to the desired organ. Besides, the recombinant protein may be made to express as a passive precursor which needs to be cleaved proteolytically in order to show its biological activity. Plants producing pharmaceuticals need to be segregated from those crops meant for human consumption for which regulatory mechanisms should be adopted for the safety of human health. Though cultivation of transgenic crops expressing recombinant proteins in greenhouses evades regulatory and safety issues, this reduces the advantages of biofarming in terms of economy and mass production of affordable medicines for supply to developing countries. Several other levels of safety can be adopted such as physio-geographical isolation, different seasons for planting and use of male sterile and transplastomic plants (Ma et al. 2003).

11.12 Conclusion

Plants are becoming popular as expression platforms for production of biopharmaceuticals of medical importance. From a technical point of view, it is feasible to produce a wide variety of antibodies in plants. Plants have several benefits such as low production cost, quick scalability, the presence of nonhuman pathogens and the capacity of plants for proper folding and assembling of proteins which make them a preferred choice as expression platforms over the conventional systems. Time may come when, in near future, plants might overtake other production systems of molecular farming due to these safety and economic advantages, though challenges exist in terms of enhancement of yield, improvement of glycosylation, bottlenecks in production, biosafety issues and industry inertia. Nevertheless, biofarming is a great boon to the pharmaceutical industry for supply of affordable medicines particularly to the developing countries.

References

Ball JM, Graham DY, Opekum AR et al (1999) Recombinant Norwalk virus-like particles given orally to volunteers: phase I study. Gastroenterol 117:40–48

Barta A, Sommengruber K, Thompson D et al (1986) The expression of a napoline synthase human growth hormone chimeric gene in transformed tobacco and sunflower callus tissue. Plant Mol Biol 6:347–357

Bevan MW, Flavell RB, Chilton MD (1983) A chimaeric antibiotic resistance gene as a selectable marker for plant cell transformation. Nature 304:184–187

Blanas E, Carbone FR, Allison J et al (1996) Induction of autoimmune diabetes by oral administration of auto antigen. Science 274:1707–1709

Chong DK, Langridge WHR (2000) Expression of full length bioactive anti-microbial human lactoferrin in potato plants. Transl Res 9:71–78

Chong DK, Roberts W, Arakawa T et al (1997) Expression of the human milk protein β-casein in transgenic potato plants. Transl Res 6:289–296

Christou P (1996) Transformation biotechnology. Trends Plant Sci 1:423–431

Chu L, Robinson DK (2001) Industrial choices for protein production by large scale cell culture. Curr Opin Biotechnol 12:180–187

Conrad U, Fiedler U (1994) Expression of engineered antibodies in plant cells. Plant Mol Biol 26:1023–1050

Daniell H, Streatfield SJ, Wyckoff K (2001) Medical molecular farming: production of antibodies, biopharmaceuticals and edible vaccines. Trends Plant Sci 6:219–226

Daniell H, Khan MS, Allison L (2002) Milestone in the chloroplast genetic engineering: an environmentally friendly era in biotechnology. Trends Plant Sci 7:84–91

Dohutia C (2014) The wonders of pharming. Horizon in The Assam Tribune. p 2

Domansky N (1995) Organ-specific expression of hepatitis B surface antigen in potato. Biotechnol Lett 17:863–866

Fischer R, Emans N (2000) Molecular farming of pharmaceutical proteins. Transl Res 9:279–299

Fischer R, Stoger E, Schillberg S et al (2004) Plant-based production of biopharmaceuticals. Curr Opin Plant Biol 7:125–128

Fraley RT, Rogers SG, Horsch RB et al (1983) Expression of bacterial genes in plant cells. Proc Natl Acad Sci USA 80:4803–4807

Franken E, Teuschel U, Hain R (1997) Recombinant proteins from transgenic plants. Curr Opin Biotechnol 8:411–416

Gelvin SB (2003) Agrobacterium mediated plant transformation: the biology behind the "gene-jockeying tool". Microbiol Mol Biol Rev 67:16–23

Gidding G (2001) Transgenic plants as protein factories. Curr Opin Biotechnol 12:450–454

Gidding G, Allison G, Brooks D et al (2000) Transgenic plants as factories for biopharmaceuticals. Nat Biotechnol 18:1151–1155

Goldstein DA, Thomas JA (2004) Biopharmaceuticals derived from genetically modified plants. Q J Med 97:705–716

Hiatt A, Cafferkey R, Bowdish K (1989) Production of antibodies in transgenic plants. Nature 342:76–78

Hood EE (2002) From green plants to industrial enzymes. Enzyme Microb Technol 30:279–283

Hood EE, Witcher DR, Maddock S et al (1997) Commercial production of avidin from transgenic maize: characterization of transformant, production, processing, extraction and purification. Mol Breed 3:291–306

Horn ME, Woodard SL, Howard JA (2004) Plant molecular farming: systems and products. Plant Cell Rep 22:711–720

Houdebaine LM (2000) Transgenic animal bioreactors. Transgenic Res 9:305–320

Huang Z, Dry I, Webstar D et al (2001) Plant-derived measles virus hemagglutinin protein induces neutralizing antibodies in mice. Vaccine 19:2163–2171

Jefferson RA, Kavanaugh TA, Bevan MW (1987) GUS fusions: β-glucuronidase as a sensitive and versatile gene fusion marker in higher plants. EMBO J 6:3901–3907

Kapusta J, Modelska A, Figlerowicz M et al (1999) A plant derived edible vaccine against hepatitis B virus. FASEB J 13:1796–1799

Kusnadi AR, Hood EE, Witcher DR et al (1998) Production and purification of two recombinant proteins from transgenic corn. Biotechnol Prog 14:149–155

Lal P, Ramachandran VG, Goyal R et al (2007) Edible vaccines: current status and future. Indian J Med Microbiol 25:93–102

Larrick JW, Thomas DW (2001) Producing proteins in transgenic plants and animals. Curr Opin Biotechnol 12:411–418

Leben M, Johansson S, Osek J (1993) Large-scale production of vibrio cholera toxin B subunit for use in oral vaccines. Biotechnology (N Y) 11:1574–1578

Lee RWH, Strommer J, Hodgins D et al (2001) Towards development of an edible vaccine against Bovine Pneumonic pasteurellosis using transgenic white clover expressing a Mannheimia haemolytica A1, Leukotoxin 50 fusion protein. Infect Immun 69:5786–5793

Ma JKC, Hiatt A, Hein M et al (1995) Generation and assembly of secretory antibodies in plants. Science 268:716–719

Ma SW, Zhao DL, Yin ZQ et al (1997) Transgenic plants expression auto antigens fed to mice to induce oral immune tolerance. Nat Med 3:793–796

Ma JKC, Drake PMW, Christou P (2003) The production of recombinant pharmaceutical proteins in plants. Genetics 4:794–805

Marquet-Blouin E, Bouche FB, Steinmetz A et al (2003) Neutralizing immunogenicity of transgenic carrot (*Daucus carota* L.)- derived measles virus hemagglutinin. Plant Mol Biol 51:458–469

Marsa L (1994) Enhancing the benefits of nature. Omni 16:14

Mason HS, Arntzen CJ (1995) Transgenic plant as vaccine production system. Trends Biotechnol 13:388–392

Mason HS, Lam DMK, Arntzen CJ (1992) Expression of hepatitis B surface antigen in transgenic plants. Proc Natl Acad Sci USA 89:11745–11749

Mason HS, Haq FA, Clement JD (1998) Edible vaccine protects mice against Escherichia Coli het labile anterotoxin (LT): potatoes expression a synthetic LT-B Gene. Vaccine 16:1336–1343

Mayfield SP, Franklin SE, Lerner RA (2003) Expression and assembly of a fully active antibody in algae. Proc Natl Acad Sci USA 100:438–442

Mishra N, Gupta PN, Khatri K et al (2008) Edible vaccines: a new approach to oral immunization. Indian J Biotechnol 7:283–294

Pant G, Sanjana WK (2014) Edible vaccines: a boon to medical science. Intl J Curr Agric Res 3:76–80

Richter L, Mason HS, Arntxen CJ (1996) Transgenic plants created for oral immunization against diarrheal diseases. Travel Med 3:52–56

Rudzik R, Claney RL, Perey YE et al (1975) Repopulation with IgA-containing cells of bronchial and intestinal lamina propria after transfer of homologous Peyer's patch and bronchial lymphocytes. J Immunol 144:1599

Ruggiero F, Exposito JY, Bournat P et al (2000) Triple helix assembly and processing of human collagen produced in transgenic tobacco plants. FEBS Lett 469:132–136

Scheller J, Guhrs KH, Grosse F, Conrad U (2001) Production of spider silk proteins in tobacco and potato. Nat Biotechnol 19:573–577

Schillberg S, Fischer R, Emans N (2003) Molecular farming of antibodies in plants. Naturwissenschften 0:145–155

Schwartz JR (2001) Advances in Escherichia coli production of therapeutic proteins. Curr Opin Biotechnol 12:195–201

Shah CP, Trivedi MN, Vachhani UD, Joshi V (2011) Edible vaccine: a better way for immunization. Int J Curr Pharma Res 3:53–56

Sijmons PC, Dekker BM, Schrammeijer B et al (1990) Production of correctly processed human serum albumin in transgenic plants. Biotechnology (N Y) 8:217–221

Singh BD (2009) Biotechnology- expanding horizons. Kalyani Publishers, Ludhiana

Stoger E, Sack M, Perrin Y et al (2000) Practical considerations for pharmaceutical antibody production in different crop systems. Mol Breed 9:149–158

Tacket CO, Mason HS, Losonsky G et al (1998) Immunogenicity in humans of a recombinant bacterial antigen delivered in transgenic potato. Nat Med 4:607–609

Veluthambi K, Gupta AK, Sharma A (2003) The current status of plant transformation technologies. Curr Sci Ind 84:368–380

Wang L, Goshnick MW, Copllel RL (2004) Oral immunization with a combination of plasmodium yoelii merozoite surface proteins 1 and 4/5 enhances protection against lethal malaria challenge. Infect Immun 72:6172–6175

Washam C (1997) Biotechnology creating edible vaccines. Ann Intern Med 127:499

Webster DE, Thomas DE, Strugnell RA et al (2002) Appetizing solutions: an edible vaccine for measles. Med J Aust 176:434–337

Wilde CD, Peeters K, Jacobs A et al (2002) Expression of antibodies and Fab fragments in transgenic potato plants: a case study for bulk production in crop plants. Mol Breed 9:271–282

Witcher DR, Hood EE, Petersen D et al (1998) Commercial production of bglucuronidase (GUS): a model system for the production of proteins in plants. Mol Breed 4:301–312

Intellectual Property Rights Issues for Herbal Products

12

Nidhi Sandal and Avinash Kumar

Abstract

The entire edifice of intellectual property rights system is based upon incentivising innovations by providing legally created private monopoly rights albeit for limited period and on certain conditions. Post TRIPS era, most of our Indian legal IP instruments have been amended, and even new legal instruments have been created in order to comply with provisions of the agreement. The present Indian law has adequate provisions for the protection of innovations in the area of herbals. During the past decade, there has been a spurt of herbal products which include health supplements and medicines for hypertension, obesity, arthritis, diabetes, neurological disorders, etc. (Liu et al. Life Sci 73:1543–1555, 2003; Modak et al. J Clin Biochem Nutr 40:163–173, 2007; Brown and Gerberg, J Sychiatr Pract 7:75–91, 2001). The bent of the global market towards herbal product is the driving force behind the R&D of big pharma companies towards the development of new herbal products. In coherence with the booming industry and extensive R&D work in the field of herbals, the role of intellectual property rights also becomes very important. Lots of innovation is taking place in R&D and all this needs to be properly protected through appropriate legal routes. Patents, copyrights, designs, trademarks and geographical indications are the types of IPRs that play an instrumental role in the legal protection of various aspects of herbal products and processes. Apart from these IP rights, the Protection of Plant Varieties and Farmers' Rights Act, 2001, was enacted to provide protection of plant varieties developed by plant breeders and farmers, so as to ultimately encourage development of new varieties of plants. It envisages facilitating the growth of seed industry ensuring high-quality seeds and planting material to the farmers. The Biodiversity Act 2002 provides

N. Sandal • A. Kumar (✉)
Defence Research and Development Organisation,
Rajaji Marg, New Delhi, 110 011, India
e-mail: avinash@hqr.drdo.in

© Springer India 2016

Vijay Veer, R. Gopalakrishnan (eds.), *Herbal Insecticides, Repellents and Biomedicines: Effectiveness and Commercialization*, DOI 10.1007/978-81-322-2704-5_12

provisions for the conservation of biological diversity, sustainable use of its components and fair and equitable sharing of benefits arising art of the new use of biological resources and knowledge. For the effective implementation of the Act, National Biodiversity Authority (NBA) was established for the effective regulation of related activities. The Act makes specific provisions that no person can apply for intellectual property rights, in India or abroad, for any invention based upon research or information on a biological resource obtained from India without seeking prior approval from the National Biodiversity Authority. This chapter discusses in detail the intellectual property rights application for protection of innovations in herbals, special provisions and guidelines of the acts and also briefly describes related regulatory requirements.

12.1 Introduction

Recently, there has been a noticeable shift in the public perception and attention from synthetic to herbal medicine. Scarcity in the area of new molecular entities, public awareness about the side effects of synthetic drugs, abundance of medicinal plants in our country or the traditional method for the cure of diseases could be the plausible factors for the shift of this attention and consequent direction of research to the area of herbals. It has been estimated that total value of the world market for herbal products stands at around $83 billion, and Europe accounts for over 50 % of the total (Dennis 2013). Due to the emergence of herbal drug companies and growing market for the herbal products, a lot of research is going on for herbals/herbal formulations. A recent PubMed search (done in September 2013) using the keyword *herbal medicinal products* (HMPs) gave rise to 30,917 hits, with about 2700 of them published in 2013 (Pelkonen et al. 2014).

India is no exception to this trend, particularly because India is endowed with a variety of recognised indigenous system of medicine, viz. Ayurveda, Siddha, Unani, homeopathy, Yoga and naturopathy catering to the health requirements of people. Charaka Samhita is one of the most ancient, comprehensive and authoritative works of Ayurveda. It is considered the original reference book of holistic Ayurvedic medicine. India's traditional system of medicines has originated from the fact that India is a nation having mega-diversity and exceptionally blessed with biological diversity due to its unique geographical location. Its natural ecosystems vary from colder Himalayan regions to the desserts in the northwest region as well as from sea coast to the green forests particularly in central and northeast India. Therefore, our traditional system of medicine provides herbal cure to both acute and chronic diseases of cardiovascular system, neurological disorders, endocrinological diseases and others.

It is interesting to note that cases related to legal protection of herbal-related innovations have successfully brought the public focus on the ill effect of wrong grant of legal protection on herbal-related innovations. In India, the post-2005 era has seen a heightened level of public debate and discourse about the very desirability of very strong patent rights particularly in the area of herbals. This debate was further fuelled when the USA granted patents on the uses of turmeric. A US patent No. 5, 401,504 was granted to two non-resident Indians Suman K. Das and Harihar P. Cohly on the use of turmeric for healing of wound. Subsequently, USPTO cancelled the patents when a re-examination was filed by the Council of Scientific and Industrial Research (CSIR), India, New Delhi, on the grounds of prior art (TKDL 2015). This particular case proved to be a historical one as for the first time any US patent claiming traditional knowledge originating from a developing was successfully challenged. A patent granted on fungicidal effect of extracts of neem seeds in Europe was also revoked by the EPO in similar fashion. In the case of basmati rice, the applicant RiceTec had to amend the claims to exclude the well-known traditional Indian basmati rice lines. These cases of biopiracy triggered a heightened level of public interest as well as awareness about the pitfalls as well as opportunities related to legal protection of innovations in the area of herbal drug. The entire debate related to patent protection in the area of traditional knowledge has been catapulted to the centre stage with focus also shifted to the desirability of carving out appropriate procedures of the legal protection of innovations using traditional resources. Rising cases of misappropriation of traditional knowledge coupled with rapid erosion of biodiversity and a concern for right of local communities holding the traditional knowledge have also raised worldwide public concern.

A total of 557 published Indian applications and 210 PCT applications have been filed by Indians during 2001–2010 (Sahoo et al. 2011). Interestingly, most of the individual inventors for these applications are herbal practitioners (doctors, vaidyas or hakims). Sahoo et al. 2011 performed a study on the patent applications and grants by Indian applicants in herbal drugs during 2001–2010. Their analysis shows that CSIR has the maximum numbers of applications not only in India but also in the USA and EU. China has a heritage of about 2000 years in the area of herbal medicine. This is further supported by the abundance in traditional knowledge and biological resources available in their country. Like other streams of research, China is pursuing its research rigorously in the area of herbal. In China alone, approximately 100,000 herbal formulae and over 11,000 individual medicinal plants have been documented, which are generally hailed as rich natural resources for developing new drugs, including new lead compounds and new types of multicomponent drugs (Wang et al. 2008; Kuhn and Wang 2008 & http://www.who.int/mediacentre/factsheets/fs134/en/).

Herbal medicinal research offers a very high potential of innovations to the researchers which, in turn, provides opportunities for patenting as well. Herbal medicines include herbs, herbal materials, herbal preparations and finished herbal products that contain as active ingredients parts of plants, or other plant materials, or combinations (Wang et al. 2008). The issues related to protection of knowledge, innovations and practices of traditional and indigenous medicine have also found echo at the forefront of international developments as well. India has played a major role in catapulting the entire issue of protection of traditional knowledge at the global stage resulting in WIPO setting up Intergovernmental Committee (IGC) as well as Doha Ministerial Declaration in 2001. Doha Declaration established a linkage between the TRIPS Agreement and the UN Convention on Biological Diversity (CBD) for fair and equitable sharing of the benefits arising from the use of genetic resources (Guidelines for Processing of Patent Applications Relating to Traditional Knowledge and Biological Material, Indian Patent Office, 2012).

As more and more interest is being generated for herbals among researchers, authorities at national and international level are taking serious note of all these developments and are making sincere efforts to make a proper equilibrium between the availability of monopoly right and the freedom for rest of public domain. IPR regime is an inseparable aspect of R&D, and in the case of herbals as well, it is very important to encourage R&D so that new products should reach market and benefit society. However, researchers or applicants need to be given due credit and incentives like commercialising their products and enjoying the monopoly rights provided by IPRs. At the same time, governments have made special provisions in the Patents Act 1970 for herbals so as to control the misappropriation of herbals/traditional knowledge like in the case of turmeric, neem and basmati. Patents are the strongest IP right giving monopoly rights to inventors/application for 20 years from the date of filing. Therefore, the Patents Act 1970 was amended to incorporate some provisions addressing the patentability of inventions using traditional knowledge. Further, Guidelines for Processing of Patent Applications Relating to Traditional Knowledge and Biological Material have been issues in 2012 by the Controller General of Indian Patent Office to provide guidance to the examiners for examining a patent application related to herbals/traditional knowledge. Another important aspect associated with the patenting of herbals is the regulatory approval from the National Biodiversity Authority. However, other legal instruments of IPR regime including copyright, trademark and design do not have any special provision for herbal innovations, and the prerequisites for the registration of any one of these rights are unanimous for all application. This chapter is an attempt to highlight all the relevant aspects related to the legal protection of herbal innovation. The authors have oriented the chapter to give an overview of all the IP rights and how these rights can be attained and used for herbals.

12.2 Intellectual Property Rights

Intellectual property systems, world over are primarily concerned about motivating innovators by providing monopoly rights to them over their creations, albeit for limited duration, if their creations meet certain laid-down conditions. Intellectual property systems essentially provide legal mechanisms enabling innovators to stop third parties from unauthorised use of their innovations. The rationale of the intellectual property system is that the 'cost' of the monopoly rights conferred to the intellectual property holder is outweighed by the 'benefits' to the society. This is one reason that each type of intellectual property needs a specific legal instrument, which tries to strike a balance between the interest of the right holder and larger public interest. The entire edifice of the intellectual property regime is based upon the principle that innovators need to be motivated by conferring them monopoly rights for certain acts if their innovations meet laid-down criteria and conditions.

Intellectual property rights is an umbrella term comprising a variety of legal rights to protect legal human creativity which includes inventions, literary and artistic works, symbols, names, images and designs. The system grants legal rights based on certain criterion making a balance between the monopoly right and larger public interest. For example, in the case of patents, monopoly rights are granted if the invention meets the criteria of novelty, nonobviousness and industrial applicability. These rights could be to make use of the inventions and right to transfer among other rights. Each type of intellectual property right demands its own set of operating rules.

In 1994, when negotiations on the WTO Agreement on Trade-Related Aspects of Intellectual Property Rights ('TRIPS') were concluded, governments of all WTO member countries (151 countries as of August 2007) had agreed to set certain basic standards for the protection of all form of intellectual property rights in all member countries. The TRIPS Agreement (Trade-Related Aspects of Intellectual Property Rights) recognises the following rights:

- Copyright and related rights
- Trademarks
- Geographical indications
- Industrial designs
- Patents
- Layout designs (topographies) of integrated circuits
- Protection of undisclosed information

All these intellectual property rights protect legally the creations of mind including inventions, literary and artistic works and symbols, names, images and designs used in commerce. Figure 12.1 provides an overview of the IPRs and the form of IP protected therein.

A patent is a legally created monopoly right for an invention granted for to the applicant for 20 years, in exchange for disclosure of his invention. This legal right enables the applicant to exclude others from making, using, selling and importing the patented product or processing for producing that product for those purposes without his consent (Controller General of Patents Designs and Trademarks 2015). It is a territorial right and therefore it is effective only within the specific territory where it is granted. A patent is granted when three criteria of novelty, inventive step and industrial applicability are satisfied. This is the worldwide applicable criteria to grant

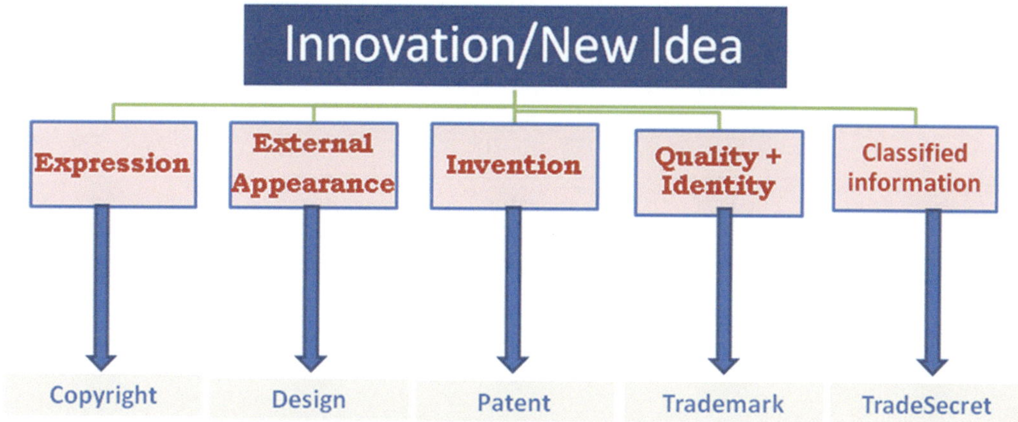

Fig. 12.1 Protecting innovations under IPR regime

a patent. However, each nation has its own legislation, and the patent is granted by their national/regional patent offices. In India, the Patents Act 1970, Section 2 (1) (i) and Section 2 (1) (j), sets out the legal requirements to obtain a patent in India. Once a patent is granted, the owner of the patent has legal rights to exclude others from making, using, selling or importing the invention in a country in which the patent has been granted for a period of 20 years from the date of filing.

For obtaining patents in several countries, the applicant has to apply in each county of interest, and the application undergoes examination as per the laws and rule of that country. Prior to 1883, the applicants had to file patent in each country of interest on the same day so as to maintain the priority of their invention. This was very difficult and cumbersome process. After a diplomatic conference in Paris in 1880, a convention called 'Paris Convention for the Protection of Industrial Property' was signed in 1883 by 11 countries. As per this convention, the applicant of the member country can file a subsequent application within 6 months (for industrial designs and trademarks) or 12 months (for patents and utility models) from the first filing in any of the member country. This convention provides 12 months time to the applicants so that they may choose the country and complete the process of filing their application in different countries. At present, there are 173 member countries of the Paris Convention.

There is another route to file a patent application in several countries, i.e. under Patent Cooperation Treaty (PCT). In the year 1970, the World Intellectual Property Organisation implemented this treaty. An applicant by filing a single international patent application through Patent Cooperation Treaty effectively can ensure priority in each of its contracting states. At present, there are 147 members of this treaty. India became its member in the year 1998. This route provides several advantages of the Paris Convention. A PCT application can be filed within 12 months from the date of filing a patent application in any one of the member country. Under PCT, there are several examining/search authorities which provide a preliminary examination within 16 months of the priority date of an application. This examination provides useful information to the applicants for deciding the countries for national phase entry. Figure 12.2 shows the flow chart for filing a patent application in several countries through PCT route.

Copyright protects the expression of ideas. It provides legal protection to the creators under the Copyright Act 1957, for original literary and artistic works, which include computer programs, multimedia and electronic databases apart from literary works, illustrations, photographic works, musical works, drawings, paintings, cinematographic works, sculpture, etc. The owners of the copyrighted works can stop others from using

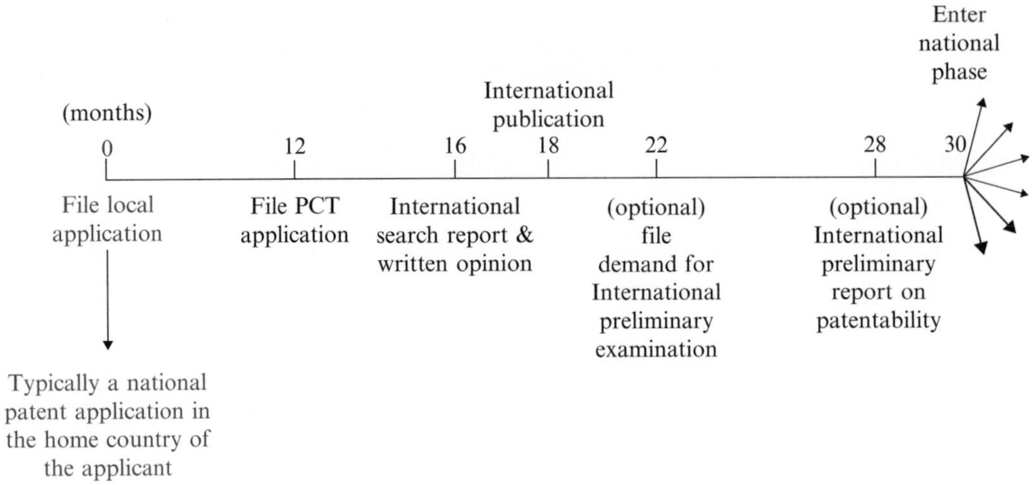

Fig. 12.2 The PCT system

such copyrighted works without their authorisation thus providing them incentives and motivation. Some of the rights, enjoyed by the owners of copyrighted works, include right to reproduction, right to communicate to the public, right to public performance, right to translation, right to adaptation, etc.

According to the definition provided in the Intellectual Property Office, India, 'A Design refers to the features of shape, configuration, pattern, ornamentation or composition of lines or colours applied to any article, whether in two or three dimensional (or both) forms. This may be applied by any industrial process or means (manual, mechanical or chemical) separately or by a combined process, which in the finished article appeals to and judged solely by the eye. Design does not include any mode or principle of construction or anything which is mere mechanical device' (Information Booklet for applicants for Registration of Design, *The Patent Office, Intellectual Property Office, Kolkata* 2010). In India, design protection is provided under the new Designs Act 2000.

According to the definition provided in the Intellectual Property Office, India, 'A trade mark (popularly known as brand name) is essentially a visual symbol which may be a word signature, name, device, label, numerals or combination of colours used by one undertaking on goods or services or other articles of commerce to distinguish it from other similar goods or services originating from a different undertaking'. It is provided under the Trade Marks Act 1999 that goods and services are classified according to the International Classification of Goods and Services.

In addition to the above-mentioned forms of intellectual property rights, i.e. patents, designs, trademark and copyrights, there are three more IPRs which provide monopoly rights to creators in the different field. These rights, although may not be of direct relevance to the innovations in herbals, cover the Plant Variety Protection and Farmers' Rights, geographical indications and the layout designs (topographies) of integrated circuits. The Plant Variety Protection and Farmers' Rights Act provides legal protection to new plant varieties including seed after fulfilment of certain conditions. A geographical indication is a sign used on goods that have a specific geographical origin and possess qualities, a reputation or characteristics that are essentially attributable to that place of origin. Layout designs (topographies) of integrated circuits are a field in the protection of intellectual property which provides legal protection to two- or three-dimensional layout or topography of an integrated circuit (IC or 'chip'), i.e. the arrangement on a chip of semiconductor devices such as transistors and passive electronic components such as resistors and interconnections. The entire gamete of IP protection has been designed to provide legal protection to all possible creation of humans; therefore, it covers different types of rights providing protection to different types of innovations after fulfilling the requisite criteria for the grant. However, for the purposes of this paper, only those IPRs and their provisions relevant for the herbals have been discussed in detail.

12.3 Patent Protection for Herbals

A patent is a set of exclusive rights granted by a government to an inventor or applicant for a limited amount of time (normally 20 years from the filing date) at the cost of making a complete disclosure of the details of his invention. It is essentially a negative right, which exclude others from using the patented invention. A patent must disclose the details of the invention so that a person skilled in the art must be able to reproduce the patented invention without undue experimentation (Section 64 (1) (h), Indian Patent Act 1970). It should clearly address the following questions:

- What was the problem?
- What was the available solution(s)?
- Why available solution(s) did not solve your problem?
- What was your idea/approach?
- How did you design and carry out experimental work?
- What were your results and data?
- How did it solve your problem?
- What other possible problems can it solve in related area?

A patent document is a technolegal document drafted to address the above aspect of the invention, but any innovation must meet the three criteria of the patentability which includes novelty, inventiveness and industrial applicability (Fig. 12.3). These criteria are universally accepted for the grant of patent in any country. Apart from these criteria, each country has their own laid-down laws which define the patentable invention, and the patents are examined according to these laws. Like in India, the Patents Act do not grant patents for the method of use, method of treatment or new use of a known substance, but in the USA or Europe, the new use of known substance is a patentable subject matter. The patent system of each country is guided by the patent law of their country. In India, a patent is granted after a thorough examination for ascertaining the novelty, inventiveness and industrial applicability of each patent application. Apart from this, there are some sections of the Act which categorise some inventions as not patentable inventions, and for herbals or inventions related to traditional knowledge, the Patents Act 1970 has some special provisions.

12.3.1 The Patents Act 1970

Indian law has adequate provisions for the protection of inventions related to herbals. Any patent application relating to herbals having novelty, inventive step and industrial applicability under Section 2 (1) (i) and Section 2 (1) (j) is patentable as per the Patents Act 1970.

> Section 2 (1) (j) : "invention" means a new product or process involving and inventive step and capable of industrial application. (Section 2 (1) (i), Indian Patent Act 1970)
> Section 2 (1) (ja): "inventive step" means a feature of invention that involves technical advance as compared to the existing knowledge or having economic significance or both and that makes the invention not obvious to a person skilled in the art. (Section 2 (1) (j), Indian Patent Act 1970)

12.3.1.1 Novelty

> An "invention" means a **new** product or process involving and **inventive step** and capable of **industrial application** (Section 2 (1) (i), Indian Patent Act 1970)

An invention is new (novel) if it has not been anticipated by publication in any document anywhere in the world or used in the country or prior claimed in an application for patent in India or form part of the knowledge, oral or otherwise, available within any local or indigenous community in India or elsewhere before the date of filing of patent application or date of priority, that is,

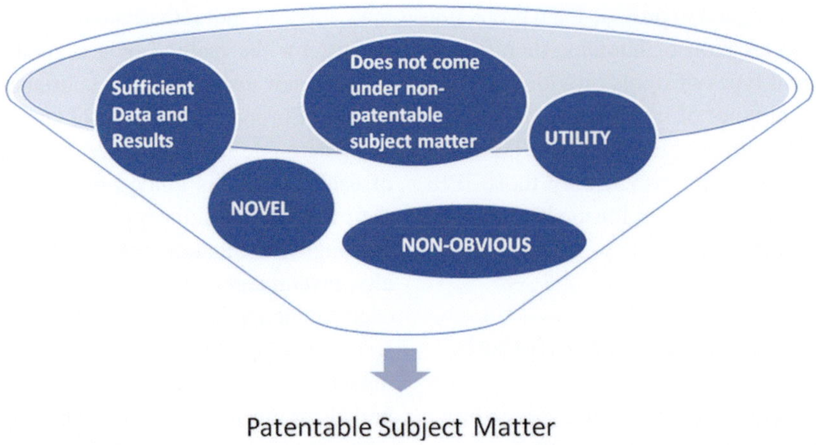

Fig. 12.3 Criteria to patentability

the subject matter has not fallen in the public domain or that it does not form part of the state of the art. As per the Patents Act 1970, novelty is

> any invention or technology which has not been anticipated by publication in any document or used in the country or elsewhere in the world before the date of filing of patent application with complete specification, i.e. the subject matter has not fallen in public domain or that it does not form part of the state of the art. (Section 2(1)(l)of the Act)

Before we proceed, it would be better to acquaint with the concept of prior art. "Prior art" is the information that was known before the date of filing a patent application. If it is public **anywhere** in the world, it is prior art (Fig. 12.4). It includes patent literature and non-patent literature comprising

- Books
- Journal/related magazines
- Abstract books
- Indexes
- Proceedings of conferences
- Catalogues
- News (printed, telecasted on TV or radio)
- Conferences, seminars, workshops
- Public prior use

It is evident to ascertain that the patentability of an invention, novelty and inventiveness of an invention is examined over the existing prior art documents. Prior art search should also be conducted by the inventors or applicants before filing a patent application. However, during the examination of a patent application, the examiners carry out extensive prior art search to check the novelty and inventiveness. Let us understand the concept of novelty with an example of a patent application disclosing a process for extracting the aqueous extract of leaves of *Gomphostemma niveum* which has at least 30 % alkaloids. In this invention, the essential features of the patent include

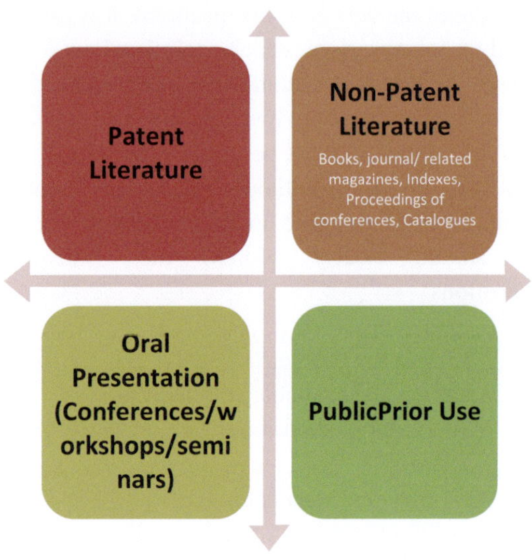

Fig. 12.4 Prior art documents

- Process for extraction
- Aqueous extract
- Leaves
- *Gomphostemma niveum*
- At least 30 % alkaloids

During prior art search, if a prior art comprising the above-listed five features of the invention is found in **one** single prior art document, then only the novelty of the above invention is destroyed, and the document is referred as 'novelty destroying document' (Fig. 12.5). Thus, to destroy novelty,

- *The prior art should disclose the invention either in explicit or implicit manner.*
- *Mosaicing of prior art documents is not followed in the determination of novelty.*
- *A generic disclosure in the prior art may not necessarily take away the novelty of a specific disclosure. For instance, a metal spring may not take away the novelty of a copper spring.*
- *A specific disclosure in the prior art takes away the novelty of a generic disclosure. For instance, a copper spring takes away the novelty of a metal spring* (MPEP, Novelty, 08.03.02).

12.3.1.2 Inventive Step

An 'inventive step' is one which makes the invention 'nonobvious to a person skilled in the art'. In other words, if the invention is obvious to the person skilled in the art, it cannot be said to involve an inventive step. As per the Patents Act 1970 and the Patents Amendment Act 2005 (which came into effect retrospectively, January 1, 2005), the definition of Inventive step was further revised to read as

> "inventive step" means a feature of an invention that involves technical advance as compared to the existing knowledge or having economic significance or both and that makes the invention not obvious to a person skilled in the art. (Section 2(1)(ja) of the Act)

Thus, the standard of inventive step has evolved to include economic significance of the invention apart from already existing criteria for determining inventive step.

Manual of Patent Office Practice and Procedure Published by the Indian Patent office provides detailed examination guidelines on inventive step. Although it must be understood that the manual has no legal binding on the examination of any patent application in case of conflict between manual and patent laws and rules, the patent laws and rules would prevail. However, the manual does provide an indication about the manner in which the examination of any patent application shall be carried out in the entire branch of patent offices. General principle of MPEP states that an invention is patentable only if it involves one or more inventive steps. In relation to the determination of patentability, an examiner first conducts an enquiry as to the novelty of the claimed invention and then proceeds to conduct an enquiry on whether the claimed invention involves one or more inventive steps (MPEP General Principle (08.03.03.01)). Then the guidelines suggest the steps to determine inventive step (MPEP Determination of Inventive step (08.03.03.02)).

(a) For determination of inventive step, all or any of the prior art(s) revealed during the search process to perform an enquiry as to whether such prior art(s) discloses the claimed invention is relied upon.
(b) Publications existing on the date of filing of complete specification would be considered as a prior art.

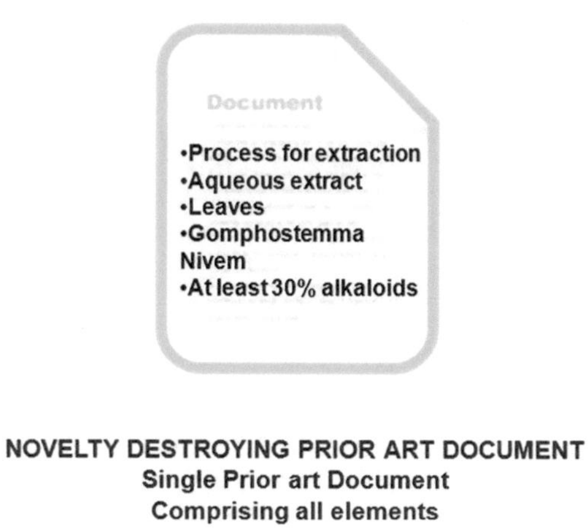

Fig. 12.5 Novelty destroying document

(c) However, Indian applications filed before but published on or after the date of filing of complete specification of the instant application are considered as a prior claiming.
(d) Invention as a whole shall be considered. In other words, it is not sufficient to draw the conclusion that a claimed invention is obvious merely because individual parts of the claim taken separately are known or might be found to be obvious.
(e) If an invention lies merely in verifying the previous predictions, without substantially adding anything for technical advancement or economic significance in the art, the inventive step is lacking.
(f) For the purpose of establishing obviousness of the invention, citing a mosaic of prior arts is permissible, provided such prior art is enabling.
(g) If the invention is predictable based on the available prior art, merely requiring workshop improvement by a person skilled in the art, the inventive step is lacking.

With respect to the previous example of *Gomphostemma niveum* which describes a process for extracting the aqueous extract of leaves of *Gomphostemma niveum* which has at least 30 % alkaloids, the inventive step over the prior art will be established if one of the elements of the invention is not available anywhere in a prior art. Let us consider a hypothetical prior art document, D1, which discloses the process for extracting an aqueous extract of leaves of *Gomphostemma niveum*. It does not mention about the alkaloidal content of the final extract. There is another prior art document D2 which discloses the aqueous extract of leaves of *Gomphostemma niveum* having 10 % alkaloids. When these two documents are read subsequently, it appears that the D1 discloses the subject matter of invention, and in view of D2, it can be deduced that the extract of D1 and D2 anticipates the presence of alkaloids in any aqueous extract of leaves of *G. niveum*. In such case, the inventive step has to be highlighted which may include the duration of extraction process or the temperature or pH of the extraction process which is yielding at least 30 % alkaloids in the final extract because 10 % alkaloids are already reported. In such a case, the exact process parameter resulting in substantial increase of alkaloids is the inventive step of the invention.

12.3.1.3 Industrial Applicability

In order for an invention to be patentable, an invention must be capable of industrial application. Industrial application in relation to patentability means that the invention is capable of being made or used in an industry. The specification explains the industrial applicability of the disclosed invention in a self-evident manner. Usually industrial applicability is self-evident. A specific utility should be indicated in the specification supported by the disclosure.

Thus, novelty, nonobviousness, industrial applicability and utility form the essential requirements of patentability. These conditions have been universally accepted as the essential prerequisites of patentability. Apart from this, Section 3 of the Patents Act 1970 elaborates on what are not inventions as per law. For the purpose of this chapter, it is important to discuss what are not inventions as per the Patents Act 1970.

12.3.2 Inventions Not Patentable

The Indian Patent Law excludes certain categories of invention from patent rights. It can be inferred that monopoly rights in these very categories are not considered to be in wider public interest. Under the Patents Act 1970, the inventions listed from Section 3 (a) to 3 (p) are not inventions and hence are not considered to be patentable. Specifically, *Section 3* (b), (c), (d), (e), (h), (i), (j) and (p) are of relevance with respect to the patent applications related to herbals (*Section 3*, Indian Patent Act 1970).

> 3 (b) An invention, the primary or intended use or commercial exploitation of which would be contrary to public order or morality or which causes serious prejudice to human, animal or plant life or health or to the environment is not an invention.
>
> 3 (c) the mere discovery of a scientific principle or the formulation of an abstract theory or discovery of any living thing or non-living substance occurring in nature;

3 (d) the mere discovery of a new form of a known substance which does not result in the enhancement of the known efficacy of that substance or the mere discovery of any new property or new use for a known substance or of the mere use of a known process, machine or apparatus unless such known process results in a new product or employs at least one new reactant.

3 (e) a substance obtained by a mere admixture resulting only in the aggregation of the properties of the components thereof or a process for producing such substance;

3 (h) a method of agriculture or horticulture;

3 (i) any process for the medicinal, surgical, curative, prophylactic, diagnostic, therapeutic or other treatment of human beings or any process for a similar treatment of animals to render them free of disease or to increase their economic value or that of their products.

3 (j) plants and animals in whole or any part thereof other than micro- organisms but including seeds, varieties and spices and essentially biological process for production or propagation of plants and animals;

3 (p) an invention which, in effect, is traditional knowledge or which is an aggregation or duplication of known properties of traditionally known component or components.

For example, an invention: a method of adulteration of food. The intended use or commercial exploitation of which is found to be injurious to public, animal or plant life or health, such as a method of adulteration of food.

Among others, Section 3 (d) was introduced in the Patents Act 1970, in the year 2005, in order to comply with the requirements of TRIPS. However, this has become an extremely powerful tool for the examiners to reject the patent applications filed on the trivial improvements. The main objective of this section is to prevent applicants from obtaining patents on already known medicines which are just a mere increment or trivial improvement of the known substances and also a refusal to the patent on discovery of new form or new use of known drugs. This particular section prevents applicants indulging in evergreening of their patent rights. This section shot into prominence when a patent application filed by Novartis before the Chennai Patent Office related to drug name GLIVEC which was slightly a different version of their 1993 patent for anti-leukaemia drug was rejected under Section 3 (d). Subsequently, the applicant challenged the constitutionality of Section 3 (d) before the High Court at Madras. The case went up to the Supreme Court of India, and in 2013, the Supreme Court of India upheld the constitutionality of Section 3 (d) and the patent remained rejected (Kant 2009). This section is also very important for the inventions related to herbals or traditional knowledge as the researchers use known herbs or other biological resources; therefore, a patent application related to herbals must not fall under this section. To summarise, the non-patentable inventions of herbals under this section, the following inventions are not considered patentable:

- Extracts per se
- New property or use of a known herb or herbal extract
- Extracts of different parts of herbs
- Mere method of extraction for a known herb
- Combination of herbal extract without significant property
- Formulations of extracts/herbs without a substantial improvement

Other than Section 3 (d), the Patents Act 1970 also has a unique provisions, like Section 3 (p) incorporated in 2002 with the implementation of the Patents (Amendment) Act 2002. Section 3 (p) states that an invention is not patentable where 'an invention is traditional knowledge or which is an aggregation or duplication of known properties of traditionally known component or components'. Traditional knowledge, being knowledge already existing, is not patentable, for example, the antiseptic properties of turmeric for wound healing or the pesticidal and insecticidal properties of *neem*. The examiner conducts an investigation by using various resources like Traditional Knowledge Digital Library (TKDL) and other references to decide as to whether the claimed subject matter falls within the purview of this provision.

Section 3 (b), (c), (e), (h), (i) and (j) also debar certain inventions from patentability criteria. Section 3 (b) states that an invention causing serious prejudice to human, animal or plant life or health or to the environment is not an invention, like a combination of herbs/herbal extract imparting properties that may cause serious health diseases or a herb that may destroy the crops or a method of extraction that makes the extract poisonous, etc. Section 3 (c) describes non-patentable inventions which fall under the category of discovery/abstract theory/scientific principle. For example, finding of a new herb occurring freely in nature is a discovery and not an invention.

Section 3 (e) is also an important section of the Patents Act 1970, which describes that mere admixture resulting only in the aggregation of the properties of the components thereof or a process for producing such substance is not an invention. An admixture resulting in synergistic properties is not considered as mere admixture, e.g. a soap, detergent, lubricant and polymer composition, etc., and hence may be considered to be patentable. A mere aggregation of known herbs/herbal extracts which does not result in any improvement and serves as comprising the properties of the constituents separately is considered a non-patentable invention. For patentability, the final product which is produced by admixing two or more ingredients or a process of producing such substances should satisfy the requirement of add-on feature which may include

- Synergistic effect: for example, combination of senna leaf extract with isabgol for enhanced laxative effect
- Improved stability: addition of a buffer in an extract to maintain the pH of the extract stable for a month or addition of a chemical to prevent sedimentation
- Decrease in side effects: combination of fennel seed extract with antihistamines (Allegra and Benadryl) so as to overcome the side effect of increased acid reflux

Section 3 (h) states that a method of agriculture or horticulture is not an invention. For example, a method of producing a plant, even if it involved a modification of the conditions or a method of producing mushrooms or a method for cultivation of algae, etc., is not patentable. Likewise, the method of improving the cultivation of herb is also not patentable. For the benefit of the society, such inventions are categorised as non-patentable inventions. Section 3 of the Patents Act 1970 has been drafted very meticulously so as to keep a balance between the interest of inventors and the public. Section 3 (i) and 3 (j) are also relevant for understanding the patentability of herbal inventions as Section 3 (i) excludes process for the medicinal, surgical, curative, prophylactic, diagnostic, therapeutic or other treatment of human beings or any process for a similar treatment of animals to render them free of disease or to increase their economic value or that of their products as an invention. For example, a medicinal method is not patentable under this section. A process of administering medicines orally, or through injectables, or topically or through a dermal patch or the order of administering two drugs or herbal extract(s) to obviate side effects or a method of treatment using herbal formulation, etc. is not patentable. Further, Section 3 (j) excludes (a) plants in whole or in part, (b) animals in whole or in part, (c) seeds, (d) varieties and species of plants and animals and (e) essentially biological process(es) for production or propagation of plants and animals, are not patentable. Section 3 clearly demarcates the inventions that cannot be considered patentable under the Patents Act 1970. This helps the researchers to understand the concept and purpose of granting a patent especially in the case of herbals as herbal medicines are considered as household items in India, for example, using fennel seeds as carminative agents and stem of *Glycyrrhiza glabra* or tulsi leaves for cough. Patent applications related to herbals must be examined with a special consideration, notably keeping in view the vast traditional knowledge that is getting transferred from one generation to another in India. Also, patenting is important for herbals; therefore it's an additional responsibility of the patent offices to grant a patent for herbal inventions considering all these above-mentioned aspects of herbals.

12.3.3 Other Relevant Provisions of the Patents Act 1970 for Herbal Patent Applications

The Patents Act 1970 has been essentially designed to ensure that only patents with technical advance are granted in the hands of applicants for asserting monopoly. Various provisions under the Act have been placed at different levels to ascertain the technical advancement of the patents. Right from the filing to prosecution and then grant, a patent application is examined thoroughly in view of the patent act and the guidelines as discussed above. There is another provision in the Act that empowers any person to challenge the validity of a patent. The pre-grant opposition (under Section 25 (1)) and post-grant opposition (under Section 25 (2)) are designed to ensure that only valid and enforceable patents are granted. Pre-grant opposition under Section 25 (1) can be filed by any person after the pre-grant publication of the patent application by way of lodging an opposition to the controller based on specific grounds (the Patents Act 1970). Post-grant opposition under Section 25 (2) can be filed by any interested person before the expiry of a period of one year from the date of publication of grant of a patent again based on specific grounds. These provisions together enable any person/ interested person to file opposition against the applied patent/granted patent to stop the grant or invalidate a wrongly granted patent, respectively.

It may be mentioned that among other grounds of opposition, Section 25 (1)(j) and Section 25 (1)(k), quoted below, are extremely relevant for getting patent rights in the area of herbals.

> Section 25: (1) Where an application for a patent has been published but a patent has not been granted, any person may, in writing, represent by way of opposition to the Controller against the grant of patent on the ground—
>
> (j) that the complete specification does not disclose or wrongly mentions the source or geographical origin of biological material used for the invention;
>
> (k) that the invention so far as claimed in any claim of the complete specification is anticipated having regard to the knowledge, oral or otherwise, available within any local or indigenous community in India or elsewhere,

Even after the grant of patent, any interested person can file opposition within a year of notification of grant of a patent on some of the specific grounds under Section 25 (2) which also include grounds mentioned above.

> Section 25 (2): At any time after the grant of patent but before the expiry of a period of one year from the date of publication of grant of a patent, any person interested may give notice of opposition to the Controller in the prescribed manner on any of the following grounds, namely:—
>
> (j) that the complete specification does not disclose or wrongly mentions the source and geographical origin of biological material used for the invention;
>
> (k) that the invention so far as claimed in any claim of the complete specification was anticipated having regard to the knowledge, oral or otherwise, available within any local or indigenous community in India or elsewhere,

Another relevant section under the Patents Act 1970 is Section 64 – revocation of patents. As per this section, any person interested or the Central Government may make a petition on any of the grounds, specified for revocation of patent under Section 64 of the Patents Act, before the Appellate Board. A patent may also be revoked by the High Court on a counterclaim in a suit for infringement of patent. There are various grounds for revocation before the elaborated in Section 64 that may be used to revoke a granted patent. Section 64 (p) and (q) incorporated via the Patents Amendments Act 2002 are related to herbal invention.

> (p) that the complete specification does not disclose or wrongly mentions the source or geographical origin of biological material used for the invention;
>
> (q) that the invention so far as claimed in any claim of the complete specification was anticipated having regard to the knowledge, oral or otherwise, available within any local or indigenous community in India or elsewhere.

All the sections of the Indian Patents Act 1970, presented at the time of formation of the

Act or incorporated during the amendments, are in coherence with each other and substantiate the fact that patenting an invention using traditional knowledge of herbals of India should be protected under the Act, but under no circumstances, any invention should be protected which violates these section. The spirit of the patent system in India is to benefit the society from the research and development in herbals and also granting monopoly rights to researches for a limited period of time (Table 12.1).

Indian Patent Office has also issued guidelines for the examiner and the controllers to examine the inventions related to the herbals/traditional knowledge. The guiding principles of this document not only help the examiners but also inventors to understand the importance of patenting activity in herbal thereby guiding inventors to draft their applications in such a way that the grant of their application should not become an impediment for public interest.

12.3.4 Guidelines for Processing of Patent Applications Relating to Traditional Knowledge and Biological Material

It is important that the innovation related to herbals should be provided legal protection under the law, but it is also the prerogative of the government to protect the biological diversity of India. Keeping in view importance of patenting in traditional knowledge, Guidelines for all Examiners and Controllers to be followed, while examining any patent application related to traditional knowledge, were issued in the year 2012. For the patent applications relating to traditional knowledge (TK), these guidelines very explicitly describe how to judge novelty and inventive step. These guidelines focus on the circumstances under which an invention should be considered patentable. The threshold for patentability has been clearly described in these guidelines with the help of six guiding principle.

Table 12.1 Patentable herbal invention in India

Patentable inventions in India
Novel formulation
Novel combinations involving selection of specific items/ingredients, specific proportions
Novel combinations that show synergy/antagonisms, better stability, better absorption/bioavailability
Novel combinations with explicit inventive steps like the addition of a chemical stabilising the formulation or enhancing penetration through the skin or increasing the rheological properties, etc.
Uniquely standardised to provide specific quality which is responsible for activity, e.g. ratios of components, etc.
Unique delivery devices like inhalation delivery devices
Combination of processes and compositions

Guiding Principle 1: If the subject-matter as claimed relates to extracts/alkaloids and/or isolation of active ingredients of plants, which are naturally/inherently present in plants, such claims cannot be considered as novel and/or inventive when use of such plants is pre-known as part of teachings of Traditional Knowledge.

Guiding Principle 2: Combination of plants with known-therapeutic effect with further plants with the same known-therapeutic agents wherein all plants are previously known for treating the same disease is considered to be an obvious combination.

Guiding Principle 3: In case an ingredient is already known for the treatment of a disease, then it creates a presumption of obviousness that a combination product comprising this known active ingredient would be effective for the treatment of same disease.

Guiding Principle 4: Discovering the Optimum or Workable Ranges of Traditionally known ingredients by Routine experimentation is not inventive.

Guiding Principle 5: In case multiple ingredients are known to have the same therapeutic activity as per traditional knowledge, taking out one single component out of them cannot be considered as inventive.

Guiding Principle 6: In case individual ingredients are already known for the treatment of a disease as a part of Traditional Knowledge, then it is obvious that a combination product comprising these known ingredients with further plants with the same known therapeutic effect would be more effective than each of the medicinal plants when applied separately (additive effect).

(Guidelines for Processing of Patent Applications Relating to Traditional Knowledge and Biological Material, 2012, Indian Patent Office)

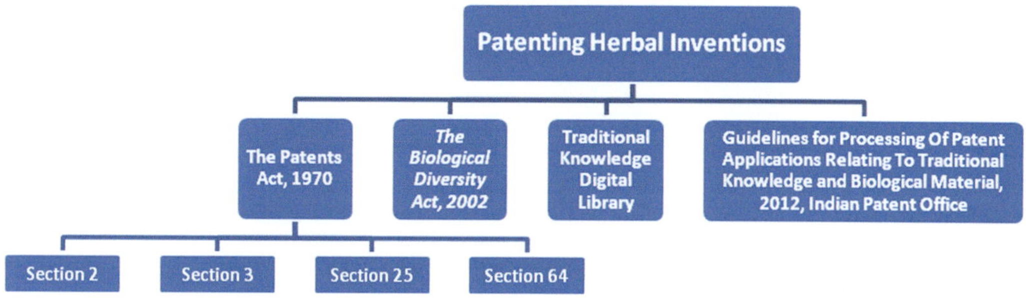

Fig. 12.6 Considerations for patenting herbal inventions

All these guiding principles very explicitly describe the inventions which are not patentable. The guiding principles are in coherence with the Patents Act 1970, and they just provide guidance to the examiners as well as patent applicants (Annexure I). Apart from these guiding principles, the guidelines also highlight that it is imperative to obtain NBA permission under the Biological Diversity Act 2002, for filing any patent application-related TK. The Biological Diversity Act 2002 provides very clearly that

> no person shall apply for any intellectual property right, by whatever name called, in or outside India for any invention based on any research or information on a biological resource obtained from India without obtaining the previous approval of National Biodiversity Authority before making such application;

> if a person applies for a patent, permission of the National Biodiversity Authority may be obtained after the acceptance of the patent but before the sealing of the patent by the patent authority concerned;

The National Biodiversity Authority shall dispose of the application for permission made to it within a period of 90 days from the date of receipt thereof. All the relevant sections of the Patents Act 1970 have been crafted to take care of biological resources of India, and few provisions are inspired by the Biodiversity Act 2002. The Indian Patent Law complements Section 6 (1) of the Biological Diversity Act 2002 by making it mandatory for the applicant of a patent to submit a declaration under Form I (Application for Grant of Patent) of the Patents Rules 2003 to the effect that 'the invention as disclosed in the specification uses the biological material from India and the necessary permission from the Competent Authority shall be submitted by me/us before the grant of patent to me/us'. This is one of the most important aspects of filing and prosecution of patents covering herbal inventions, and it would be a good idea to discuss this in detail for thorough understanding of the readers/researchers (Fig. 12.6).

12.3.5 National Biodiversity Authority

Patents together with access and benefit sharing are a critical component of conserving biodiversity. This was acknowledged in the objectives of the United Nations Convention on Biodiversity (CBD) to conserve biodiversity (Art 1) together with the recognition that patents and other forms of intellectual property should support the CBD's objectives. The access and benefit-sharing objectives of the CBD have now been implemented in India, and the National Biodiversity Authority constituted under Biodiversity Act 2002 is the nodal centre to obtain permissions related to IPRs in the area of traditional knowledge.

12.3.5.1 The Biological Diversity Act 2002

India is one of the 12 mega biodiversity countries of the world and accounts for 7–8 % of the recorded species. The biodiversity legislation regulates access to biological resources. As mentioned in the introduction, the Act provides for conservation of biological diversity, sustainable use of its components and fair and equitable sharing of benefits arising art of the new use of biological resources and knowledge. The Act established National Biodiversity Authority for the regulation of related activities. It also established State Biodiversity Boards and contains important provisions so as to stop indiscriminate use, misappropriation as well as granting of monopoly rights on biological resources (NBA 2015). The following three-tier structures at the national, state and local level have been created:

- National Biodiversity Authority (NBA)
- State Biodiversity Boards (SBB)
- Biodiversity Management Committees (BMCs)

It provides that any foreign national or corporation can obtain any biological resources occurring in India or knowledge associated thereto for research or commercial purposes only after taking the approval of the National Biodiversity Authority. Not only that, it also provides according to the province of Section 4 that even the result of research related to biological resources of India cannot be transferred to any foreign individual or corporate without the approval of National Biodiversity Authority. However, it does make exception for publication of research papers as well as for certain collaborative research projects.

Section 4 of the Act makes specific provisions that no person can apply for intellectual property rights, in India or abroad, for any invention based upon research or information on a biological resource obtained from India without seeking prior approval from the National Biodiversity Authority. In the following situations, NBA's permission is required:

- For commercialization of research results when the source material used for research belongs to countries biodiversity
- When research results have to be shared with foreigners
- When a foreigner/institution wants access to the country's biodiversity for undertaking research

However, the following are exempted from above:

- Local people and community of the area for free access to use biological resources within India
- Growers and cultivators
- Vaidyas and hakims
- Normally traded commodities
- Collaborative research with approval of the Central Govt.

The National Biodiversity Authority, an autonomous body created in 2003, performs the role of regulatory as well as advisory body on the matters related to biological resources. The entire set of responsibilities, mandated under the Biological Diversity Act (2002), is performed in a decentralised manner with NBA advising the Central Government on matters related to conservation, sustainable use and benefit sharing.

The State Biodiversity Boards advise the concerned state governments on issues related to biodiversity. The local-level Biodiversity Management Committees (BMCs) are responsible for documentation of biological diversity, preservation of habitats, conservation of domesticated stocks, breeds of animals and microorganisms, etc. The National Biodiversity Authority has been able to create State Biodiversity Boards in a number of states apart from creating about 30,000 BMCs.

Section 6 (1) makes it mandatory to seek prior approval from NBA for filing any IPR application, whether in India or abroad on a biological resource obtained from India. The applicant has to fill Form III required to obtain an application for intellectual property right and submit at NBA for approval. This is a detailed form which requires information about the biological material used for the invention (Annexure I). Effective January 1, 2005, it has become mandatory for a patent applicant to furnish a declaration in Form I, to be submitted to the patent office along with patent specifications for seeking patent rights, to the effect that the applicant would be submitting the necessary permission from the competent authority before grant of patent, in case the specification uses the biological material from India.

12.3.6 Traditional Knowledge Digital Library

As the purpose of this chapter is to provide a holistic view about the patenting in herbals/traditional knowledge, it is important to update the readers about Traditional Knowledge Digital Library (TKDL). The genesis of TKDL can be traced to the legal battle fought by CSIR in US Patent Office for re-examination of US Patent Number US 5401504, granted to two US-based Indians for wound healing properties of turmeric. This was a joint project initiated by the Department of Ayurveda, Yoga and Naturopathy, Unani, Siddha and Homoeopathy (AYUSH) (erstwhile Department of Indian Systems of Medicine and Homoeopathy, ISM&H) and National Institute of Science Communication and Information Resources (NISCAIR) (erstwhile National Institute of Science Communication, NISCOM) in order to prevent misappropriation of disclosed traditional knowledge (TK). It serves as a more easily accessible non-patent literature database that deals with traditional knowledge subject matter (Gupta, 2005). It integrates multidisciplinary skills in traditional knowledge, classification expertise, International Patent Classification, information technology and language expertise in French, German, Spanish and Japanese. Therefore, project team represents above skill set.

At present, apart from the Indian Patent Office, the following seven patent offices are using TKDL, and negotiations are underway with the New Zealand IP Office for signing the access agreement.

(i) Japan Patent Office (Apr 2011)
(ii) United Kingdom Patent & Trademark Office (Feb 2010)
(iii) Canadian Intellectual Property Office (Sep 2010)
(iv) German Patent and Trade Mark Office (Oct 2009)
(v) United States Patent and Trademark Office (Nov 2009)
(vi) Intellectual Property, Australia (Jan 2011)
(vii) European Patent Office (Feb 2009)

12.3.6.1 TKRC and IPC Concordance

One of the major factors for the success of TKDL is the unique classification system called the Traditional Knowledge Resource Classification (TKRC) on which TKDL is based and which makes the use of TKDL easy and effective in carrying out prior art searches. Traditional knowledge documentation lacked a classification system. Therefore, a modern classification based on the structure of International Patent Classification (IPC) was evolved. This has been attempted for Ayurveda and has been named as Traditional Knowledge Resource Classification (TKRC). The TKRC like the IPC has a system of classification based on hierarchical system of language-independent symbols for retrieving non-patent literature on Indian systems of medicine. TKDL concentrates only on the aspect of defensive protection which just prevents others from claiming any form of intellectual property protection over traditional knowledge and does not recognise or confer any rights on the knowledge holders.

12.3.6.2 TKDL Database

TKDL database is essentially a dynamic database covering more than two lakh formulations collected from Ayurveda, Unani, Siddha and Yoga texts, continuously updated. The entire information is provided in a standard format. It also provides modern names to the plants, diseases and processes and establishes linkage between traditional and modern knowledge. Over the years, TKDL has been successfully used for the cancellation/withdrawal of a number of patent applications filed in the USA, European Patent Office, etc. For example, formulations on Indian Systems of Medicine appear in the form of a text, which comprises the following main components:

- Name of the drug
- Origin of the knowledge
- Constituents of the drug with their parts used and their quantity
- Method of preparation of the drug and usage of the drugs
- Bibliographic details

TKDL gives modern names to plants (e.g. *Curcuma longa* for turmeric), diseases (e.g. fever for jwar) or processes, mentioned in the literature related to Indian Systems of Medicine, and establishes relationship between traditional knowledge and modern knowledge. The change that TKDL has brought in has been quite impressive. As of August 2011, 53 patent applications of the pharma companies of the USA, Great Britain, Spain, China, etc. had been either set aside or withdrawn or cancelled or declared as dead patent applications on the basis of third-party observations submitted by the TKDL team based on the information present in the TKDL database.

12.4 Copyright

Indian state provides the strongest possible protection to the creators of copyrightable works through the Indian Copyright Act amended from time to time. Works protected under Copyright Act are as follows:

(i) Literary, dramatic and musical work. Computer programs
(ii) Artistic work
(iii) Cinematographic films including soundtrack and video films
(iv) Record on any disc, tape, perforated roll or other device

The general rule is that copyright lasts for 60 years. In the case of original literary, dramatic, musical and artistic works, the 60-year period is counted from the year following the death of the author. In the case of cinematograph films, sound recordings, photographs, posthumous publications, anonymous and pseudonymous publications, works of government and works of international organisations, the 60-year period is counted from the date of publication.

India has a very strong and comprehensive copyright law based on Indian Copyright Act 1957 that was amended in 1981, 1984, 1992, 1994 and 1999. The amendment in 1994 was a response to technological changes in the means of communications like broadcasting and telecasting and the emergence of new technology like computer software. The 1999 amendments have made the Copyright Act fully compatible with Trade-Related Aspects of Intellectual Property Rights (TRIPS) Agreement. With these amendments, the Indian copyright law has become one of the most modern copyright laws in the world.

Herbal innovations falling under any one of the categories defined above can be protected through copyright. It may include labels, monographs, pamphlets, papers, product information leaflets of herbal products, etc. Unlike patents, there are no special guidelines or provisions in the Act for the innovations related to herbals. The copyright protects the form of expression rather than the subject matter of the writing. However, it is equally important to take copyright protection wherever applicable as it involves commercial interests of the stakeholders.

International Scenario The 1886 Berne Convention first established recognition of copyrights among sovereign nations, rather than merely bilaterally. Under the Berne Convention, copyrights for creative works do not have to be asserted or declared, as they are automatically in force at creation: an author need not ‘register’ or ‘apply for’ a copyright in countries adhering to the Berne Convention. The regulations of the Berne Convention are incorporated into the World Trade Organization’s TRIPS Agreement (1995), thus giving the Berne Convention effectively near-global application.

12.5 Design

Industrial design right is intellectual property right that protects the visual design of objects that are not purely utilitarian. An industrial design consists of the creation of a shape, configuration or composition of pattern or colour or combination of pattern and colour in three-dimensional form containing aesthetic value. An industrial design can be a two- or three-dimensional pattern used to produce a product, industrial commodity or handicraft. An industrial design is registrable under the Designs Act 2000, if it meets the following prerequisites:

The design should be new or original, not previously published or used in any country before the date of application for registration. The novelty may reside in the application of a known shape or pattern to new subject matter.

The design should relate to features of shape, configuration, pattern or ornamentation applied or applicable to an article.

The design should be applied or applicable to any article by any industrial process.

The features of the designs in the finished article should appeal to, and are judged solely by, the eye. This implies that the design must appear and should be visible on the finished article, for which it is meant.

Significantly distinguishable from known design or a combination of known designs.

Not comprise or contain scandalous or obscene matter.

Not be contrary to public order or morality.

The Locarno Agreement provides an internationally agreed classification system based upon the functionality of the goods under the design registration. Overall there are 32 classes which are further divided into subclasses. The assigned classification must correspond to the functionality of the item under consideration for design registration. Normally, the name of the article should be such that it is common/familiar in the trade or Industries. The name of the article as mentioned in the application form should correspond with the representation of the article as filed.

The design right is initially granted for 10 years which could be further extended by another 5 years by paying a one-time extension fee of Rs 2000. Industrial design protection is largely associated with the external appearance influencing commercial value of the products like in the case of formulations, shape of tablets, bottles used for dispensing medicines and shape of outer packaging box. There are no special provisions for herbals for design registration, and the herbal innovations under this IPR regime are examined and registered like any other design from any field.

International Protection The Hague Agreement facilitates filing of design application in several countries. An applicant file a single international application with national office, a party to the Hague Agreement governed by World Intellectual Property Office, and can seek protection in all the member countries of the agreement. The design rights historically originated in the United Kingdom in 1787 with the Designing and Printing of Linen Act and have expanded from there (Table 12.2).

12.6 Trademark

A **trademark** or **trade mark** (represented by the symbol ™) or **mark** is a distinctive sign or indicator of some kind which is used by an individual, business organisation or other legal entity to identify uniquely the source of its products and/or services to consumers and to distinguish its products or services from those of other entities. A trademark could be typically a name, word, phrase, logo, symbol, design, image or a combination of these elements. There are a lot of trade-

Table 12.2 What herbal innovations can be protected under the IPRs

Herbal innovations	
Copyright	Labels, monographs, pamphlets, papers, product information leaflets of herbal products, etc.
Design	Shape of tablets, bottles used for dispensing medicines, shape of outer packaging box, etc.
Trademark	Logo, symbol, design, image or a combination of these elements
Geographical indications of goods	Goods from a specific part of the country having distinctiveness and quality
Protection of Plant Varieties and Farmers' Rights	New plant varieties
Trade secret (No registration possible)	Process, formula, business information, design, instrument, pattern, compilation of information, etc.

marks related to herbals products, like GANDHAM, a herbal bath soap; Nature's gold crème, Nature's Fruit Bleach by Nature's Essence Private Limited; NIKHAR soap; etc. Trademark protection adds value to the product by providing it a desired identity through name, logo, etc. It is equally important for herbal products as there is a boom in the herbal market, and trademark facilitates trading of products.

The owner of a registered trademark may commence legal proceedings for trademark infringement to prevent unauthorised use of that trademark. The owner of a common law trademark may also file suit, but an unregistered mark may be protectable only within the geographical area within which it has been used or in geographical areas into which it may be reasonably expected to expand.

The Trade Marks Registry was established in India in 1940 and presently it administers the Trade Marks Act 1999 and the rules thereunder. It acts as a resource and information centre and is a facilitator in matters relating to trademarks in the country. The main function of the Registry is to register trademarks, which qualify for registration under the act and rules.

The duration of protection afforded to a 'mark' varies from country to country and registrations are issued for finite periods of time. However, because of the fundamental purposes of marks – namely, avoiding public confusion, encouraging competition and protecting the owners' goodwill – registrations may be renewed and thus extend indefinitely as long as the marks are used.

The initial registration of a trademark in India is for a period of ten years but may be renewed from time to time for an unlimited period by payment of the renewal fees.

International Protection It is important to note that although there are systems which facilitate the filing, registration or enforcement of trademark rights in more than one jurisdiction on a regional or global basis (e.g. the Madrid and CTM systems), it is currently not possible to file and obtain a single trademark registration which will automatically apply around the world. Like any national law, trademark laws apply only in their applicable country or jurisdiction, a quality which is sometimes known as 'territoriality'.

The major international system for facilitating the registration of trademarks in multiple jurisdictions is commonly known as the 'Madrid System'. Madrid System provides a centrally administered system for securing trademark registrations in member jurisdictions by extending the protection of an 'international registration' obtained through the World Intellectual Property Organization. This international registration is in turn based upon an application or registration obtained by a trademark applicant in its home jurisdiction.

The primary advantage of the Madrid System is that it allows a trademark owner to obtain trademark protection in many jurisdictions by filing one application in one jurisdiction with one set of fees and make any changes (e.g. changes of name or address) and renew registration across all applicable jurisdictions through a single administrative process. Furthermore, the 'coverage' of the international registration may be extended to additional member jurisdictions at any time.

12.7 The Geographical indications (GI) of Goods (Regulation and Protection) Act

A geographical indication (GI) is a name or sign used on certain products which corresponds to a specific geographical location or origin (e.g. a town, region or country). The use of a GI may act as a certification that the product possesses certain qualities or enjoys a certain reputation, due to its geographical origin.

In December 1999, the parliament passed the Geographical Indications of Goods (Registration and Protection) Act 1999. This Act seeks to provide for the registration and better protection of geographical indications relating to goods in India. India, as a member of the World Trade

Organization (WTO), enacted the Geographical Indications of Goods (Registration and Protection) Act 1999 which has come into force with effect from September 15, 2003.

Under Section 1 (e) of Act, GI is defined as:

> 'Geographic Indication' in relation to goods, means an indication which identifies such goods as agricultural goods natural goods or manufactured goods as originating or manufactured in the territory of a country or a region or locality in that territory, where a given quality reputation or other characteristic of such good is essentially attributed to its geographical origin and in case where such goods are manufactured goods, one of the activities of either the production or of processing or preparation of the goods concerned takes place in such territory, region or locality as the case may be.

A few of the agricultural goods registered as GI are provided in the following:

1. Navara rice: Certificate No.40 dated November 20, 2007

   Registered proprietor: Navara Rice Farmer's Society, Karukamanikalam, near Chittur, Kerala

   Medicinal rice used in Ayurveda treatment

   Varieties covered: Two black glumed and golden yellow glumed varieties of Navara rice
2. Palakkadan matta: a popular rice variety

   Bold red rice with a unique taste because of its special geographical area and peculiar weather of Eastern wind

   Registered proprietor: Palakkad Matta Farmers Producer Company Ltd.

   Varieties covered: 10 – Aryan, Aruvakkari, Chitteni, Chenkazhama, Chettadi, Thavalakanna, Eruppu, Poochamban, Vattan Jyothy and Kunjukunj. However, more rice varieties with matta properties cultivated in Palakkad can be added to this list after detailed examinations.

Any association of persons or producers or any organisation or authority established by or under any law for the time being in force representing the interest of the producers of the concerned goods, who are desirous of registering geographical indication in relation to such goods, can apply in writing to the Registrar.

The application for registering geographical indication should include the various requirements and criteria as specified in Rule 32 (1) which are:

(i) The reason to designate the good as a geographical indication
(ii) The class of goods
(iii) The territory
(iv) The particulars of appearance
(v) Particulars of producers
(vi) An affidavit of how the applicant claims to represent the interest
(vii) The standard benchmark or other characteristics of the geographical indication
(viii) The particulars of special characteristics
(ix) Textual description of the proposed boundary
(x) The growth attributes in relation to the GI pertinent to the application
(xi) Certified copies of the map of the territory
(xii) Special human skill involved, if any
(xiii) Number of producers
(xiv) Particulars of inspection structures, if any, to regulate the use of geographical indication

Registration of a GI enables producers to stop unauthorised use by others thereby boosting exports and their economic prosperity. Currently, there are 235 registered GI in India out of which 50 belongs to agricultural category including Malabar pepper from Kerala, Coorg Green Cardamom from Karnataka, Naga Mirchi from Nagaland, Guntur Sannam Chilli from Andhra Pradesh, etc. (Geographical Indications Registry India, 2015).

International Protection The TRIPS Agreement essentially stipulates the following obligations on the part of member countries in relation to the protection of GIs. There are, in effect, two basic obligations on WTO member governments relating to GIs in the TRIPS Agreement:

1. **Article 22 of the TRIPS Agreement** says that all governments must provide legal opportunities in their own laws for the owner

of a GI registered in that country to prevent the use of marks that mislead the public as to the geographical origin of the good. This includes prevention of use of a geographical name which although literally true 'falsely represents' that the product comes from somewhere else.

2. **Article 23 of the TRIPS Agreement** says that all governments must provide the owners of GI the right, under their laws, to prevent the use of a geographical indication identifying wines not originating in the place indicated by the geographical indication. This applies *even where the public is not being misled*, where there is no unfair competition and where the true origin of the good is indicated or the geographical indication is accompanied by expressions such as 'kind', 'type', 'style', 'imitation' or the like. Similar protection must be given to geographical indications identifying spirits.

12.8 The Plant Variety Protection and Farmers' Rights

The purpose of providing legal protection to new plant varieties is to encourage the plant breeders for their innovation. The rights provided to the plant breeders over their new plant varieties, for a limited period of time, motivate them to invent new plant varieties in the larger public interest. The Plant Variety Protection and Farmers' Rights Act 2001 was enacted in India to protect the new plant variety. Rules for the same were notified in 2003.

12.8.1 Objectives of Plant Variety Protection and Farmers' Rights Act

(i) To stimulate investments for research and development both in the public and the private sectors for the development of new plant varieties by ensuring appropriate returns on such investments

(ii) To facilitate the growth of the seed industry in the country through domestic and foreign investment which will ensure the availability of high-quality seeds and planting material to Indian farmers.

(iii) To recognise the role of farmers as cultivators and conservers and the contribution of traditional, rural and tribal communities to the country's agro biodiversity, by rewarding them for their contribution through benefit sharing and protecting the traditional right of the farmers.

(iv) More importantly this act provides safeguards to farmers by giving farmers rights while providing for an effective system of protection of plant breeders' rights. The Act seeks to safeguard researchers' rights as well. It also contains provisions for safeguarding the larger public interest. The farmer's rights include his traditional rights to save, use, share or sell his farm products of a variety protected under this Act, provided the sale is not for the purpose of reproduction under a commercial marketing arrangement.

12.8.2 Varieties Registrable under the Plant Variety Act

1. A new variety if it conforms to the criteria of novelty, distinctiveness, uniformity and stability
2. An extant variety if it conforms to criteria of distinctiveness, uniformity and stability

12.8.3 Definition of Novelty, Distinctiveness, Uniformity and Stability

Novelty Plant variety is novel if on the date of filing of the application for registration for protection, the propagating or harvested material of such variety has not been sold or otherwise disposed of, by or with the consent of breeder or his successor, for the purpose of exploitation of such variety.

In India earlier than one year, or outside India, in the case of tree or vines, earlier than six years, or in any other case, earlier than four years, before the date of filing such application, provided that a trial of a new variety which has not been sold otherwise disposed of shall not affect the right to protection.

Distinctiveness New plant variety will be considered distinct if it is clearly distinguishable by at least one essential characteristic from any other variety whose existence is a matter of common knowledge in any country at the time of filing of the application.

Uniformity New plant variety will pass uniformity test if subject to the variation that may be expected from the particular features of its propagation it is sufficiently uniform in its essential characteristics.

Stability New plant variety will be considered stable if its essential characteristics remain unchanged after repeated propagation or, in the case of a particular cycle of propagation, at the end of each such cycle.

Compulsory Plant Variety Denomination After satisfying the above four essential criteria, every applicant shall assign a single and distinct denomination to a variety with respect to which he is seeking registration: in the case of trees and vines, eighteen years from the date of registration of the variety and, in the case of extant varieties, fifteen years from the date of the notification of that variety by the Central Government under Section 5 of the Seeds Act 1966. In other cases, it is fifteen years from the date of registration of the variety.

Initially the certificate of registration shall be valid for nine years in the case of trees and vines and six years in the case of other crops and may be revived and renewed for the remaining period on payment of fees as may be fixed by the rules.

At present, the Protection of Plant Varieties and Farmers' Rights Authority, India, has issued a list of 88 crops/species for which seeds can be submitted to the authority for testing (Protection Of Plant Varieties And Farmers' Rights Authority, India 2015). These crops/species include isabgol, menthol mint, brahmi, coriander, almond, walnut, grapes, etc.

International Protection Under the TRIPS Agreement, it is obligatory on part of a member to provide protection to new plant variety either through patent or an effective sui generis system or a combination of these two systems. India was therefore under an obligation to introduce a system for protecting new plant variety. India opted for sui generis system and enacted the New Plant Variety Protection and Farmers' Rights Act. However, in many countries such plants can be protected through patent and UPOV Convention.

UPOV is an abbreviation of Union pour la Protection des Obtentious Vegetals (Union for protection of new varieties of plant). It is an international convention which provides a common basis for the examination of plant varieties in different member states of UPOV for determining whether a plant variety merits protection under UPOV or not.

12.9 Trade Secret

Trade secret can also be a useful legal vehicle for protecting innovations related to herbals, when dealing with outsiders' improper acquisition, disclosure and use of relatively secret information. Trade secret is unique among all the other legal instruments of legal protection as it is limited and fragile. It does not apply to publicly available, reverse-engineered or independently developed information. Broadly, a **trade secret** can be a formula, practice, process, design, instrument, pattern or compilation of information which is not generally known or reasonably ascertainable, by which a business can obtain an economic advantage over competitors or customers. In some jurisdictions, such secrets are referred to as 'confidential information' or 'classified information' (WIPO 2015).

The precise language by which a trade secret is defined varies by jurisdiction (as do the particular types of information that are subject to trade secret protection). However, there are three factors that, although subject to differing interpretations, are common to all such definitions. A trade secret is information that:

- Is not generally known to the public
- Confers some sort of economic benefit on its holder (where this benefit must derive *specifically* from its not being generally known, not just from the value of the information itself)
- Is the subject of reasonable efforts to maintain its secrecy

A company can protect its confidential information through non-compete and non-disclosure contracts with its employees (within the constraints of employment law, including only restraint that is reasonable in geographic and time scope). The law of protection of confidential information effectively allows a perpetual monopoly in secret information unlike patent which has only 20-year term. The lack of formal protection, however, means that a third party is not prevented from independently duplicating and using the secret information once it is discovered.

Trade secrets are by definition *not* disclosed to the world at large. Instead, owners of trade secrets seek to keep their special knowledge out of the hands of competitors through a variety of civil and commercial means, not the least of which is the use of non-disclosure agreements (NDA) and non-compete clauses. An employee may be required to sign an agreement for not revealing his or her prospective employer's proprietary information, in exchange for the opportunity to be employed by the holder of secrets. Often, the employee will also sign over rights to the ownership of own intellectual works produced during the course (or as a condition) of their employment. Similar agreements are often signed by other companies with whom the trade secret holder is engaged, e.g. with the trade secret holder's vendors, or third parties in licensing talks or involved in other business negotiations. Trade secret protection *can*, in principle, extend indefinitely, and this may offer an advantage over patent protection, which lasts only for a specifically limited period of time. Coca-Cola, the most famous trade secret example, has no patent for its formula and has been very effective in protecting it for many more years than the twenty years of protection that a patent would have provided. The relationship between intellectual property law, secrecy and disclosure with respect to herbals has important consequences. In the case of herbals, it is all the more important that society as a whole should be benefited from the disclosure of commercially valuable information. If any property of a part of tree has been found, then society has an interest in encouraging the disclosure of this knowledge to other entities that can improve upon it and bring it to the larger public. Trade secret should be used as an effective mechanism for protection of herbal innovations.

12.10 Conclusion

Herbal innovations are important for the society, and to encourage research in the area of herbals, it is imperative that the IPR regime should provide adequate protection to the innovations made by inventors. Herbals have a huge potential to fill the gap arising due to the lack of new molecular entities. Researchers have been looking at the herbals and their activities for clues for development of new molecular entities. IPR regime had been instrumental in promoting research and motivating researchers for their innovations, and there is no reason that herbals should be an exception for the same. However, any IPR regime must make a proper equilibrium between the availability of monopoly right and the freedom for rest of public domain. In the case of herbals, some of the provisions of IPR regimes, particularly patent regime, have been designed to do the same. It must be understood thoroughly that herbal innovations are treated on equal footing with other innovations for granting any IPR. Herbal innovations can be protected but one needs to find the right combination of IPRs and the IPR tools and use them. Patents provide the strongest monopoly right to the herbal inventions. It will be instructive to highlight grant of a patent on an invention related to the processing, extraction, composition and use of extracts of a plant *Plectranthus amboinicus* filed by a Taiwanese Company, the Development Center for Biotechnology (patent application number 1556/KOL/2007). The Indian Patent Office upheld during a pre-grant opposition proceeding, where some of the claims were objected by the opponent (CSIR). Obviously, the doors of even the strongest possible monopoly right for herbals are not totally closed in India or elsewhere. To summarise, patents are granted for herbal innovations in case:

- The formulation is novel.
- Novel combinations involve selection of specific items/ingredients and specific proportions.
- Novel combinations show synergy/antagonisms, better stability and better absorption/bioavailability.
- Novel combinations have explicit inventive steps like the addition of a chemical stabilising the formulation or enhancing penetration through skin or increasing the rheological properties, etc.
- Uniquely standardised to provide specific quality those are responsible for activity, e.g. ratios of components, etc.
- Uniquely delivered like inhalation delivery devices.
- Patents are grantable for combination of processes and compositions.

However, unlike patents, innovations related to herbal products protected under trademark laws, copyrights law, design act, GI, plant varieties, etc. are treated equally with other innovations. There are no special provisions or guidelines under any of these above-mentioned IPRs for herbals thereby strengthening the researchers to protect their innovations.

References

Brown RP, Gerbarg PL (2001) Herbs and nutrients in the treatment of depression, anxiety, insomnia, migraine, and obesity. J Psychiatr Pract 7:75–91

Controller General of Patents Designs and Trademarks (2015) http://ipindia.nic.in/ipr/patent/patents.htm. Accessed 10 Mar 2015

Dennis J (2013) Nutraceuticals World. International perspectives on herbs and botanicals. http://www.nutraceuticalsworld.com/issues/201307/view_features/internationalperspectivesonherbsbotanicals/. Accessed 15 Apr 2015

Geographical Indications Registry, India (2015) http://ipindia.nic.in/girindia/. Accessed 10 Apr 2015

Guidelines for Processing of Patent Applications Relating to Traditional Knowledge And Biological Material (2012) Indian Patent office, India

Gupta VK (2005) Traditional knowledge digital library Asia-Pacific database on Intangible Cultural Heritage (ICH). In: Asia-Pacific Cultural Centre for UNESCO (ACCU), Bangkok, 13–16 Dec 2005

Information Booklet for applicants for Registration of Design, The Patent Office, Intellectual Property Office, Kolkata 2010

Kant A (2009) Section 3(d): 'New' Indian perspective. J Intellect Prop Rights 14:385–396

Kuhn T, Wang Y (2008) Artemisinin an innovative cornerstone for anti-malaria therapy. Prog Drug Res 66:383–422

Liu JC, Hsu FL, Tsai JC et al (2003) Antihypertensive effects of tannins isolated from traditional Chinese herbs as non-specific inhibitors of angiotensin converting enzyme. Life Sci 73:1543–1555

Modak M, Dixit P, Londhe J et al (2007) Indian herbs and herbal drugs used for the treatment of diabetes. J Clin Biochem Nutr 40:163–173

National Biodiversity Authority (2015) http://www.nbaindia.org/introduction.htm. Accessed 10 Mar 2015

Nutraceuticals World (2015) http://www.nutraceuticalsworld.com. Accessed 15 Mar 2015

Pelkonen O, Xu Q, Fan TP (2014) Why is research on herbal medicinal products important and how can we improve its quality. J Tradit Complement Med 4:1–7

Protection of Plant Varieties and Farmers' Rights Authority, India (2015) http://www.plantauthority.gov.in/. Accessed 10 Apr 2015

Sahoo N, Manchikanti P, Satya Hari Dey SH (2011) Herbal drug patenting in India: IP potential. J Ethnopharmacol 137:289–297

TKDL (2015) Traditional Knowledge Digital Library, India. http://www.tkdl.res.in/tkdl/langdefault/common/Biopiracy.asp?GL=Eng. Accessed 10 Mar 2015

Wang L, Zhou GB, Liu P, Song JH, Liang Y, Yan XJ et al (2008) Dissection of mechanisms of Chinese medicinal formula realgar-indigo naturalis as an effective treatment for promyelocytic leukemia. Proc Natl Acad Sci U S A 105:4826–31

WIPO (2015) World Intellectual Property Organisation, Geneva. http://www.wipo.int/sme/en/ip_business/trade_secrets/trade_secrets.htm. Accessed 10 Apr 2015

13 Regulatory Approval of Botanical Products Including Herbal Drugs, Food, and Insecticides for Commercialization

P. Chattopadhyay and Vijay Veer

Abstract

India, with its enormous topographic and climatic diversity, is home to a large number of herbs with medicinal properties. The great herbal healers of India were pioneers in the use of herbs as medicine, and the modern medical research is greatly benefitted by their findings. After post-Vedic era the traditional medicine system of India was renamed as Ayurveda. The modern Ayurvedic system is more scientific in all respects including safety, toxicity, dose accuracy, clinical evidences, efficacy, etc. Botanicals are also explored as food supplements and nutraceuticals, and now the concept of value-added specialized foods has arisen. Food supplements from plants, algae, fungi, or lichens have become widely available in the Indian market. Many countries including India regulate the use of botanicals as dietary supplements, foods, and medicines, and these should fulfill the specific requirements for consideration and regulatory clearance. Botanical insecticides are now widely accepted as attractive alternatives as compared to the synthetic chemical insecticides as they are less toxic and safer to health and environment. The chapter elaborates the regulatory process for botanical products including herbal drugs, food, and insecticides in India.

13.1 Introduction

India is blessed with vast topographic and climatic diversity and thus has a large number of diversified herbs, which have the unique medicinal property. The therapeutic use of plants has started from 400 before the Common Era (BCE) onward and coming out with spirituality of the communities of thinkers including the Buddha and others. The Rig Veda (1500 BC) is one of the

P. Chattopadhyay (✉) • Vijay Veer
Defence Research Laboratory, (Defence Research and Development Organisation),
Tezpur, 784 001, Assam, India
e-mail: chattopadhyay.drdo@gmail.com;
vijayveer50@yahoo.com

© Springer India 2016
Vijay Veer, R. Gopalakrishnan (eds.), *Herbal Insecticides, Repellents and Biomedicines: Effectiveness and Commercialization*, DOI 10.1007/978-81-322-2704-5_13

important documents on rational use of plants as a medicine and emphasizes about medicinal knowledge of botanicals. The great herbal healers Maharishi Charaka and Maharishi Sushruta were pioneers in the use of herbs as medicine. The modern medical research is benefitted by their findings including the concepts of ailments of human body. The *Charaka Samhita* and *Sushruta Samhita* are the richest books of Indian herbs used as medicine. After post-Vedic era the traditional medicine system of India was renamed as Ayurveda which means the "complete knowledge for long life." The modern Ayurvedic system is more scientific in all respects including safety, toxicity, dose accuracy, clinical evidences, efficacy, etc. (Lachure 2012; Lodha and Bagga 2000; Verpoorte and Mukherjee 2003; Warude and Patwardhan 2005).

Botanicals are also explored as food supplements and nutraceuticals, and now the concept of value-added specialized food has arisen. Interestingly, sweets for diabetics, cholesterol-lowering food supplements, and other categories of foods are available in the market. Food supplements from plants, algae, fungi, or lichens have become widely available in the Indian market. The added value of the use of ginkgo, garlic, St. John's wort, and ginseng in food supplements is well claimed. A variety of claims regarding possible health benefits are mentioned in the labels, and these are being sold as over the counter (OTC) in pharmacies, specialist shops, shopping malls, online shops, etc. Also, such foods are promoted through advertisement in the electronic media. Many countries including India regulate the use of botanicals as dietary supplements, foods, and medicines, and these should fulfill the specific requirements for consideration and regulatory clearance. The Drugs and Cosmetics Act, 1940, strongly regulated herbal products and claims should not be under the property of drugs. The safety of the herbal supplements in food is a great concern, and the toxicity and botanical contaminants should be controlled at the time of cultivation, storage, or processing. The contaminants including heavy metals such as lead or mercury, herbicides, pesticides, or microbial metabolites including the toxins such as mycotoxins are a great problem during the cultivation, processing, or storage. Modern analytical techniques are available for the detection of the heavy metal contaminants and toxins. Many countries have adopted numerous methodologies including those in the manufacturing, licensing, and trading to ensure their safety, efficacy, and quality. Indian legislations have strongly controlled the use of traditional botanical products in food supplements or as medicinal products with the claiming (Anonymous 2002; Bhatt and Bhatt 1996; Burton 2003; CDSCO 2001; Ernst 2002; FICCI-Ernst and Young 2015; HADSA 2015; Kalra 2003; Kumar 2007; Lachure 2012; Lodha and Bagga 2000; NIN 2015; Stein 2002; Straus 2002; FDA 2015; Verpoorte and Mukherjee 2003; Warude and Patwardhan 2005.

Another important segment is the use of botanicals as insecticides. Botanical insecticides are accepted as attractive alternatives as compared to the synthetic chemical insecticides as they are less toxic and safer to health and environment. Bioactive compounds derived from plants effective against arthropod pests are required. Botanical insecticides are susceptible to light and moisture and are metabolized into less toxic or nontoxic compounds. Sometimes the synergism is quite common in botanicals although only limited literature is available on the synergism of toxicity. An example is the synergism between nicotine and rotenone which are commonly used as insecticides. In inhalation and dermal exposure, nicotine produced skin reactions and rotenone potentiated the toxicity. Another very common botanical insecticide is pyrethrum, which is used for the control of fleas and mosquitoes, but toxicities are also observed with its use.

The use of botanicals as insecticides has a great future for the underdeveloped and developing countries like India because of their low cost. Moreover, organic foods are very widely accepted because of less toxicity of the botanical insecticides and thus have high commercial values. Thus, botanicals used as insecticides should be used in a controlled manner to overcome the toxicities and under purview of the Insecticides Act and Rules when they are claimed as insecticides (Fig. 13.1).

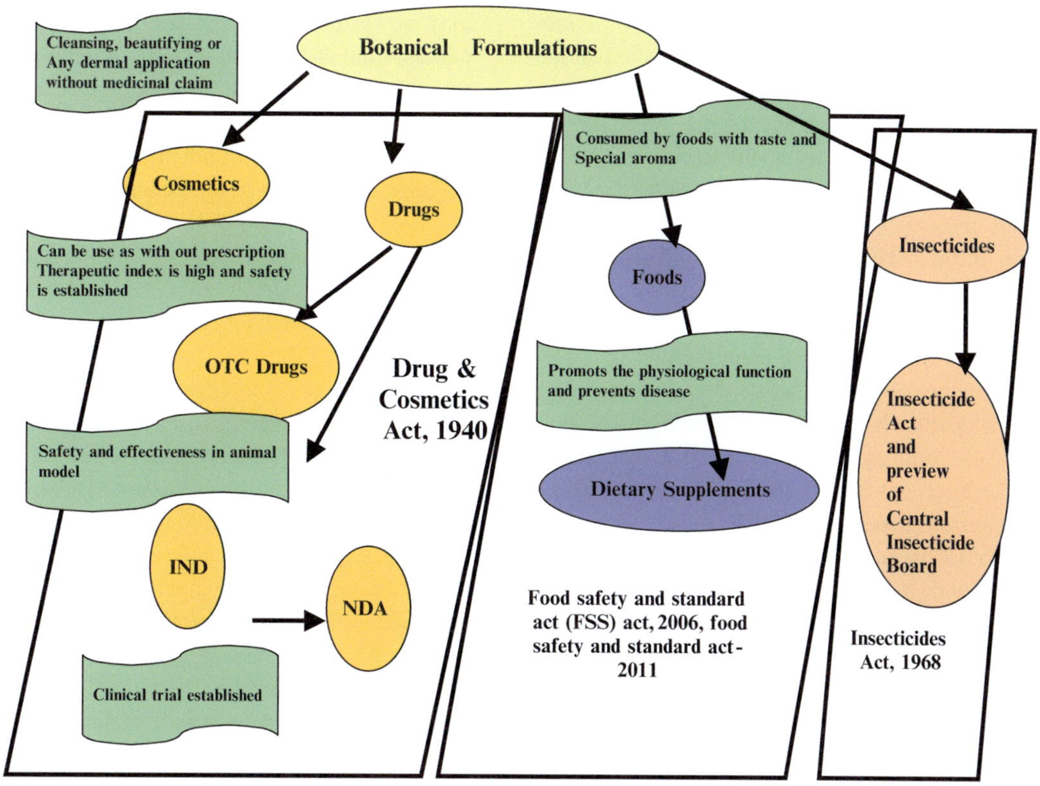

Fig. 13.1 Regulation process of herbals in India

13.2 Herbals as Foods

In the Indian market, it is very common that the botanicals are sold in the name of food or health supplements under various nomenclatures including "dietary food supplements," "nutritional supplements," "food supplements," and "health supplements." It is very difficult to differentiate these, and a narrow gap exists between "food," "drugs" or "drugs/medicines," and "nutritional supplements." Botanical ingredients may be classified into drugs as well as food supplements, and moreover all ingredients coming under the category of the different nomenclature are treated as nutritional supplements. In the context of the Indian regulation, most of the herbals are used as food and nutraceuticals with claim to cure or prevent disease. Sometimes it is very difficult to understand that the food is really acting as medicine. Prior to 2006, the Indian manufacturers were constrained to only the Prevention of Food Adulteration Act, 1954, and followed the rules framed under it. After 2006, implementation of the Food Safety and Standards (FSS) Act, the use of herbal as a food became restrictive, and the act almost stopped legally the manufacturing and trade including selling, import, distribution, or export of genetically modified articles of food, novel food, irradiated food, special dietary using foods, functional foods, nutraceuticals, proprietary foods, health supplements, or any specialized category foods and such other articles of food until notified by the central government in accordance with the FSS Act or rules laid norms. Implementation of the Food Safety and Standards Act, 2011, sometimes known as licensing and registration of food or FSS regulation, also strengthens the FSS Act, and more demarcation was formed between the so-called dietary food supplement, food supplements, nutritional supplements, and health supplements. The act is prerogative and directive to mandate the compulsory registration by any food business firms or manufacturer or any means operator with the Food Safety and Standards Authority of India (FSSAI). FSSAI controls the safety of the food for the human use and controls the unethical claims such as controlling of disease or enriching the vitamins, minerals, enzymes, etc. It is true that a very thin line exist between food or health supplement and drugs. In most of the cases, it is observed that nutritional enrichment was done with vitamins or minerals. Sometimes the manufacturer or trader accepts the food safety laws rather than the drugs and cosmetics law, which are more rigorous so as to avoid running the business with the risk of prosecution in case of incorrect categorization. The various aspects of herbals as food as per the available literature (Anonymous 2002; FICCI-Ernst and Young 2015; FTC 2015; HADSA 2015; Kalra 2003; NIN 2015; Stein 2002; FDA 2015; WHO 2005) are discussed in the chapter.

13.2.1 Section 22 of the FSS Act

This section provides categorization of functional foods, nutraceutical, health supplement, and drugs:

1. Foods can be formulated or processed for use in a particular physical or physiological condition or specific disease and disorder conditions to meet the necessary requirements of the particular condition. In view of this, these foods may differ from the natural food in the following aspects:
   - (i) Plant extract in water, ethyl alcohol, or hydroalcoholic extract, single or in combination or concentrate form.
   - (ii) Mineral, vitamin, or enzyme supplementation within the permissible limits for Indians.
   - (iii) Processed from animal origin.
   - (iv) A dietary substance may be allowed to humans as supplement that increases the total dietary intake.
2. Claiming in the label as "food for special dietary uses or functional foods" or "health supplements or nutraceuticals similar to such foods" and not under the conventional foods may be formulated in the form of powders, tablets, granules, capsules, jelly, liquids, and other dosages.
3. It is mandatory that in any circumstance, the product does not carry within the drug definition in clause (b) and Ayurvedic, Siddha, and Unani drugs as defined in clauses (a) and (h) of Section 3 of the Drugs and Cosmetics Act, 1940 (23 of 1940), and rules. Under any circumstances the food supplement cannot be introduced intravenously or by any means of parenteral route.
4. The product cannot consider the mitigation or healing of any unspecific disease or any disorder or condition. The claim cannot be for prevention in the categories of disease. However, the claim may be considered under the regulations made under FSSA only in certain health benefit or promotional claims.
5. Schedule of the Narcotic Drugs and Psychotropic Substances Act, 1985, and Schedules E and EI of the Drugs and Cosmetics Rules, 1945, should be strictly followed, and no enlisted drug should be incorporated.

13.2.2 Regulatory Authority for Nutraceuticals in India

The Food Safety and Standards Authority of India (FSSAI) is the apex body to regulate the nutraceutical market in India (FSSAI 2015). The Food Safety and Standards Authority (FSSA) is centralized to the governing of foods including the name of the special dietary uses or functional foods or nutraceuticals or health supplements or food enriched with nutrients including the minerals, proteins, or vitamins.

13.2.3 Functions of FSSAI

- (i) Controlling the registration and license for manufacturing of the foods
- (ii) Controlling the import of foods from the foreign origin including no objection certificates from the FSSAI and Institutional Animal Ethics Committee (IAEC) from Directorate General of Foreign Trade (DGFT)

13.2.4 Definition of Controlling of Foods under Section 22(1) of the FSS Act, 2006

Under the Indian law, the authorization of any person to produce or manufacture or distribute, sell, or import any foods in the category of functional foods or nutraceuticals or health supplements or special dietary use is under the FSS Act or FSS regulation. Sometimes conflicts of the food as health supplements and drugs arise. The Ministry of Health and Family Welfare in exercise of power conferred under Sections 22 and 92 of the FSS can act to decide the product status.

Health supplements are enriched with more nutrients including minerals, proteins, metals, vitamins or derived compounds, enzymes or amino acids, and other dietary substances or whole or part of plants or botanical sources from the plant or animal origin. It is strictly monitored that the claim should not violate the Drugs and Cosmetics Act, 1940. Usually, the special foods should be used for the known pharmacological/physiological functions without claiming any alternation of the disease conditions or pharmacological or physiological processes of normal body. The special dietary supplements are intended to supplement the normal diet of a person, and the composition of these foodstuffs differs significantly from the composition of ordinary foods of comparable nature.

13.2.5 Safety Issues of Dietary Food Supplements under Section 98 and Rule 37A of the FSS Act, 2006

Section 98 of the FSS Act regulates the safety of the special dietary foods and it is clearly mentioned in the act as:

> Notwithstanding the repeal of the enactment and orders specified in the second schedule, the standards, safety requirements and other provisions of the Act and the rules and regulations made there under and Orders listed in that schedule shall continue to be in force and operate till new standards are specified under this Act or rules and regulations made there under…

The Prevention of Food Adulteration Rules (PFA Rules), 1955, regulates the manufacture of health food supplements. However, the PFA Rules are not implemented to health and nutritional supplements, which are required to enforce for the quality control and safety issues.

13.2.6 Nutshell of Rule 37A of the Prevention of Food Adulteration Rules, 1955, for the Manufacturing of Proprietary Foods

(a) Separate license is required for manufacturing of each proprietary food product.
(b) A composite license is required to manufacture Indian traditional snacks and sweets by all the sweet or biscuit manufacturer.
(c) Label should contain the name of the food and category of the food.
(d) In the process of manufacture of proprietary food products, any tobacco and nicotine or related product should not be incorporated.
(e) If the food contains allergenic and/or hypersensitive ingredients, as identified under the rules, the label must mention it under clause (24) of the sub-rule of Rule 42.
(f) Proprietary food products should not contain any adulterated food additives or food and/or category of food.
(g) As per the rule (37A), the proprietary food should be mandatorily complying with the Prevention of Food Adulteration (PFA) Act.

13.2.7 Necessary Steps and Requirements for Approval

Registration as manufacturing unit falls under the state license authority including the tax, fire certification, and pollution certification. Further for industry act certification under the state, the applicant should prepare in the following way for the approval of the FSS Act. Complete form of part "A" along with self-attested declaration form are required for the registration. In case of import license, form B of Schedule 2 and whereas for the manufacturing form C of Schedule 2 along with self-attested declaration form are required. The following documents are also required as mentioned below:

(a) Plant layout/blueprint of the processing unit.
(b) List of directors or the responsible persons.
(c) Fire safety certificate.
(d) List of equipments and machinery.
(e) Address proof with photograph.
(f) Food category desired to be manufactured as per the FSS Act or PFA Rules.
(g) Name and address of the responsible person with authorization.
(h) Quality control or analysis report.
(i) Affidavit of possession of manufacturing premises.
(j) Any partnership deed/affidavit/memorandum and article copy should be submitted as per the factory act.
(k) No objection certificate of the manufacturer in case of importing the food.
(l) Safety management system or certificate complying with PFA Rules.
(m) Source of materials with certificate.
(n) Analysis of pesticide residue certificate or report of groundwater intended to use in the manufacturing.

13.3 Botanicals as Insecticides

Botanicals are also used as insecticides in the agriculture practice since at least two millennia in India. It is well documented that India was using botanicals as insecticides for more than 150 years. In the mid-1930s–1950s, the discoveries of major classes of synthetic chemical insecticides, viz., organochlorines, organophosphates, carbamates, and pyrethroids, led to reduced use of botanicals as insecticides in agriculture. However, due to the concept of organic farming and with awareness on the ill effects of the synthetic insecticides, botanicals have gained acceptance again.

Unforeseen poisoning of applicators, farmworkers, and consumers is reported after the use of insecticides. The environmental impact of the insecticides is also reported which destroyed the nonspecific targets including fish, birds, and other wildlife. The long persistent effects were also seen like extensive groundwater contamination and pollination imbalance by natural ways, and there is occurrence of resistance to pesticides.

13.3.1 Herbal Insecticides in India

In organic farming or in high-value commercial agriculture of products like mango, tea, and grapes, the herbal insecticides are used for economical excellence and commercial demand. In India, the commonly used herbal insecticides include pyrethrum, rotenone, neem, and essential oils and the limited used of nicotine. The other herbal products are being studied for insecticidal activities including garlic oil and *Capsicum* oleoresin. Governments responded to the problems associated with toxicities and untargeted toxicities of synthetic insecticides by banning or severely restricting the use of the most dangerous synthetic insecticides and creating policies to replace the synthetic insecticides by the other safer insecticides. The herbal insecticides have the potential to replace the synthetic insecticides.

13.3.2 Regulation of Herbal Insecticides in India

The import, manufacture, sale, transport, distribution, and use of insecticides are regulated by the Insecticides Act, 1968, which came into force on 1 March 1971. This act regulates the claim as insecticides of a compound or its formulation to prevent risk of poisoning to human beings or animals or nontarget species or the environment and the matters connected therewith. The herbal insecticide claiming as insecticide also comes under the purview of the Insecticides Act. The Central Insecticides Board (CIB) under the Ministry of Agriculture is the controlling authority and advises the central and state governments on matters on implementation of the act. The herbal pesticides should also be registered by the Pesticide Registration Committee. Before the registration of the pesticides, the herbal product should be scheduled as insecticides for preview under the Insecticides Act (CIBRC 2015).

13.3.3 Process of Registration

The requisite data are to be submitted along with specified forms to the Central Insecticides Board for consideration of the Registration Committee (CIBRC 2015). The data are evaluated by the expert committees for category claiming as insecticides, herbicides, and fungicides. The certificate along with approved label is issued by the Registration Committee. The certificate is essential for the manufacturing and import of the particular herbal pesticides. According to the act, enforcement is executed by the appointed inspectors. It is important that the government encourages the industry to participate in the responsibilities for producing safe and efficient pesticides under the act.

13.3.4 Data Requirement

The data in the following four categories are required for registration (Fig. 13.2):

1. Chemistry
2. Toxicology
3. Bioefficacy
4. Packaging and labeling

13.3.4.1 Chemistry

In this part the following data are required:

1. Chemical composition: The % w/w, the active chemical composition, and synergism if any are to be mentioned. The total formulation is to be disclosed in % w/w.
2. Physicochemical properties: Physical properties – density, refractive index, or any other physical property – are required to be mentioned.
3. Method of analysis along with undertaking for product quality: The established analytical tools are to be mentioned for quality control.
4. Identification and quantification of impurities: The impurity and the analytical method for identification are required.
5. Shelf life data: The storage conditions and the shelf life maintaining the activity are also required to be mentioned.

13.3.4.2 Toxicity

The various toxicological tests with the proposed formulation are to be performed as per the OECD guidelines, which are accepted by the Central Insecticides Board. The following tests are required:

1. Acute oral toxicity in rat and mice, acute dermal toxicity
2. Acute inhalation toxicity
3. Primary skin irritation
4. Irritation to mucous membrane
5. Mutagenicity
6. Toxicity to birds, fish, and honeybees
7. Human toxicity information from foreign countries

13.3.4.3 Bioefficacy

Three agroclimatic and two location bioefficacy data with the proposed formulations are required.

13.3.4.4 Packaging and Labeling

Similar to the synthetic insecticides, herbal insecticides are also required to comply with the label and leaflets as per Insecticides Rules, 1971. According to the act, type packaging, manner packaging, and compatibility with the packaging material are required to be mentioned. Also the details of primary packaging, secondary packaging, and transport packaging are to be mentioned.

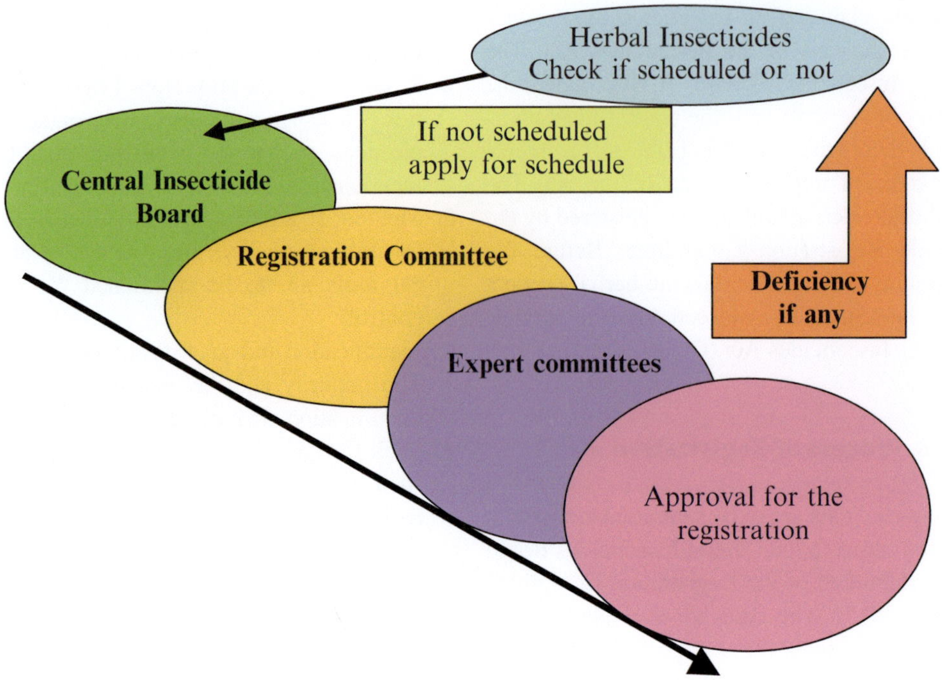

Fig. 13.2 Registration process of the herbal insecticide

It is also required to be mentioned in the label the methods of storage, use, and disposal of the product along with safety or cautionary statement and hazard if any associated with the herbal insecticides. In the formulations, the container content compatibility data is not required if kerosene is used as a solvent. If solvent is other than kerosene, container content compatibility data is required to be submitted. If the herbal is not originated from India, complete data is to be submitted under Section 3 of Insecticides Act for approval for the import.

13.4 Herbal Medicine

Human civilization started using many herbs for curing diseases and these are being used as medicine since long time. The science of alternative medicine, Ayurveda, originated in ancient India. The ancient scholars Atreya and Agnivesa in 800 BC practiced with herbal medicine. Later on further research and applications were consolidated by Charaka who wrote *Charaka Samhita*. It contains the principles and practices and is counted as the standard textbook for almost 2000 years and was translated into many languages. The book covered the basic physiology like digestion, metabolism, and human immunity including the etiology of the diseases. The Indian practitioners contributed a lot to surgery also. The great medical theoretician and practitioner Sushruta, in the book *Sushruta Samhita*, mentioned at least seven branches of surgery including excision, scarification, puncturing, exploration, extraction, evacuation, and suturing. Rhinoplasty (plastic surgery) and ophthalmology (ejection of cataracts) were mentioned perhaps for the first time in India. The book also mentioned the anatomical positions of the human dead body with the anatomical locations.

Various aspects of herbal and traditional medicines including quality control, safety issues, and regulatory process are described in the chapter as per Bhatt and Bhatt (1996), Burton (2003), CDSCO (2001), EMEA (1999), Ernst (2002), FTC (2015), Kumar (2007), Lachure (2012), Lodha and Bagga (2000), Marcus and Grollman (2002), Stein (2002), Straus (2002), US FDA (2000, 2015), Verpoorte and Mukherjee (2003),

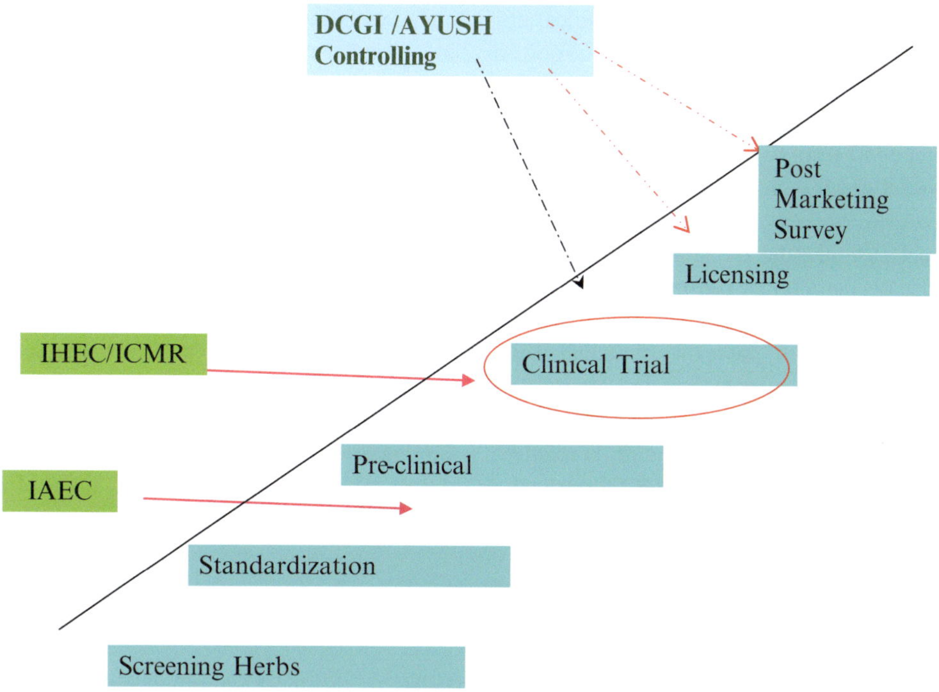

Fig. 13.3 Regulation of herbal drug in India

WAHO (2008), Warude and Patwardhan (2005), WHO (1999, 2000, 2004a, b, c, d, e, 2005, 2007, 2009) (Fig. 13.3).

13.4.1 Classification of Herbal Medicine and the Regulations

It is true that Indians are using plant medicines since long time and the birth of Ayurveda took place in India. However, all extracts or compounds isolated from a plant are not described in traditional system or ancient literature. A large number of medicinal plants are available in India and each has specific therapeutic importance. Most of the Indian forests are full with the different medicinal plants and aromatic plants having importance in perfumery industries. The practicing with the herbal medicine is also very old and traditional. The word "Ayurveda" originated from the Sanskrit words "ayur" for age or life and "Veda" which means knowledge. Thus, Ayurveda means the inner of science of life or longevity. Thus the herbal medicines are derived from the rich traditions of ancient civilization imparting the scientific heritage. The safety issues associated with the use of herbal medicine is an important aspect. Although a number of herbals are being used for the treatment of various diseases, the safety issue is not well proven and detailed scientific data are not available. These are classified into different categories as the following:

1. The use of plants and extract reported in the ancient literature like Ayurveda, Siddha, or Unani literature or plants that are used by traditional system physicians. This type is classified as complementary alternative medicine (CAM) and is in the purview of AYUSH and is regulated by AYUSH. The plant as a medicine will be registered as Ayurvedic medicine, and the manufacturing license will be provided by the state government.
2. The isolated compound from a traditionally described plant is considered as new chemical entity (NCE), and the regulatory aspects are the same as that of the allopathic medicine as per the norms of the Drug Controller General

of India (DCGI). The safety and clinical efficacy data are required for registration or approval.

3. If the traditional medicine or any isolated compound from plants, which are not evaluated clinically, and its proper therapeutic effect are not reported in the texts of traditional systems, the preparation methodology and consideration of the treatment as a new substance are different. This type of compounds is under the new chemical entity (NCE), and the toxicological data are required for the acute, subacute, and chronic toxicity. Further all the clinical evaluation is required. These compounds are under DCGI purview.
4. If the plant or any related preparations are not described traditionally or the method of extraction is different, then such type of compounds/formulations falls under the purview of DCGI, and all regulatory requirements are mandatory before being evaluated clinically.

13.4.2 Legal Framework in India

The drug-related issues are usually dealt under the Drugs and Cosmetics Act (DCA), 1940, which is amended time to time. The laws are implemented for both import and export of the products. The laws are implemented and enforced by the state and central governments. The Drug Controller General of India (DCGI) has the prime responsibility to permit the clinical trials in human volunteers for a new drug or NCE after satisfying the toxicity data. The DCGI also regulates the introduction and approval of new drugs and import licenses of new drugs. The state governments act as an enforcing agency of DCA and central directives and grant permission for setting up new manufacturing facilities and manufacturing license and also monitor the sale of drugs.

13.4.3 Problems of Herbal Drugs and Quality Control

Most of the herbal drugs frequently fail to meet the standards of consistency in composition and biological activity. The other problems are associated with the identification of plants, variability in genetic origins, agroclimatic conditions, change of harvesting procedures and extraction process, and alternations of pharmacologic or therapeutic effects. These problems are overcome with the help of chromatographic techniques to identify the marker compounds for the standardization and with control over the batch-to-batch consistency. However, quality assurance is still an important issue. The safety of herbal drugs is an important issue as most of the herbal drugs are adulterated with a number of herbals and contaminated with toxic metals, pesticides, and fumigation agents. Herbal medicines are contaminated with microorganisms and metabolites of microbial toxins during longtime storage.

13.4.4 Preclinical Toxicity Studies

Safety is the most important issue in any drug development. Nonclinical studies or Phase 0 or preclinical research is the determining step to ascertain the product safety and further development studies. This step in the drug development determines the probability of toxic reaction and proposed mechanism. Preclinical toxicology studies are mandatory prerequisite before administration of drugs to human beings so as to assess the adverse toxicity of the drugs.

Moreover, preclinical toxicology studies are to understand the adverse effect of the herbal drugs at the conditions of the dosage forms, amount, interaction, physiologic conditions, etc. Further, preclinical study highlights the limit and quantity of dosage forms, adverse reaction, target organ toxicity, or any side effects. The selection of safe dose is also derived from the preclinical toxicity, and preclinical research helps clinician to select secure dose/s during clinical studies and later on to conclude the therapeutic amounts.

Safety categories are classified according to the periods of use as follows:

Category 1: Safety established since long time by users.

Category 2: Safety is covered by well-established documentation under specific conditions of use.

Category 3: Uncertain safety and the safety data are required to be established, and this class is sometimes identical to that of any new substance.

As per AYUSH guidelines, toxicities are classified into two major categories – acute and long-term toxicities or subacute toxicity. India being a signatory to WTO, the safety aspect as per international regulatory guidelines of WHO and OECD is to be followed for commercialization in the international market.

13.4.5 Acute Toxicity Testing

Acute toxicity tests are indicative of the dose level where mortality may be observed, and this toxicity level is known as threshold toxicities or lethal doses (LD_{50}). Ideally for toxicity evaluation of herbal drugs, two species are used, and the dose is adjusted accordingly to the human dose. In most of the cases, 10 times of the therapeutic human dose is considered for the highest level of safety determination.

13.4.6 Subacute Toxicity Testing

The subacute toxicity tests are also done in two species with a minimum of 90 days of study at the presumed therapeutic dose (TD), average dose (TD × 5), and the highest dose (TD × 10) of the herbal. The dose-response character (DRC) should be determined for the conclusion of the toxic changes or "no observable effect level" (NOEL).

The toxicity observations are usually as follows:

1. Changes in general behavior including the food intake and weight are noted during study periods.
2. Changes in hematology.
3. Biochemical changes.
4. Changes in urine.
5. Changes in organ weight.
6. Gross examination of organs.

The preclinical toxicity of NCE molecules is under the purview of DCGI, is more extensive, and requires more parameters, which include:

1. 14–28-day repeated-dose toxicity studies
2. 90-day repeated-dose toxicity studies
3. 180-day repeated-dose toxicity studies
4. Systemic toxicity studies in single or multiple dose
5. Reproductive toxicity studies including male fertility study, female fertility study, teratogenicity study, and perinatal study
6. Genotoxicity including gene mutation in bacteria, cytogenetic changes in mammalian cells, chromosomal damage, and carcinogenicity assay
7. Dermal toxicity studies including acute, subacute, sensitization assay, photoallergy, or dermal phototoxicity
8. Inhalation toxicity assay
9. Ocular toxicity assay

13.4.7 Evidence of Clinical Efficacy

The regulatory authority in India is the Ministry of Health and Family Welfare and controlled by the Central Drugs Standard Control Organization (CDSCO). CDSCO is the responsible authority for device controlling and regulatory measures to control drugs, diagnostics, and medical devices in the implemented standards and norms. CDSCO has the equal power to prosecute the violators of rules or norms related to drugs.

The use of drug to volunteers for evaluation is known as clinical trial. CDSCO is also the regulatory authority of clinical trial of new drugs and importing of new drugs. Since India is a big country having the maximum variety in population including race and ethnicities, a number of multinationals are putting the final destination to India as the clinical trial hub. The Indian government after realizing the potentiality of clinical trial, put clinical trial in the judiciary framework by amending of the Drug and Cosmetics Rules, 1945, known as Schedule Y. The schedule directed the guidelines and laid norms and requirements for clinical trials in

India. In Schedule Y, the patients' right and ethical issues are considered foremost, and it implemented the proper informed consent, consideration of ethical issues by forming institutional ethical committee, and other regulatory parameters like financial involvement between the sponsor and volunteer, insurance, clinical trial registry, and strong monitoring. However, clinical trial is almost the same in the nomenclature of herbal drugs or modern medicines. The herbal drug clinical trials are classified as follows:

Phase 1: The aim of Phase 1 is to assess the safety testing in healthy human volunteers and 20–80 volunteers are recruited. The study period is usually 1 year.

Phase 2: In this study period, the safety and efficacy testing is the prime concern and 100–300 volunteers are recruited. The study period is usually 2 years.

Phase 3: In this phase the safety and effectiveness are assessed at a larger scale and 1000–3000 volunteers are recruited. The study period is usually 3 years. Sometimes a multicentric trial is done.

Phase 4: In this phase post-marketing surveillance is executed and the regulatory authority is responsible. The study period is several years and the drugs can be withdrawn in this phase if found harmful.

Sometimes the safety data of herbal drugs are known and are used for a long time. In this case, the herbal drug should be registered by conducting the clinical trial. The data required for the clinical trial are as follows:

A. Description of product and documentation
   1. Description of herbs used for a long time
   2. Current use and commercial significance
B. Chemistry, manufacturing, and controls
   1. Raw material
   2. Drug substance or product details
      (a) Quantitative description
      (b) The composition or quantitative description and quality control
C. Labeling
D. Environmental assessment or claim
E. Pharmacology
F. Bioavailability
G. Clinical safety and use considerations

A randomized clinical trial is the most accepted design of the clinical trial for manual therapies, though other designs of clinical trial such as randomized, double blind, and placebo-controlled (or dose-response) are also accepted. In herbal drugs, the clinical trial in equivalence designs is not considered. Herbal drugs are quite complicated in the dose selection, which are based upon the response and traditional methodologies for the dose selection, etc.

13.4.8 Adverse Drug Reaction (ADR) Monitoring

Adverse drug reactions are commonly associated with the drug-drug interactions, drug-food interaction, and physiological compatibility of the patients. Extreme adverse drug reactions may lead to death. In allopathic medicine adverse drug reaction is mandatory and report should be made directly to DCGI or US FDA. It is commonly believed that the herbal drugs are safer as compared to other synthetic or modern medicines. But there are number of cases reported in which the relevant interactions with other herbal drugs produce serious adverse effects. Several reports are there with the interaction of St. John's wort causing serum with cyclosporine and anti-retroviral agents to alter the bioavailability, while consumption of garlic is reported to increase the bioavailability of anticoagulant activities.

13.4.9 Filing for Herbal Drug Approval

The regulatory aspects of herbal drug are quite different from the modern medicine or allopathic drugs. The process for AYUSH category drugs is somewhat easier than the herbal drugs in NCE categories.

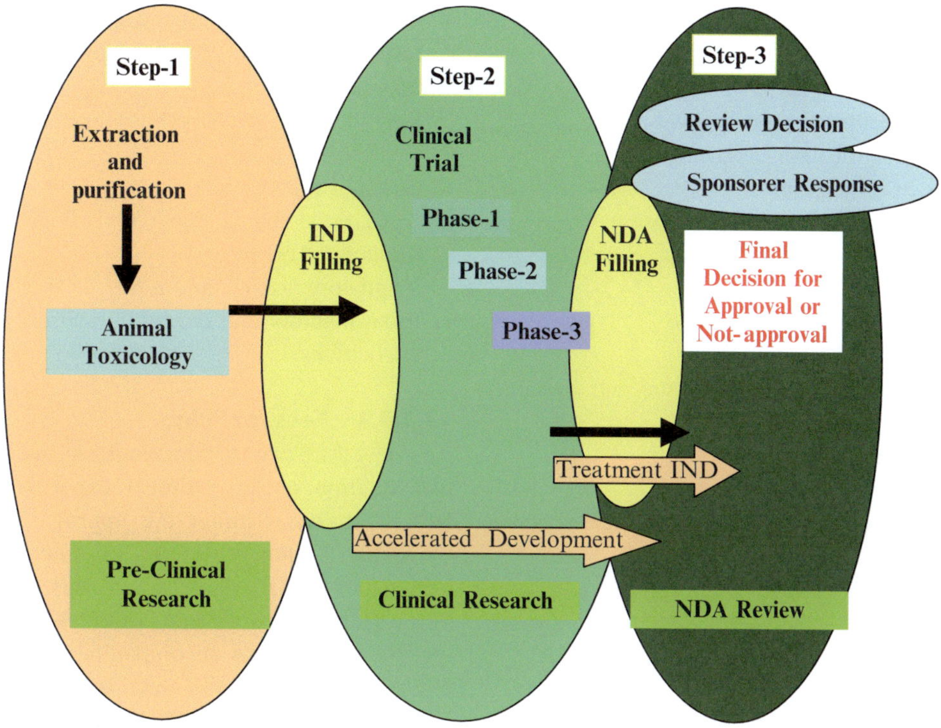

Fig. 13.4 Filing process for herbal drug approval

13.4.9.1 Requirements for Investigational New Drug (IND) Applications of Botanicals

Clinical trial of botanicals, which are being used currently or were used traditionally for a long time, can be initiated with small knowledge on new chemistry, manufacturing, and controls (CMC) or toxicological data. Since the drug is in use for a long time and it seems that the safety is already established, finally pilot trial is required. In most cases, the sponsor is permitted for randomized, parallel, dose-response study for the evaluation of the better products (Fig. 13.4).

The following data are required for application of IND approval:

(a) Cover page
(b) Tables and contents
(c) Introduction and general investigational plan
(d) Brochure
(e) Protocol
(f) Chemistry, manufacturing, and controls
(g) Pharmacology and toxicological data
(h) Previous uses of the product and human experience or products currently used in other countries

13.4.9.2 Requirements for New Drug Approval (NDA) Applications of Botanicals

After IND approval the botanicals are entered to the clinical trials, and the data required for the clinical trial were already mentioned. If successful in the clinical trial, the drug can be filed as NDA application and can be marketed exclusively as a new drug.

The data required for NDA approval are as follows:

1. Description of product and documentation if used previously or human data available
2. Chemistry, manufacturing, and controls
   (a) Raw material including source, collection process, and seasonal variation (must

comply with the Indian Biological Diversity Act)
   (b) Details of herbal
   (c) Identification qualitatively
   (d) Identification chemically
   (e) Specifications
   (f) Process of manufacturing
   (g) Quality control methods
   (h) Methodology of test
   (i) Standards and reference
   (j) Closure and containers
   (k) Stability data and closure and container interaction data
   (l) Label of container
3. Botanical drug product
   (a) The composition and qualitative description
   (b) Specifications of acceptance
   (c) Process of manufacturing
   (d) Quality control tests including the test methods, containers, and closure
4. Placebo
5. Labeling
6. Claims of categorical exclusion

Additional data are also required along with the clinical data for NDA approval as follows:

(a) Botanical raw materials
(b) Process of manufacturing
(c) Variation and consistency batch-to-batch process
(d) Established specifications including the marker compound concentration or the pharmacological active content
(e) Established test procedures and analytical methods
(f) Established reference standard
(g) Stability-indicating analytical data including the product and the container
(h) The data of preclinical and clinical studies are required to compare the similarities and/or differences with other marketed products or mimic molecules
(i) Facilities of manufacturing and testing
(j) Claim for categorical exclusion

Preclinical safety assessment:

1. Toxicity studies in repeat dose
2. Data of nonclinical pharmacokinetics/toxic-kinetics
3. Reproductive toxicology
4. Genotoxicity studies
5. Carcinogenicity assay
6. Any special pharmacology/toxicology studies for targeted organ
7. Regulatory considerations
8. Bioavailability and drug-drug and drug-food interactions

13.4.9.3 Fees for Filing

Any new drug cannot be considered in clinical investigation or any clinical experiment by any institution without the approval of the Licensing Authority defined in clause (b) of Rule 21 of the Drugs and Cosmetics Act. The Phase 1 clinical trial can be initiated after filling out the Form 44 accompanied by a fee of 50,000 along with the data required under Schedule Y. For the Phase 2 trial, the additional fee of 25,000 is required and the data of Phase 1.

Permission of Phase 3 trial requires an additional fcc of 25,000, and the data emerging from Phase 2 and Phase 1 are necessary. After being satisfied with the results of clinical trials, license is permitted by the regulatory authorities within Form 45 or Form 45-A or Form 46 or Form 46-A with the conditions as necessary which are stated therein. No additional fees are required for the manufacturing and import of the drugs after successful completion of clinical trials by the applicant.

13.4.9.4 Licenses for Manufacturing of ASU Drugs

1. Chapter IV-A and Chapter V of the Drugs and Cosmetics Act, 1940, are directive of ASU drugs. The provisions of ASU drugs are also mandatory to be fulfilled as mentioned in parts XVI, XVI-A, XVII, and XVIII of the Drugs and Cosmetics Rules, 1945.
2. Schedule T: Grant of manufacturing license.
3. Application forms: 24-D application for the grant/renewal of license for manufacturing.

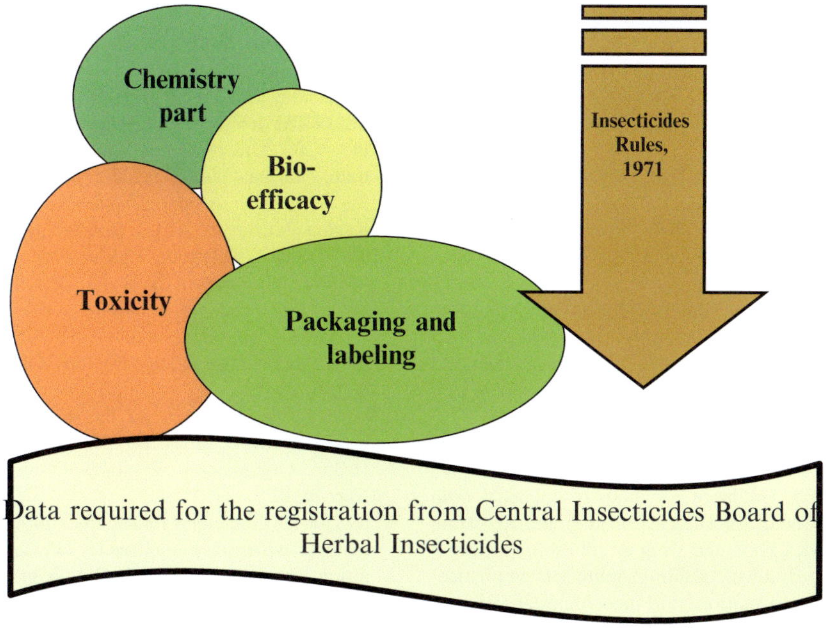

Fig. 13.5 Data required for the registration process of herbal insecticides

Usually completed pro forma is sent to AYUSH, Delhi, or respective state branch office.

13.4.9.5 Time Frame for Granting License

The granting of license is systematic and timely response is obtained (Fig. 13.5):

1. If the conditions are satisfied, the time limit to grant licenses is fixed to 90 days after the receipt of application.
2. Scrutiny of the application and any shortcoming are informed within 15 days of the receipt of application.
3. Inspections of the premises of the applicant firm are conducted within 21 days of receipt of complete documents.
4. Disposal of application within 180 days includes all respect, i.e., grant of license or the rejection of the application.

References

Anonymous (2002) Problems with dietary supplements. Med Lett Drugs Ther 44:84–86

Bhatt AD, Bhatt NS (1996) Indigenous drugs and liver disease. Ind J Gastroenterol 15:63–67

Burton B (2003) Complementary medicines industry in crisis after recall of 1546 products. BMJ 326:1001

CDSCO (2001) Guidelines for clinical trials on pharmaceutical products in India – good clinical practices. Central Drugs Standard Control Organisation, Directorate General of Health Services, Ministry of Health & Family Welfare, Government of India, New Delhi

CIBRC (2015) Central Insecticides Board and Registration Committee. http://www.cibrc.nic.in. Accessed 28 Mar 2015

EMEA (1999) Report from the Adhoc working group on herbal medicinal products 1997/1998 EMEA/HMPWG/25/99. European Agency for the Evaluation of Medicinal Products, London. http://www.ema.europa.eu/docs/en_GB/document_library/News_article/2009/11/WC500015693.pdf. Accessed 28 Mar 2015

Ernst E (2002) Heavy metals in traditional Indian remedies. Eur J Clin Pharmacol 57:891–896

FICCI-Ernst and Young (2015) Nutraceuticals critical supplement for building Healthy India. http://plethico-fd.com/reports%5CNutraceuticals_Final_Report.pdf. Accessed 28 Mar 2015

FSSAI (2015) Food Safety and Standards Authority of India. http://www.fssai.gov.in

FTC (2015) Federal Trade Commission. https://www.ftc.gov. Accessed 28 Mar 2015

HADSA (2015) Health Foods and Dietary Supplements Association http://www.hadsa.com

Kalra EK (2003) Nutraceutical: definition & introduction. AAPS Pharmsci 5(2):Article 25. doi:10.1208/PS/0500225

Kumar GS (2007) Regulatory roadmap for herbal medicine. Business Horizons, New Delhi
Lachure PS (2012) Exploration of some medicinal plants used by tribals from Digras region of District – Yavatmal Maharashtra, India. Int J Sci Res Publ 2:1–11
Lodha R, Bagga A (2000) Traditional Indian system of medicine. Ann Acad Med Singapore 29:37–41
Marcus DM, Grollman AP (2002) Botanical medicines – the need for new regulations. N Engl J Med 347:2073–2076
NIN (2015) National Institute of Nutrition, Hyderabad, India. http://www.ninindia.org. Accessed 28 Mar 2015
Stein MC (2002) Are herbal products dietary supplements or drugs? An important question for public safety. Clin Pharmacol Therap 71:411–413
Straus SE (2002) Herbal medicines – what's in the bottle? N Engl J Med 347:1997–199
US FDA (2000) Guidance for industry botanical drug products. Centre for Drug Evaluation and Research, United States Food and Drug Administration. http://www.fda.gov/downloads/drugs/guidancecompliance-regulatoryinformation/guidances/ucm070491.pdf. Accessed 28 Mar 2015
US FDA (2015) United States Food and Drug Administration. http://www.fda.gov. Accessed 28 Mar 2015
Verpoorte R, Mukherjee PK (eds) (2003) GMP for botanicals. Business Horizons Pharmaceutical Publishers, New Delhi
WAHO (2008) Traditional medicine workshop report. Situational analysis of the level of development of traditional medicine in the ECOWAS member States. West African Health Organization
Warude D, Patwardhan B (2005) Botanicals: quality and regulatory issues. J Sci Ind Res 64:83–92
WHO (1999) Essential drugs in the WHO African region: situation and trend analysis. World Health Organization Regional Committee for Africa, Windhoek, Namibia. http://apps.who.int/iris/bitstream/10665/1926/1/AFR-RC49-11%20EN.pdf?ua=1. Accessed 28 Mar 2015
WHO (2000) Promoting the role of traditional medicine in health systems: a strategy for the African region. WHO Regional Office for Africa, Ouagadougou, Burkina Faso. www.afro.who.int. Accessed 28 Mar 2015
WHO (2004a) WHO model legal framework for the practice of TM: a bill for traditional health practitioners in the WHO African region, Tools for institutionalizing traditional medicine in health systems of countries in the WHO African region. World Health Organization Regional Office for Africa, Brazzaville
WHO (2004b) WHO model code of ethics for traditional health practitioners in the WHO African region, Tools for institutionalizing traditional medicine in health systems of countries in the WHO African region. World Health Organization Regional Office for Africa, Brazzaville
WHO (2004c) Guidelines for the registration of traditional medicines in the WHO African region. World Health Organization Regional Office for Africa, Brazzaville
WHO (2004d) Guidelines for documenting data on ethnomedical evidence, Guidelines for clinical study of traditional medicines in the WHO African region. World Health Organization Regional Office for Africa, Brazzaville
WHO (2004e) Guidelines for the regulation of herbal medicines in the south-east Asia region developed at the Regional Workshop on the Regulation of Herbal Medicine, Bangkok, 24–26 June 2003, World Health Organization, Regional Office for South-East Asia, New Delhi. http://apps.searo.who.int/PDS_DOCS/B0560.pdf?ua=1. Accessed 28 Mar 2015
WHO (2005) National policy on traditional medicine and regulation of herbal medicine, Report of WHO global survey, World Health Organization, Geneva. http://whqlibdoc.who.int/publications/2005/9241593237.pdf. Accessed 28 Mar 2015
WHO (2007) Guidelines on good manufacturing practices (GMP) for herbal medicines. World Health Organization, Geneva
WHO (2009) Handbook of good laboratory practices (GLP) quality practices for regulated Non-clinical research and development. Special programme for research and training for tropical diseases. World Health Organization, Geneva

MIX
Papier aus verantwortungsvollen Quellen
Paper from responsible sources
FSC® C105338

If you have any concerns about our products,
you can contact us on
ProductSafety@springernature.com

In case Publisher is established outside the EU,
the EU authorized representative is:
Springer Nature Customer Service Center GmbH
Europaplatz 3, 69115 Heidelberg, Germany

Printed by Libri Plureos GmbH
in Hamburg, Germany